Modelling Aspects of Water Framework Directive Implementation

Modelling Aspects of Water Framework Directive Implementation

Edited by
Peter A. Vanrolleghem

Publishing
London • New York

Published by **IWA Publishing**
Alliance House
12 Caxton Street
London SW1H 0QS, UK
Telephone: +44 (0)20 7654 5500
Fax: +44 (0)20 7654 5555
Email: publications@iwap.co.uk
Web: www.iwapublishing.com

First published 2010

Typeset in India by OKS Prepress Services.

Disclaimer
The information provided and the opinions given in this publication are not necessarily those of IWA or the European Commission and should not be acted upon without independent consideration and professional advice. IWA, the European Commission and the Author(s) will not accept responsibility for any loss or damage suffered by any person acting or refraining from acting upon any material contained in this publication.

British Library Cataloguing in Publication Data
A CIP catalogue record for this book is available from the British Library

Library of Congress Cataloging-in-Publication Data
A catalog record for this book is available from the Library of Congress

ISBN 10: 1843392232
ISBN 13: 9781843392231

Contents

By Peter A. Vanrolleghem. ISBN: 9781843392231. Published by IWA Publishing, London, UK.

Authors for this volume

Anker Lajer Højberg
Department of Hydrology
Geological Survey of Denmark and Greenland (GEUS)
E-mail: alh@geus.dk

Henrik Madsen
Head of Innovation, Water Resources Department, DHI
E-mail: hem@dhigroup.com

Florian Pappenberger
European Centre for Medium-Range Weather Forecasts
E-mail: florian.pappenberger@ecmwf.int

Marco Ratto
European Commission, Joint Research Centre
E-mail: marco.ratto@jrc.ec.europa.eu

 Modelling Aspects of Water Framework Directive Implementation Volume 1. By Peter A. Vanrolleghem. ISBN: 9781843392231. Published by IWA Publishing, London, UK.

Jens Christian Refsgaard
Department of Hydrology
Geological Survey of Denmark and Greenland (GEUS)
E-mail: jcr@geus.dk

Huub Scholten
Wageningen University
E-mail: huub.scholten@wur.nl

Jeroen P. van der Sluijs
Utrecht University
E-mail: j.p.vandersluijs@uu.nl

Veronique Vandenberghe
E-mail: veronique.vandenberghe@soresma.be

Ann van Griensven
Associate Professor in Environmental Hydroinformatics
Department Hydroinformatics and Knowledge Management
UNESCO-IHE Institute for Water Education
E-mail: a.vangriensven@unesco-ihe.org

Peter A. Vanrolleghem
Canada Research Chair in Water Quality Modelling modelEAU
Département de génie civil et génie des eaux, Université Laval
E-mail: Peter.Vanrolleghem@gci.ulaval.ca

Preface

Water resources planning and management and the development of appropriate policies require methodologies and tools that are able to support systematic, integrative and multidisciplinary assessments at various scales. It also requires the quantification of various uncertainties in both data and models, and the incorporation of stakeholders participation and institutional mechanisms into the various tools and risk assessment methodologies, to help decision makers understand and evaluate alternative measures and decisions. This requirement has been explicitly recognized in the context of the European Union Water Framework Directive (WFD). Designing the Programme of Measures is one of the key elements in building up the river basin management plan as defined by the WFD. The planning process for producing the Programme of Measures consists of several consecutive steps, starting from setting the objectives to the analysis of alternative ways of reaching them. It becomes evident that this is a challenging task which needs support from appropriate information systems and modelling tools that are able to cope with the complexity of the water system and planning process. This is particularly important in the context of current discussions on climate change impacts on the water sectors and actions needed

By Peter A. Vanrolleghem. ISBN: 9781843392231. Published by IWA Publishing, London, UK.

to adapt to those changes. Since changes in the availability of water resources and occurrence of extreme events will affect not only the water supply and sanitation services but also other economic sectors, like agriculture, forestry, tourism, inland navigation, hydropower, health, etc. it is important to prepare a framework which allows policy makers to make integrated assessments.

To respond to this challenge, the European Commission has initiated in the context of the 5th EU Framework Programme for Research and Technological Development (FP5) a large CatchMod initiative. Its objective was to develop and harmonize integrated modelling tools that could be used by water authorities for planning and managing water resources in an integrated way at the scale of river basins. This very ambitious initiative with long-term objectives has been implemented trough a dedicated cluster constituted by a core group of several research projects and coordinated by a concerted action, the Harmoni-CA project. CatchMod and Harmoni-CA addressed several key issues in modelling research, including model concepts and regimes, scale issues and data availability, understanding stakeholders participation, participatory approaches and institutional mechanisms and their incorporation in the various biophysical models and risk assessment methodologies to help decision-makers understand and evaluate alternative approaches, etc. Science-policy integration is another important challenge that has been addressed by the CatchMod initiative and the Harmoni-CA project.

In many instances, the lack of communication and of clear coordination mechanism hinders the research outputs from being used or simply known by policy makers, and hinders policy research needs from being communicated to the scientific communities in a timely fashion. By initiating and facilitating the discussion between the Common Implementation Strategy Coordination Group of the WFD and the various researchers, Harmoni-CA helped to foster synergies between researchers, policy makers and water managers on different levels to achieve a common understanding of the various WFD implementation issues. It also helped consolidate knowledge generated by several EC projects (of the CatchMod cluster and beyond) on various cross-cutting and complementary research issues, with a view of providing technical support for the implementation of the WFD.

This publication is one of the concrete outcomes of the Harmoni-CA concerted action as part of a 4-volume series of Guidance Reports that guide water professionals through the implementation process of the Water Framework Directive, with a focus on the use of ICT-tools (and in particular modelling). They are supposed to be complementary to the Guidance Documents produced by the EU Directorate General for Environment. An initial volume "**Water Framework Directive: Model supported Implementation – A Water Manager's Guide**", edited by Fred Hattermann and Zbigniew W Kundzewicz, provides a practical guidance to water managers and other interested stakeholders on the model-supported implementation of the WFD. The three other volumes bring more detailed guidance regarding a range of aspects of such WFD-implementation. This particular volume 1 focuses on the "**Modelling aspects of WFD implementation**", containing four Guidance Reports: (i) Quality assurance and better modelling practices, (ii) Model calibration and validation, (iii) Sensitivity analysis, and (iv) Uncertainty analysis. The two other volumes cover "**Integrated Assessment for WFD implementation: Data, economic and human dimension**" in five additional Guidance Reports and "**Decision support for WFD implementation**" with two final Guidance Reports. These Guidance Reports were written as part of the work conducted within the different CatchMod cluster projects. Therefore, some of these reports were already finalized in 2005–2006 and report the status at the time. Evidently, research and practice have moved forward, but it is expected that the review and synthesis nature of the Guidance Reports makes that the information they contain is still very actual.

These four volumes provide a good example of disseminating research results of policy relevance. We hope that this publication will reach a broad audience, both science and practice oriented, thus helping to further strengthen the links between scientific research, policy and decision making.

Dr. Panagiotis Balabanis
Principal Scientific Officer
Directorate General for Research
Commission of European Communities

Prof. Peter A. Vanrolleghem, Editor
Current address: modelEAU
Université Laval
Québec (QC) Canada

Guidance Report I.1
Quality assurance in model-based water management: Better modelling practices

Huub Scholten and Jens Christian Refsgaard

Abstract

This Quality Assurance (QA) Guidance Report reviews state-of-the-art Quality Assurance (QA) for model-based water management and provides practical guidance on good modelling practices. Modelling credibility is compromised due to miscommunication, malpractice and overselling model capabilities. The proposed QA guidance is a more specific informal standard on top of ISO 9000. QA practices for model-based water management vary between different modelling domains and countries. QA guidelines can be classified according to how much focus is put on the dialogue between the modelling team and the water manager as Type 1 (internal technical guidelines), via Type 2 (public technical guidelines) to Type 3 (public interactive guidelines). This QA Guidance Report proposes an integrated QA approach with a modelling Knowledge Base (KB), KB-editor, MoST and an organization to maintain and improve modelling KB and supporting software.

A clear problem statement on the need for QA illustrated with examples of poor quality, an introduction of the interaction between the Water Management Process and the Modelling Process and an overview of available QA approaches in a wide range of application domains precede the QA guidance itself. Extensive appendices provide the necessary detail to the interested reader.

Keywords

Modelling credibility, ISO 9000, modelling knowledge base, guidelines, Modelling Support Tool

1. WHY QUALITY ASSURANCE FOR MODEL-BASED WATER MANAGEMENT?

1.1. Problems in modelling

Models describing water flows, water quality, ecology and socio-economics are increasingly being used to support water management functions and decisions. Stakeholders and decision makers are therefore often confronted with results from models whose underlying assumptions they, especially those without modelling background, have difficulty assessing. At the same time insufficient attention is generally given to documenting the predictive

By Peter A. Vanrolleghem. ISBN: 9781843392231. Published by IWA Publishing, London, UK.

capabilities, the uncertainties involved and the limitations of models used (Refsgaard *et al.*, 2005; Jakeman *et al.*, 2006).

As a result of this development many modelling studies have produced results that were not considered credible or understandable by the water managers and stakeholders and that therefore were not utilised. This tendency is reflected in the increasing number of guidelines for good modelling practise that are appearing in these years (Refsgaard *et al.*, 2005). A summary of reasons for the growing interest in Quality Assurance includes:

- Ambiguous terminology and a lack of understanding between key-players (modellers, clients, auditors, stakeholders and concerned members of the public);
- Bad practice (careless handling of input data, inadequate model set-up, insufficient calibration/validation and model use outside of its scope);
- Lack of data or poor quality of available data;
- Insufficient knowledge on the processes;
- Poor communication between modellers and end-users on the possibilities and limitations of the modelling project and overselling of model capabilities;
- Confusion on how to use model results in decision making;
- Lack of documentation and clarity on the modelling process, leading to results that are difficult to audit or reproduce;
- Insufficient consideration of economic, institutional and political issues and a lack of integrated modelling.

1.2. Examples of poor quality of the modelling process

The following subsections provide a number of examples of poor quality experienced in the modelling process. The examples illustrate various aspects of poor quality such as misuse, miscommunication, malpractice and careless handling of data.

1.2.1. Bloopers in modelling

All professional modellers have experienced incorrect applications of models. The following examples originate from the (Dutch) Good Modelling Practice Handbook (Van Waveren *et al.*, 1999).

Blooper: Introducing the 'ditch factor' in the calibration

"In calibration of our nutrients drainage and leaching model, the prediction was not at all in keeping with the measurements. There was a factor 2 difference each time. Upon further analysis, it became apparent that processes take place at the transition from unsaturated groundwater to surface water in particular, i.e. in the ditches and canals, which result in enormous retention. Those processes were not included in our model. We were in a hurry and therefore could not adapt the model. We introduced a 'ditch factor' of 0.5. The model then made perfect sense!"

Blooper: Water sucked out of sea and disappeared up mountain
"We had carried out calibration using a calibration tool. What we hadn't realized however, was that the tool only presented the absolute value of the discharge. The discharge presented therefore said nothing about the *direction* of the flow. Due to considerable extraction somewhere in the middle of the project area, the situation could arise whereby, after calibration, the water was 'sucked' out of the sea, flowed up the mountain and disappeared into the 'extraction point'. The absolute value of the calibrated discharges was pretty much in keeping with the measured values, however! Warning: this error/pitfall could even occur when using a 'normal' presentation tool, because the user can input the positive flow direction per branch/segment of a schematization."

Blooper: Negative concentrations
"Upon completion of a water quality model, the nitrogen concentrations (nitrate and ammonium) in the boundary layer between water and bed showed severe oscillations. Further investigation proved that this happened around the value 0, and that the model in fact calculated negative nitrogen concentrations. This was due to the integration algorithm chosen on the one hand, and the absence of a mechanism to prevent this, on the other hand. When this situation was remedied, the performance of the model became much better."

Blooper: Algae refused to grow
"We had made a model in order to observe planktonic algae into more detail. However, the algae refused to grow. Upon closer consideration, it appeared that the residence time of the lake was less than one day. Consequently, the algae did not have the chance to grow in the model. In a similar model, the algae would not grow because the initial algae concentration was set to zero. You don't have to be a biologist to see that algae will not grow, i.e. multiply when there are no algae in the first place!"

Blooper: Phosphate and phosphor are two totally different things".
"Many errors are made with units of nutrients. In (inorganic) fertilizer, a common mistake is to forget to convert P_2O_5 to P, the same applying to NO_3 to N. This can lead to gross underestimation or overestimation of the results."

1.2.2. Lack of documentation of model capabilities and limitations

The County of Copenhagen was until 2006 the authority responsible for water resources management in the county where the city of Copenhagen abstracts groundwater for most of its water supply. During the last 15 years the use of groundwater models increased steadily. Modelling was performed by consulting firms. Initially modelling was typically performed as a supporting task in connection with many studies of groundwater availability or groundwater protection. When a study was finished the model was terminated and the next study in the same area started modelling more or less from scratch, and often with another consultant. As several modelling studies were reported, it became clear that the predictions of for instance groundwater recharge to a particular aquifer varied substantially from one model study to another. Often the difference was up to a factor of two. As the studies did not report the uncertainties related to their model predictions and did not explicitly report on the limitations on the validity of the model, this obviously resulted in a considerable degree of scepticism and mistrust in modelling from the water authority side.

Another incident added significantly to this general lack of trust in modelling. In 2000 the county asked five groups of Danish consulting firms to conduct studies of the aquifer's vulnerability to pollution in a 175 km^2 area west of Copenhagen, where the groundwater abstraction amounts to about 12 million m^3/year. The key question to be answered was: "which parts of this particular area are most vulnerable to pollution and need to be protected?" The five consultants used significantly different approaches, including many different types of models ranging from very simple index based tools to sophisticated integrated hydrological models based on Modflow or MIKE SHE. The conclusions of the five consultants regarding vulnerability were strikingly different and clearly did not provide a sound basis for deciding anything about which areas should be protected. More details about this case are provided in Refsgaard *et al.* (2006).

1.2.3. Miscommunication and lack of public participation

Blooper: Model was too good

"For years, we had worked with a certain high water model, which we knew to generally predict slightly higher water levels than actually occurred in practice. In time, we produced a new model which predicted the water levels almost exactly. The consequences: a year later there was a flood, with relatively great damage. What had happened? The local water board always assumed that our old model's prediction would be a standard half meter too high. For the sake of convenience, the local water board had also deducted a half a meter from our new prediction. Now that the prediction was accurate, nobody was prepared for the resultant flood." (Example from the Dutch Good Modelling Practice Handbook; Van Waveren *et al.*, 1999.)

Motivation for developing quality assurance guidelines
The background for modelling guidelines for the Murray-Darling Basin in Australia was "a perception among the end-users that model capabilities may have been 'over-sold', and that there is a lack of consistency in approaches, communication and understanding among and between modellers and water resources managers, often resulting in considerable uncertainty for decision making" (Middlemis, 2000). Similar motivation is given in Californian guidelines (BDMF, 2000): "Unfortunately, stakeholders, decision-makers, and even their technical staff, often lose confidence in models because of (1) an inadequate understanding of modelling principles and (2) inconsistencies in the way models are developed and used and (3) a lack of model and data upkeep".

1.3. Context and objectives of this guidance document

The EU Water Framework Directive (WFD) provides European policy at the river basin scale. With requirements imposed by the WFD the trend to base water management decisions to a larger extent on model studies and to use more sophisticated models is likely to be reinforced. Therefore, it is crucial to improve the credibility of modelling as much as possible. One of the obvious means to achieve this is to improve the quality assurance of the modelling process.

It is important to note that the modelling studies typically do not address the entire water resources management decision process, such as the WFD process, but rather support certain elements of the process. The role of modelling as part of the WFD process may be illustrated schematically as in Figure 1.1. The inner circle in Figure 1.1 depicts a simplified version of the WFD planning process (EC, 2004a) with the main elements:

- *Identification* including assessment of present status, analysis of impacts and pressures and establishment of environmental objectives. Here modelling may be useful for example for supporting assessments of what are the reference conditions, for assessment of the pressures and what are the impacts of the various pressures in combination to monitoring data and expert judgement (EC, 2004b).

- *Designing* including the set up and analysis of programme of measures designed to be able in a cost effective way to reach the environmental objectives. Here modelling will typically be used for supporting assessments of the effects and costs of various measures under consideration.
- *Implementing* the measures. Here on-line modelling in some cases may support the operational decisions to be made.
- *Evaluation* of the effects of the measures on the environment. Here modelling may support the monitoring in order to extract maximum information from the monitoring data, e.g. by indicating errors and inadequacies in the data and by filtering out the effects of climate variability.

This main WFD process is a participatory process with important elements of public participation.

Figure 1.1. The role of modelling in the water resources management process within the context of the Water Framework Directive (WFD)

Modelling can be used as a tool at various stages of the WFD process, as illustrated by the four smaller circles in Figure 1.1. The typical steps and elements of a modelling process are illustrated in Figure 1.2. The most important interactions between the modelling process and the main planning process are:

- The modelling process starts with a thorough framing of the problem to be addressed and definition of modelling objectives and requirements for the modelling study (Step 1 in Figure 1.2). Water managers and stakeholders dominate this step, which basically is identical to part of the broader planning process. A participatory based assessment of the most important sources of uncertainty for the decision (WFD) process should be used as a basis for prioritising the elements of the modelling study. The uncertainty assessments made at this stage will typically be qualitative.
- The main modelling itself is composed of Steps 2, 3 and 4 of Figure 1.2. Here the link with the main planning process consists of dialogue, reviews and discussions of preliminary results involving water managers and, according to decisions made at the start of the modelling, some stakeholders. As the modelling process proceeds uncertainty assessments are typically made more quantitative (rather than qualitative) and the uncertainties assessed by the modeller are confronted with the water manager and stakeholder expectations to accuracy.
- The finalisation of the modelling study (equivalent to the last step in Figure 1.2), typically includes scenario simulations. Here the water managers and the stakeholders again have a dominant role. The decisions made at the outcome of this step on the basis of modelling results are made in the context of the main planning process. Uncertainty assessment of model predictions is a crucial aspect of the modelling results and should be communicated in a way that is accessible for the stakeholders in the further WFD process.

The objective of this document is to provide guidance on quality assurance of the modelling process related water resources management. Thus, the document does not focus on quality assurance related to the broader policy and public participation processes. The target audience for the document is professionals involved in modelling. This includes modellers themselves as well as the persons in the water manager's and stakeholders' organisations designated to interact with the modeller in the modelling process.

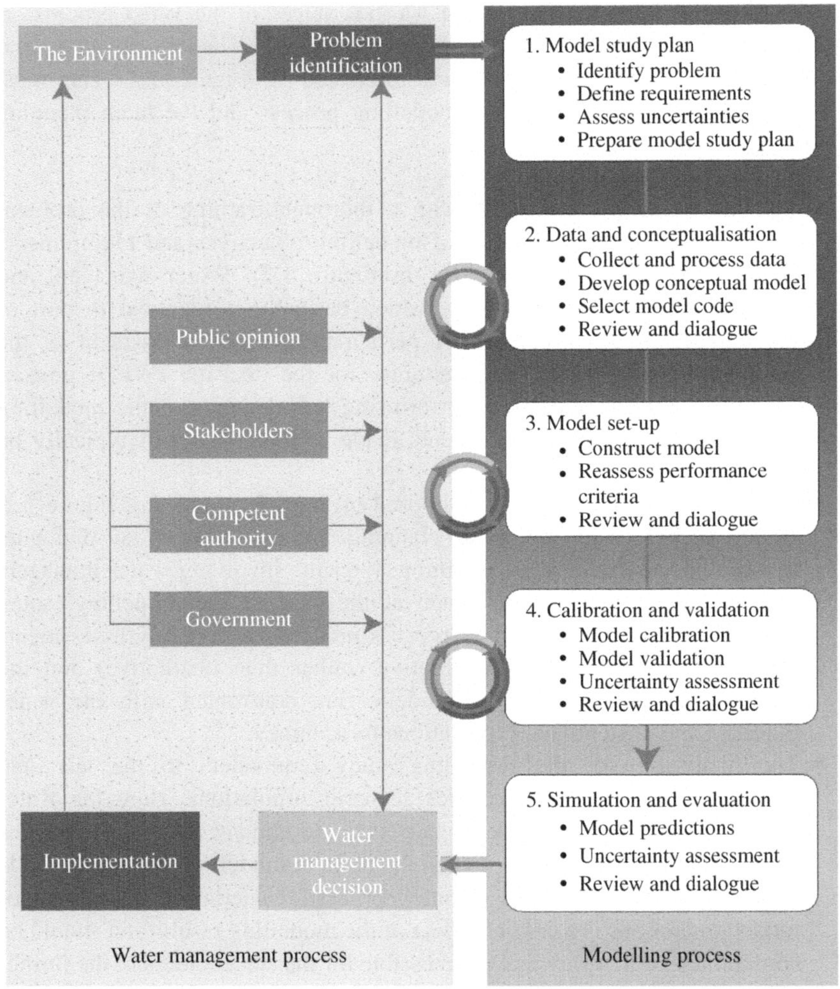

Figure 1.2. The interactions between the water management process (left) and the five steps of the modelling process (right)

2. ON QUALITY ASSURANCE

2.1. On quality

In his benchmark book on quality, Crosby (1979) uses sex as a metaphor for quality:

> *"The problem of quality management is not what people don't know about it. The problem is what they think they do know. . . .*
> *In this regard, quality has much in common with sex. Everybody is for it.* (Under certain conditions, of course.) *Everyone feels they understand it.* (Even though they wouldn't want to explain it.) *Everyone thinks execution is only a matter of following natural inclinations. (After all, we do get along somehow.) And, of course, most people feel that problems in these areas are caused by other people.* (If only they would take the time to do things right.)"

There are many reasons to adopt quality assessment as an integral part of all model-based water management studies (see Chapter 1). Quality is not the result of 'following natural inclinations', but requires extra effort in the course of model development or model use. This demands a lot of discipline and hard work.

In contrast to software development, quality assurance for developing and using simulation models is not common practice. If scientists build such models, they see a model as a research tool and hardly as a product that can be used to support decision making, scenario analysis, planning or policy making. Usually modellers develop a model without considering the possibility that others may apply it. In more technical domains (e.g. hydrology) there is a tendency to improve and evaluate model quality. In domains with less hard ('grey') models (e.g. ecology, environmental science, socio-economics), which are based on incomplete, controversial theories and hypotheses, quality assessment is not a modeller's every day routine (Scholten and Udink ten Cate, 1999).

2.2. Quality approaches in a historical perspective

In the early days of manufacturing, products of work were inspected and a decision was made whether to accept or reject it. In the beginning, these inspections were done by inspectors, which were also production workers. Later, separate inspection departments were established, with trained inspectors, standards, recording of inspection data and predefined accuracy and fault tolerances. These quality control approaches were mainly focused on *detection of defects*. In the 1920's a statistical theory was developed and applied to control quality. Statistical Process Control appeared to be effective in monitoring a process with control charts. By collecting readily available data from samples, variations in the process that may affect the quality of the end product or service can be detected and corrected before any waste is produced. A second approach focuses on *prevention of defects* and this is more related to quality assurance by controlling the process.

The lead in quality efforts switched from USA to Japan and back to the USA. From the cross-fertilization between American and Japanese approaches a new approach emerged, called TQM or Total Quality Management. The International Organization for Standardization (ISO) defines Total Quality Management (TQM) as a management approach for an organization, centred on quality, based on the participation of all its members and aiming at long-term success through customer satisfaction, and benefits to all members of the organization and to society. In Japan, TQM comprises four process steps[1], namely:

(1) *Kaizen* – Focuses on *continuous process improvement*, to make processes visible, repeatable and measurable.
(2) *Atarimae Hinshitsu* – The idea that *things will work as they are supposed to* (for example, a pen will write).
(3) *Kansei* – Examining the way the user applies the product leads to improvement in the product itself.
(4) *Miryokuteki Hinshitsu* – The idea that *things should have an aesthetic quality* (for example, a pen will write in a way that is pleasing to the writer).

TQM requires that organisations maintain this quality standard in all aspects of its business: things have to be done right the first time and defects and waste have to be avoided. We will give some more explanation in order to comprehend the scope of TQM, which is – together with ISO standards – a leading quality assurance paradigm. The term *management* in TQM includes planning, organizing, directing, control, and assurance. Furthermore, *total* in *total quality* is called *total* because it consists of 3 qualities: *quality of profit* to satisfy the needs of the shareholders, *quality of products and services* to satisfy specific needs of customers (consumer, therefore end-customer) and *quality of life* – at

[1]Wikipedia: http://en.wikipedia.org/wiki/Total_Quality_Management.

work and outside work – to satisfy the needs of employees in the organisation. Total Quality Management goes beyond ISO Standards series, which have been established after World War II.

Quality in software development got a boost by Philip B. Crosby's Quality Management Maturity Grid QMMG, an organizational maturity matrix (Crosby, 1979). QMMG's successor is Watts Humphrey's *Capability Maturity Model* or CMM (Humphrey, W.S. 1989). The *CMM* was initially funded by military research. The United States Air Force funded a study at the Carnegie-Mellon Software Engineering Institute to create an abstract model for the military to use as an objective evaluation of software subcontractors. In the 1990's attempts were made to translate CMM, developed for software engineering) to a modelling version (Scholten and Udink ten Cate, 1995, 1999).

2.3. Quality assurance and ISO

2.3.1. The International Organization for Standardization (ISO)

ISO or the International Organization for Standardization is the world's leading standardization organisation. ISO standards specify requirements for products, services, processes, materials and systems and for good practice of organizational management. ISO's members are standardisation and normalisation organisations in most countries in the world. Besides ISO there are other international standardisation organisations, e.g. CEN (Committee European the Normalisation or European Standardisation Organization), IEC (International Electrotechnical Commission) and the ITU-T (International Telecommunication Union).

ISO has a current portfolio of more than 16000 standards that provide practical solutions and achieve benefits for almost every sector of business, industry and technology. ISO's standards cover agriculture, construction, mechanical engineering, manufacturing, distribution, transport, medical devices, and information and communication technology, but also standards for services.

2.3.2. ISO standards history

Errors committed in manufacturing have been an impetus for creating quality standards. For example, in World War I, a high percentage of shells failed to explode. This was traced to different definitions of an inch by the two major UK armaments manufacturers, leading to calibration standards. In World War II, the United Kingdom had serious problems with accidental detonations in weapons factories and other manufacturing flaws. In an attempt to solve the problem, the Ministry of Defence placed inspectors in the factories to oversee the production process; to supply to the Government, a company had to write up the procedure for making their product, have the procedure approved by the Ministry and ensure their workers followed the procedure. Parallel to this, the United States developed requirements for contractors for shell, aircraft and missile suppliers during and shortly after World War II. In 1947 ISO was founded as an organization to develop and disseminate world-wide industrial and commercial standards. In 1959, the United States established Quality Program Requirements, MILQ-9858, a quality standard for military procurement, detailing what suppliers had to do to achieve conformance. By 1962, NASA had similarly developed Quality System Requirements for its suppliers. In 1963 MILQ-9858 was slightly improved (MILQ-9858 A). In 1968, NATO adopted the AQAP (Allied Quality Assurance Publication) specifications for the procurement of NATO equipment[2]. In 1979 British Standard Institute adopted these NATO standards and adjusted it for civilian use, i.e. BS 5750.

In 1987 ISO published an adapted BS 5750 as ISO 9000. New releases of ISO 9000 have been published since 1987 (e.g. 1994, 2000). ISO 9000 currently includes three quality standards: ISO 9000:2005, ISO 9001:2000, and ISO 9004:2000. ISO 9001:2000 presents requirements, while ISO 9000:2005 and ISO 9004:2000 present guidelines. In 2008 a new version of ISO 9000 will be published. All of these are process standards and do not apply to products.

2.3.3. *ISO families of standards*

The ISO 9000 and ISO 14000 families[3] of standards are widely known. ISO 9000 has become an international reference for quality requirements in business to business dealings and ISO 14000 helps organizations to meet their environmental problems. ISO 9000 and ISO 14000 families are known as *generic management system standards. Generic* means, that they can be applied to all kind of organizations, for many types of products (even if a product is actually a service), in many sectors, and for various kind of organizations, including business enterprises, public administrations, government departments and research institutes. *Management system* refers to what the organization does to manage its processes, or to activities.

In the next sections ISO 9000 will briefly be discussed (Section 2.3.4) and its merits for modelling evaluated (Section 2.3.5).

---

[2]AQAP-1: *NATO Requirements for an Industrial Quality Control System* and AQAP-2: *Guide for the Evacuation of a Contractor's Quality Control System for Compliance with AQAP-1.*

[3]See http://iso.nocrew.org/iso/en/aboutiso/introduction/index.html.

2.3.4. *ISO 9000 quality standards in general*

The ISO 9000 family of standards is initiated and maintained by ISO's TC176 (Technical Committee on *quality management and quality assurance*) and is administered by accreditation and certification bodies. ISO 9000 includes the following standards[4]:

- *ISO 9000:2005 Quality management systems – Fundamentals and vocabulary*: covers the basics of what quality management systems are and also contains the core language of the ISO 9000 series of standards.
- *ISO 9001:2000 Quality management systems – Requirements*: is intended for use in any organization which designs, develops, manufactures, installs and/or services any product or provides any form of service. It provides a number of requirements which an organization needs to fulfil if it is to achieve customer satisfaction through consistent products and services which meet customer expectations. This is the only implementation for which third-party auditors may grant certifications.
- *ISO 9004:2000 Quality management systems – Guidelines for performance improvements*: covers continual improvement. This gives advice on what you could do to enhance a mature system. This standard very specifically states that it is not intended as a guide to implementation.

ISO 9001:2000[5] includes[6]:

- A set of procedures that cover all key processes in the business;
- Monitoring processes to ensure they are effective;
- Keeping adequate records;
- Checking output for defects, with appropriate corrective action where necessary;
- Regularly reviewing individual processes and the quality system itself for effectiveness;
- Facilitating continual improvement.

Although the ISO 9000 way of working, i.e. defining and following guidelines from process definitions, monitor processes and keeping records and checking products for defects, will lead to some quality level of processes and

---

[4]Wikipedia: http://en.wikipedia.org/wiki/ISO_9000.
[5]ISO 9001:2000 includes the previous standards ISO 9002 and ISO 9003.
[6]Wikipedia: http://en.wikipedia.org/wiki/ISO_9000.

the resulting products, the basic idea is that the ISO 9000 is rather abstract and universal, i.e. applicable for a too wide range of companies and organizations. Therefore, various industry sectors have developed their own interpretations of the ISO 9000 standards, on top ISO 9000, with more specific requirements and more appropriate expert guidance. Specific guidance has been developed for a range of sectors (e.g. AS 9000 for aerospace manufacturers, PS 9000 for pharmaceutical packaging materials, QS 9000 for automotive manufacturers, TL 9000 for the Telecom Industry, TickIT for software development in UK, ISO/IEC 25000:2005 for Software product Quality Requirements and Evaluation (SQuaRE), ISO 13485:2003 for the medical industry's equivalent). These specific guidance/standards are therefore a particularization of ISO 9000 standards.

2.3.5. Are ISO 9000 quality standards sufficient for modelling?

The available quality standard, ISO 9000 and those built on top of the ISO 9000 family, are not focused on modelling and simulation. They are not sufficient for modelling and simulation, because they are not specific enough (ISO 9000) or focussed on different fields of application. The approach followed in this QA guidance document for model based water management matches with the basic ideas of ISO 9000, i.e. describing what to do (process definition), monitoring and recording what is done, comparing what is done with what should have be done and testing for satisfactory (intermediate) products and results (assessing the soundness of the conceptual model, calibrated model, validation and uncertainty analysis of scenario runs), are to some extent part of the modelling process and at the same time aiming at finding flaws in resulting products of the modelling process. The approach followed here combines testing products, e.g. model validation with a process approach.

ISO 9000 is concerned with *quality management*, i.e. what organizations do to enhance customer satisfaction by meeting customer and other relevant requirements and to improve their performance in this regard. ISO 9000 defines Quality Assurance as *'part of quality management focused on providing confidence that (the customer's) quality requirements will be fulfilled'*. ISO 9000 '*does not establish requirements for products*', as these '*can be specified by customers or by (an) organization in anticipation of customer requirements, or by regulation. The requirements for products and in some cases associated processes can be contained in, for example, technical specifications, product standards, process standards, contractual agreements and regulatory requirements*'.

Therefore, extra guidance for model-based water management, which is more specific, detailed and includes state-of-the-art expertise, fits perfectly in ISO 9000's Quality Assurance definition.

2.4. On standards and guidelines

2.4.1. Informal standards versus formal standards

International, European, and National standards are formal standards, established by consensus to provide a common set of rules or guidelines for specific products or activities. They are usually voluntary standards, but are widely adopted by producers to improve interoperability and to satisfy the perceived or explicit wishes of the customer (Packman and Old, 2005). Some standards may be enforced by laws, regulations, or directives, usually as an approved means of achieving compliance. Formal standards are usually drafted by representative committees of experts and volunteers (to allow all relevant groups to participate), and then put forward by the standards authorities for public debate. However, individual businesses, associations, organisations, or 'workshops' can develop their own informal standards, often with the support of the standards authorities. These standards may also be adopted by businesses or specified by customers. Within Europe, in order to implement a directive, the European Commission may 'mandate' the European Committee for Standardisation (CEN) to develop a voluntary 'Harmonised' standard to be adopted by each member country. A brief outline of the procedure (typically lasting three years) for drafting a formal European Standard is given in CEN (2004), and the procedure for developing an informal CEN Workshop Agreement is given in CEN (2005).

The main difference between formal standards on the one hand and informal standards or guidelines on the other hand is the way in which formal standards are publicly debated. Informal standards or guidelines are discussed within a smaller group and can best be seen as the consensus within this community.

2.4.2. ISO and modelling standards

Many commercial modelling contracts already require evidence of a good Quality Management structure such as that described in the ISO 9000 family of standards (Packman and Old, 2005). ISO 9000 does not describe *Quality Assurance Standards* per se (the phrase in the title of the 1994 edition has been dropped), but rather a generic Quality Management system that:

- Defines responsibilities;
- Analyses customer requirements;
- Defines and controls the processes to achieve those requirements;
- Manages the resources necessary;
- Assesses effectiveness within each individual process – in order to continually improve the product and increase customer satisfaction.

Packman and Old (2005) state that in all three definitions of Quality Assurance (of ISO 9000, discussed in Section 2.3.5 and of NRC (NRC, 1990) and HarmoniQuA (Scholten, 2006), discussed in Section 3.1), are management standards (defining the responsibilities, procedures and protocols that should be followed) and product standards (defining the required technical specification and relevant regulations). For a manufactured product, functionality and serviceability may be the prime product standards, based on a clear understanding of the processes, stresses and failure modes involved, and with the opportunity to test product samples to failure. A river basin model, however, is likely to be a 'one off' product, in which (1) the understanding of processes, their complexity and interactions, is incomplete and must be inferred (especially under extreme conditions), and where (2) failure is a relative and ill-defined concept. Functionality (based on proposed usage) and accuracy may be the prime product requirements, but it can be difficult to define an appropriate accuracy at the study outset, and to evaluate it subsequently.

Packman and Old (2005) remark, that numerical accuracy is a prime requirement, but it can be expressed in many ways and over many features (mean/maximum/annual values of water levels/flows/pollutants, rates of change, correlations, during rising or falling flows, etc.) and may be affected by the availability and accuracy of measurement data (system inputs and outputs, presence/absence of trends or extremes such as floods/droughts, point/spatial errors, etc.). The choice may depend on the study objectives (customer requirements) rather than any product standards, and may differ between domains in a multi-disciplinary study. In practice, a number of accuracy criteria may be specified, including some of a more subjective nature, and it may be technically or practically impossible to meet all of them – though this may only become clear as the study progresses. Alternatively, a generalised accuracy criterion could be met, but the model will still behave poorly under specific conditions of interest.

Packman and Old (2005) conclude that these aspects of model-based water management require translating product standards into procedures (1) to ensure that an appropriate model is chosen and properly applied, and (2) to test the model limitations, inaccuracies and uncertainties in order to understand and minimise their impact. These activities are already components of the ISO 9000 Quality Management System, so establishing a formal or informal (guideline-like) set of standards for model-based water management may be viewed as providing sector specific guidance on applying ISO 9000 to model-based water management. Specific guidance has already been developed for a range of sectors (see Section 2.3.4).

3. STATE-OF-THE-ART QA FOR MODEL-BASED WATER MANAGEMENT

3.1. Quality assurance defined for modelling

Quality assurance (QA) is defined by the National Research Council, USA (NRC, 1990) as *the procedural and operational framework used by an organization managing the modelling study to assure technically and scientifically adequate execution of all tasks included in the study and to assure that all modelling-based analysis is reproduced and defensible.*

In line with this within the HarmoniQuA project QA is defined as *the procedural and operational framework used by an organization managing the modelling study to build consensus among the organizations concerned in its implementation, to assure technically and scientifically adequate execution of all tasks included in the study, and to assure that all modelling-based analysis is reproducible and justifiable* (Scholten, 2006). This modification of the definition of the National Research Council (NRC, 1990) includes the organizational, technical and scientific aspects, but in addition the need to build consensus among the organizations concerned.

It is clear from these definitions that the QA for modelling includes the modelling products such as the conceptual model, the model code, the calibrated and validated site-specific model and the model scenario runs as well as the entire modelling process with its interactions between modeller, water manager and stakeholders.

3.2. Guiding principles – three different approaches

Different approaches have been used to ensure good quality in water resources modelling. Inspired by de Groof *et al.* (2005) and Hansen and Grøn (2006) the approaches may be categorised according to three different guiding principles:

- *Regulation driven*. Performance is regulated by detailed standards on how to perform a study. These regulations often combine technical guidance and policy governed instructions linked to certain legislation. The regulations are typically issued by public authorities and have a status that makes it compulsory to subscribe to them. So this principle results in a centralistic top-down approach. The advantages of such regulations are that they often provide a transparent and operational set of rules for how to perform a given task for water managers, stakeholders and consultants. The disadvantages are that to provide detailed descriptions of what to do in all possible cases they become rather resource demanding and rigid. This approach is typically adopted for work carried out within public authorities. An example of regulatory driven guidelines is the UK Flood Estimation Handbook (IH, 1999; Packman, 2002) that provides detailed prescriptions on how to use various models to perform flood estimation. Another example is the set of Common Implementation Strategy (CIS) guidance documents aiming at supporting the WFD implementation (EC 2004a; EC 2004b). These CIS guidance documents are, however, broader than modelling and not very detailed, so they leave room for many different interpretations about how to do things and should therefore be combined with other approaches to assure good quality.
- *Expertise driven*. Performance is here governed by knowledge and experience from experts. This will result in delegation of modelling tasks to modelling specialists, who have documented experience and a well proven positive track record. In its ultimate way it may result in certification of certain tasks to individuals or organisations. This is a decentralised approach where the responsibility is entrusted to specialists. The advantage of this approach is that it allows knowledge and experience to be utilised in a flexible manner. The major disadvantage is that it is subjective in nature. Different experts may have different experiences and assessments, so the procedure is usually not reproducible and transparent. This approach is typically used in connection with consultancy services, where the client fully trusts the consultant.
- *Quality driven*. Modelling is here performed according to detailed technical guidelines, including a documentation of the quality of work. Examples of such guidelines are norms and standards. Such standards can ensure that all the important elements are considered. Furthermore, it may ensure that the study is carried out within a formal framework involving the entire organisation. The major disadvantage is that the standards can be very detailed and the formal documentation of quality typically requires a lot of bureaucratic work and hence considerable resources.

3.3. Type of QA guidelines for modelling

3.3.1. Classification

QA in the modelling process has two main components: (a) QA in development of model codes and (b) QA in relation to application studies. The present guidance document focuses on the second component only. QA in model application studies includes data analyses, methodologies of good modelling practice, reviews and administrative procedures. Such QA guidelines can be classified according to how much focus is put on the consensus building process between the modeller and the water manager in the following three classes (Refsgaard *et al.*, 2005):

- *Internal technical guidelines* (*Type 1*) established and used internally by the modeller's organisation.
- *Public technical guidelines* (*Type 2*) established as public guidelines and used internally by the modeller's organisation.
- *Public interactive guidelines* (*Type 3*) established as public guidelines and based on regulation of the interaction between the modeller and the water manager throughout the modelling process.

If we refer to the three guiding principles (Section 3.2) both the internal and the public technical guidelines are typically quality driven, while the public interactive guidelines also to some extent are expertise driven.

Type 1: Internal technical guidelines

Most organisations involved in modelling studies have some kind of internal QA procedures. They usually focus on the technical aspects, i.e. to ensure that the modelling work itself is done without making unqualified judgements or errors. The betters of these are based on the modelling protocols and similar scientifically based procedures originating from the research community. These procedures are internal in nature because they have been established or adopted unilaterally by the modeller's organisation, and because they seldom deal with the interaction between modeller and end-user. Examples of Type 1 guidelines include:

- Internal QA procedures, common in many companies.
- Text books. Many textbooks contain chapters with recommended modelling protocols.
- Manuals to software packages with hints on the best way to use a model.

Type 2: Public technical guidelines
These guidelines often contain the same substance as the internal technical guides mentioned above. However, they differ in the sense that they have been prepared through a consultative and consensus building process involving many persons and organisations. They focus on the technical aspects and give no or little emphasis to the interaction between the modeller and the end-user. Examples of Type 2 guidelines include:

- Standards from American Society for Testing and Materials (e.g. ASTM, 1994).
- The USGS guidelines for evaluating of ground-water flow models (Reilly and Harbaugh, 2004).

Type 3: Public interactive guidelines
These guidelines have, like the public technical guidelines (Type 2), been established through a public consultative and consensus building process. However, they differ from the Type 2 guidelines by an additional focus on regulating the interaction between the modeller and the water manager, who often have the roles of consultant and client, respectively.

Important elements in public interactive guidelines are reviews that, in addition to QA in the sense of technical guidance, can facilitate the consensus-building process between the parties. Experience shows that such a process is crucial for the overall credibility of the modelling process. Examples of such QA guidelines include (more details on these guidelines provided in Section 4.4:

- The Dutch guidelines (Van Waveren *et al.*, 1999; Scholten and Groot, 2002).
- The Australian groundwater flow modelling guidelines established by the Murray-Darling Basin Commission (Middlemis, 2000).
- The Danish groundwater modelling guidelines (Sonnenborg and Henriksen, 2005).
- Californian guidelines prepared by Bay-Delta Modelling Forum (BDMF, 2000).

3.3.2. Development stage and prevalence of QA guidelines

Reviews of a number of existing QA guidelines (Refsgaard *et al.*, 2005) revealed significant differences in current practice, both between domains and between different countries. In some domains and some countries there has been a clear trend over the past couple of decades to move from Type 1 to Type 2 or Type 3 guidelines. Refsgaard *et al.* (2005) hypothesised that the stage of QA guideline development largely depends on the maturity of both the scientific discipline and the modelling market in the respective country or region.

3.4. Existing guidelines

Table 3.1 contains a list of some of the more recent guidelines with an indication of some of their respective characteristics.

3.5. Discussion of QA in water resources modelling

3.5.1. Key aspects in QA guidelines

Guiding principles

Quality assurance procedures are composed of a mix of the three guiding principles outlined in Section 3.2. For instance regulatory driven guidelines often prescribe use of (quality driven) standards for some of the detailed regulation. The guidelines listed in Table 3.1 are mostly quality driven, but with important expertise driven elements such as peer reviews and stakeholder interaction. As these guidelines are often used within a regulatory context, such as implementation of the WFD, the quality assurance in reality becomes a mix of all three principles: (a) they are based on quality driven guidelines providing recommendations on how to perform the technical details and how to manage the interaction between modeller, water manager and stakeholders; (b) they include subjective expertise in terms of peer reviews performed by independent external experts; and (c) they are most often carried out within the context of legislation prescribing that the work has to be carried out within the framework of certain regulatory guidelines.

A particular problem with some of the regulatory driven guidelines is that they are prescriptive with respect to methodologies including some times even which model code to use. As new knowledge appears the old methodologies and codes will inevitably become outdated. Therefore in order for such guidelines to be in accordance with state-of-the-art it is required that they regularly be updated.

Interactive guidelines

The dialogue between the different players is crucial to ensure that the output from the modelling process is understandable for stakeholders and beneficial for the client. The importance of involvement of stakeholder and public opinions is emphasised by Pahl-Wostl (2002) and Jakeman *et al.* (2006) and addressed in many of the guidelines listed in Table 3.1. In HarmoniQuA (Chapter 4), each of the five major steps (Figure 4.3) is therefore concluded with a dialogue task, in terms of either contract negotiation (first step) or reviews (last four steps). A dialogue task encourages the assessment of the present step and provides the opportunity to redefine the content of the model study plan for the next step based upon the results and findings of the present step. These dialogue steps provide flexibility to the modelling study and ensure that the tasks that have yet to be performed can be modified according to the achieved results and perceptions of modeller and client.

Peer review
According to our experience the single most important instrument to ensure a good quality of the modelling process is to perform independent reviews of the modelling process. Like in the scientific tradition, peer review of the modelling work is the best way of evaluating the quality. Peer review is based on the principle of 'expertise driven' (Section 3.2) and is subjective by nature, which may be considered a weakness as compared to the more objective 'quality driven' guidelines with step by step descriptions of what to do. However, if we were not allowing the subjective element, we would deprive the process of the knowledge of experienced professionals and such knowledge is virtually impossible to convert to guideline text, because each modelling case is unique, while the guidelines have to be generically formulated. By combining the generic guidelines and the peer review we believe that we achieve the optimal mix of quality and expertise driven QA. Most of the guidelines in Table 3.1 emphasise peer review.

Transparency and reproducibility
Transparency and reproducibility are important, especially for large studies involving use of complex models. This is recommended in most guidelines. However, to efficiently implement this requires a supporting tool which enables modelling teams, consisting of modellers, managers and auditors, to be guided through the modelling process, to monitor all modelling activities and to oversee the status of each task to perform. With an increasing tendency to reuse existing models or rebuild them with additional data, modified conceptual models (revised model structure and/or inclusion of additional processes) and improved calibration and validation tests, such functionality becomes very important. To our knowledge the only existing tool dedicated for this purpose existing today is MoST (see Section 4.5).

Table 3.1. Key characteristics of some selected QA modelling guidelines

Title used in this report	Developer	Domain	Technical or Interactive	Peer review (Note 1)	How is uncertainty accounted for			Model validation (Note 1)	Format	Supporting tools	Reference
					Where (Note 2)	Qualitative aspects?	Model structure (Note 3)				
ASTM	American Society for Testing and Materials	Groundwater (other domains also covered, but not reviewed here)	Technical	–	**	No	*	*	A number of separate guides	None	E.g.: ASTM (1992) ASTM (1994) ASTM (1996)
NL GMP	More than 10 organisations in NL	Generic + all water domains	Interactive	–	**	No	–	**	Report	None	Van Waveren *et al.* (1999)
Australian	Murray-Darling Basin Commission, Australia	Groundwater flow	Interactive	***	***	No	**	**	Report	Scoreboard and templates	Middlemis (2000)
Bay-Delta	Bay-Delta Modeling Forum, California	Generic (based on surface water experience)	Interactive	***	*	No	*	**	Report	None	BDMF (2000)
German	German Geological Society	Groundwater	Interactive	–	–	No	*	**	Report	None	German Geological Society (2001)

(*continued*)

Table 3.1. *Continued*

Title used in this report	Developer	Domain	Technical or Interactive	Peer review (Note 1)	How is uncertainty accounted for			Model validation (Note 1)	Format	Supporting tools	Reference
					Where (Note 2)	Qualitative aspects?	Model structure (Note 3)				
CREM	US EPA	Generic	Interactive	***	***	Yes	**	**	Report	None	Pascual *et al.* (2003)
USGS	US Geological Survey	Groundwater	Technical	–	*	No	*	–	Report	None	Reilly and Harbaugh (2004)
DK-GW	GEUS	Groundwater	Interactive	***	***	Yes	***	***	Report	None	Sonnenborg and Henriksen (2005)
MoST	HarmoniQuA (EU RTD)	Generic + all water domains	Interactive	***	***	Yes	***	***	Built into software tool	Comprehensive (MoST)	Refsgaard *et al.* (2005); Scholten *et al.* (2007)
Jakeman EMS	Jakeman *et al.*	Generic	Interactive	*	**	Yes	**	**	EMS Position Paper	None	Jakeman *et al.* (2006)

Note 1 – peer reviews/validation

– No/not considered
* Mentioned/vaguely described
** Accounted for explicitly or implicitly
*** Strong emphasis

Note 2 – where/how is uncertainty considered

– Not at all
* Vaguely, but no explicit consideration
** At the end/explicit
*** Throughout the study/comprehensive

Note 3 – model structure uncertainty

– Not considered at all
* No, but focus on conceptual models
** Considered
*** Strong emphasis

Accuracy criteria

Establishment of accuracy criteria for a modelling study is a very important, but difficult, issue. Modellers often establish numerical accuracy criteria in order to classify the goodness of a given model (e.g. Henriksen *et al.*, 2003; Scholten and Van der Tol, 1998). These attempts are very useful in making the performance more transparent and quantitative, but do not provide an objective means to decide what the optimal accuracy criteria really should be in a given case. According to Refsgaard and Henriksen (2004) no universal accuracy criteria can be established, i.e. it is generally not possible from a natural scientific point of view to tell when a model performance is good enough. Such acceptance criteria will vary from case to case depending on the socio-economic context, i.e. what is at stake in the decisions to be supported by the model predictions. An appropriate question may be: how do we translate the 'soft' socio-economic objectives to 'hard-core' model performance criteria? This is obviously a challenge that can not be solved by natural science alone, but needs to be addressed in a much broader context including aspects of economy, stakeholder interests and risk perception.

Performance statistics must comprise quantifiable and objective measures. However numerical measures can not stand alone. Often expert opinions are necessary supplements.

Uncertainty assessments

Quality assurance and uncertainty assessments are two aspects that are very closely linked (Refsgaard *et al.*, 2007). Initially, the manager has to define accuracy criteria from a perception of which uncertainty level he/she believes is suitable for a particular case (see *accuracy criteria* above). Subsequently, as the modelling study proceeds, the dialogue between modeller and manager has to continue with the necessary trade off between modelling accuracy and the cost of the modelling study. In the uncertainty assessments it is very important to go beyond the traditional statistical uncertainty analysis. Thus, e.g. aspects of scenario uncertainty and ignorance should generally be included and in addition the uncertainties originating from data and models often need to be integrated with socio-economic aspects in order to form a suitable basis for the further decision process (e.g. Van Asselt and Rotmans, 2002). Uncertainty assessment is included in all guidelines included in Table 3.1. However, most of them only consider uncertainty assessment as something that needs to be done late in a modelling study together with the scenario simulations. The guidelines which most strongly emphasise that uncertainty should be on the agenda from the very first day and throughout a modelling study are HarmoniQuA's modelling guidelines.

Assessment of uncertainty due to errors in the model structure is a particularly difficult task and is most often neglected. One way of evaluating this source of uncertainty is through the establishment of alternative conceptual models (Refsgaard *et al.*, 2006). Among the guidelines listed in Table 3.1 above this aspect is most strongly emphasised in HarmoniQuA's modelling guidelines.

Model validation
Although experience shows that models generally perform poorer in validation tests against independent data than they do in calibration tests, model validation is in our opinion a neglected issue, both in many modelling guidelines and in the scientific literature. Maybe many scientists have not wanted to use the term validation due to the scientific philosophically related controversies, but in any case many scientists are not advocating the need for model validation. One of the unfortunate consequences of this 'lack of interest' is that not much work has been devoted to developing suitable validation test schemes since Klemes (1986). In our opinion further development of suitable testing schemes, particularly for non-linear models and for applications comprising extrapolations beyond the calibration data basis, is a major future challenge.

Supporting software tools
Modelling is a complex process and QA is far from trivial. Therefore there is an obvious need for software tools to support QA of the modelling process. Such support is first of all required to document what is going on in the modelling process, so that the transparency and reproducibility can be improved. To our knowledge the only existing software tool dedicated for this purpose existing today is MoST.

3.5.2. Organisational requirements for QA guidelines to be effective

Modelling studies involve several parties with different responsibilities. The key players are modellers and water managers, but often reviewers, stakeholders and the general public are also involved. To a large extent the quality of the modelling study is determined by the expertise, attitudes and motivation of the teams involved in the modelling and QA process.

QA will only be successful if all parties actively support its use. The attitude of the modellers is important. NRC (1990) characterises this as follows: "most modellers enjoy the modelling process but find less satisfaction in the process of documentation and quality assurance". Scholten and Groot (2002) describe the main problem with the Dutch Handbook on Good Modelling Practice as "they all *like* it, but only a few *use* it". The water manager, however, has a particular responsibility, because he/she has the power to request and pay for adequate QA in modelling studies. Therefore, QA guidelines can only be expected to be used in practice if the water manager prescribes their use. It is therefore very important that the water manager has the technical capacity to organise the QA process. Often, water managers do not have individuals available with the appropriate training to understand and use models. An external modelling expert should then be sought to help with the QA process. However, this requires that the manager is aware of the problem and the need.

4. MODELLING KNOWLEDGE BASE AND SUPPORT TOOL

4.1. QA in the HarmoniQuA approach

The HarmoniQuA project[7] aimed at developing a common, generic methodology, terminology and guidelines to support multidisciplinary model-based water management by developing a modelling Knowledge Base (modelling KB), building a Modelling Support Tool (MoST), testing these products, extending the project results to the wider professional community and raising awareness to use the developed methodology and tool to facilitate quality assurance and to make modelling projects more transparent (Refsgaard and Henriksen, 2004; Refsgaard *et al.*, 2005; Scholten, 2006; Scholten *et al.*, 2007)[8]. The HarmoniQuA approach is built on two pillars. The first pillar is a body of explicit knowledge of modelling expertise, shared by a substantial part of the modelling community. It is organised in a modelling Knowledge Base with prescriptive guidance on how to model. The second pillar comes from the human dimension and consists of human intellect, expertise and practices within the modelling team.

The general concept of modelling support in the HarmoniQuA context is summarised in Table 4.1 and Figure 4.1.

---

[7]HarmoniQuA is an acronym for *Harmonizing Quality Assurance in model-based catchment and river basin management*. From 1 January 2002 to 31 December 2005, this research project was supported by the European Commission under the Fifth Framework Programme and contributed to the implementation of the Key Action "Sustainable Management and Quality of Water" within the Energy, Environment and Sustainable Development Programme. Contract n°: EVK1-CT-2001-00097.

[8]See for a complete list of publications: http://harmoniqua.wau.nl/public/Product/papers.htm.

Table 4.1. General concept of modelling support for water management, developed in HarmoniQuA

Modelling support item	*Resulting product*
1. Develop modelling guidance for model-based water management, based on consensus expertise (knowledge) of a representative group of experts and professionals.	Consensus knowledge
2. Organise this modelling expertise in a Knowledge Base to facilitate updating and maintenance.	Knowledge Base
3. Provide the consensus knowledge, stored in the Knowledge Base, as guidance in modelling projects.	Modelling guidance
4. Monitor what actually is done in modelling projects and store this for reuse by others on the internet/LAN.	Modelling journal
5. Help managing modelling projects.	Tool to support managing model-based water management projects
6. Help generating reports for various purposes and audiences.	Reports of model-based water management projects

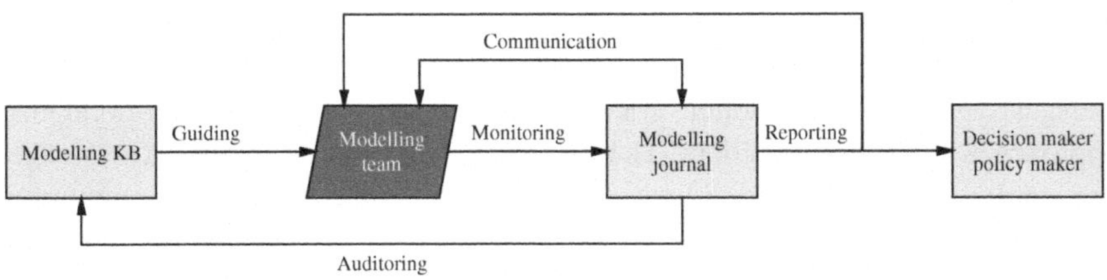

Figure 4.1. Overall set-up of modelling support for water management, developed in HarmoniQuA

The integrated approach, proposed in this chapter assumes that applying the outlined approach will reduce several of the modelling problems, summarised in Chapter 1, resulting in:

(1) Support for better communication and mutual understanding of all persons involved with a glossary for model-based water management and a clear, well, defined modelling process definition, useful to provide guidance;
(2) Guidance for better modelling practices to avoid malpractice with regard to data handling, model set-up, calibration, validation and scope of applicability;

(3) Sufficient documentation and transparency of modelling projects;
(4) A more structured project from a client (water manager) perspective, as its process is well-defined and widely accepted;
(5) Support of integrated model-based water management projects, including socio-economic aspects;
(6) Modelling projects that can easily be audited and reconstructed.

In addition to the above outlined approach, the following typical setting of modelling projects is assumed:

(1) Water management problems to be solved with models are typically *multidisciplinary* (or multi-domain)[9];
(2) Instead of a single problem solver or modeller, modelling projects are carried out by *modelling teams*, consisting of:
   a. *Water manager*, i.e. client;
   b. One or more *modellers*;
   c. *Auditor* (internal, external);
   d. *Stakeholders*;
   e. *Public*.
(3) *Team cooperation* is partly in real and partly virtual, using the internet or a LAN;
(4) In this team cooperation the interaction (in real or virtual) between team members and especially between water manager and modellers is emphasised.

4.2. Design considerations of modelling KB and MoST

The modelling Knowledge Base (KB) and Modelling Support Tool (MoST) are designed according to a set of functional requirements in interaction with modelling experts, knowledge engineers and software developers. Details of the requirements and how they are aimed to be met in the design are given in Appendix A.

The KB should support cooperation in multidisciplinary teams. This has been realized in different ways, including serving various user types, various (water management) domains, as part of its functional architecture (see Figure 4.2). MoST can optimally be used in a setting, in which a (distributed) team, consisting

[9]*Multidisciplinary* means here belonging to more than one discipline; *multi-domain* refers to water management domains, e.g. hydrodynamics, groundwater, precipitation-runoff, flood forecasting, surface water quality, biota (ecology), socio-economics.

of problem owner/(water) manager, modeller(s), auditor, stakeholders and interested members of the public cooperate in a multidisciplinary model-based water management project (i.e. covering more than one water management domain). What team members do is stored in a modelling journal, which is shared by all team members and stored on a project server (on Internet or on a LAN). A server application is responsible for managing modelling journals, enabling that all team members see and work in a single modelling journal, shared by all team members. MoST acts in this way as a client application. The KB is available at a central server to facilitate maintenance and updates. MoST supports updating the local versions of the guidelines from this central server (Scholten *et al.*, 2006).

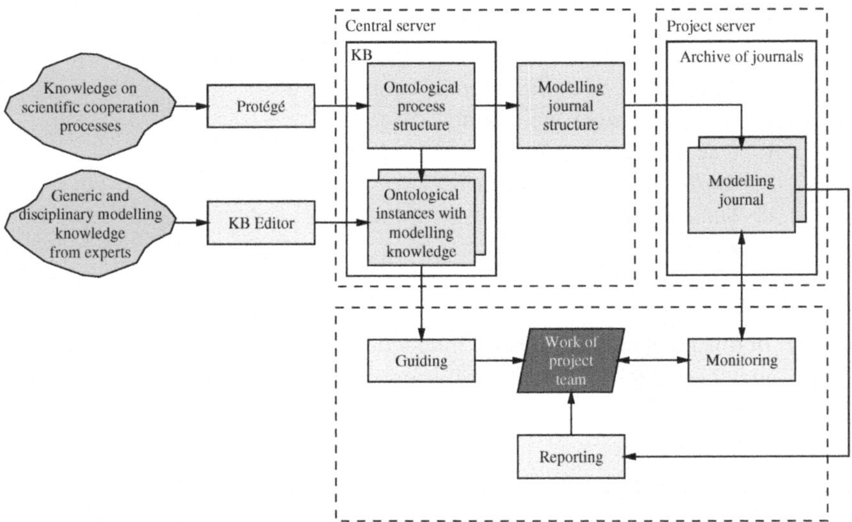

Figure 4.2. Sketch of the client-server architecture with the Modelling Support Tool, MoST, the ontological modelling KB, outlining MoST's major functions, being guiding (from the modelling KB, monitoring and reporting. Modelling journals are located on a project server (from Scholten *et al.*, 2006)

4.3. Modelling KB in MoST

4.3.1. MoST and Quality Assurance (QA)

As defined in HarmoniQuA: *"Quality Assurance (QA) is the procedural and operational framework used by an organisation managing the modelling study to build consensus among the organisations concerned in its implementation, to assure technically and scientifically adequate execution of all tasks included in the study, and to assure that all modelling-based analysis is reproducible and justifiable"*. This modification of the NRC (1990) definition includes the organisational, technical and scientific aspects, but also the need to build consensus among the organisations concerned (see Section 3.1).

4.3.2. Structure and guiding principles of the Knowledge Base (KB)

The modelling process as defined by HarmoniQuA is central to MoST and the structure of its Knowledge Base (KB). The modelling process has been decomposed into five steps. The five-step system is shown in the flowchart (Figure 4.3). Each step includes several tasks. Each task has an internal structure i.e. name, definition, explanation, interrelations with other tasks, activities, activity related methods, references, sensitivity/pitfalls, task inputs and outputs.

The KB contains knowledge specific to seven domains (groundwater, precipitation-runoff, river hydrodynamics, flood forecasting, water quality, ecology, and socio-economics), and forms the heart of the tool. A computer based journal is produced within MoST where the water manager and modelling team record the progress and decisions made during a model study according to the tasks in the flowchart. This modelling journal can be used when auditing the model study to judge its quality.

The most important QA principles incorporated in the KB are:

- The five modelling steps conclude with a formal *dialogue* between the modeller and manager, where activities and results from the present step are reported, and details of plans for the next step (a revised work plan) are discussed.
- *External reviews* are prescribed as the key mechanism of ensuring that the knowledge and experience of other independent modellers are used.
- The KB provides *public interactive guidelines* to facilitate dialogue between modellers and the water manager, with options to include auditors, stakeholders and the public.
- There are many *redo-loops*, some technical involving only the modeller, and others that may require a decision before doing costly additional work.
- The KB allows *performance and accuracy criteria* to be updated during the modelling process. In the first step the water manager's objectives and requirements are translated into performance criteria that may include qualitative and quantitative measures. These criteria may be modified during the formal reviews of subsequent steps.
- Emphasis is put on *validation schemes*, i.e. tests of model performance against data that have not been used for model calibration.
- *Uncertainties* must be explicitly recognised and assessed (qualitatively and/or quantitatively) throughout the modelling process.

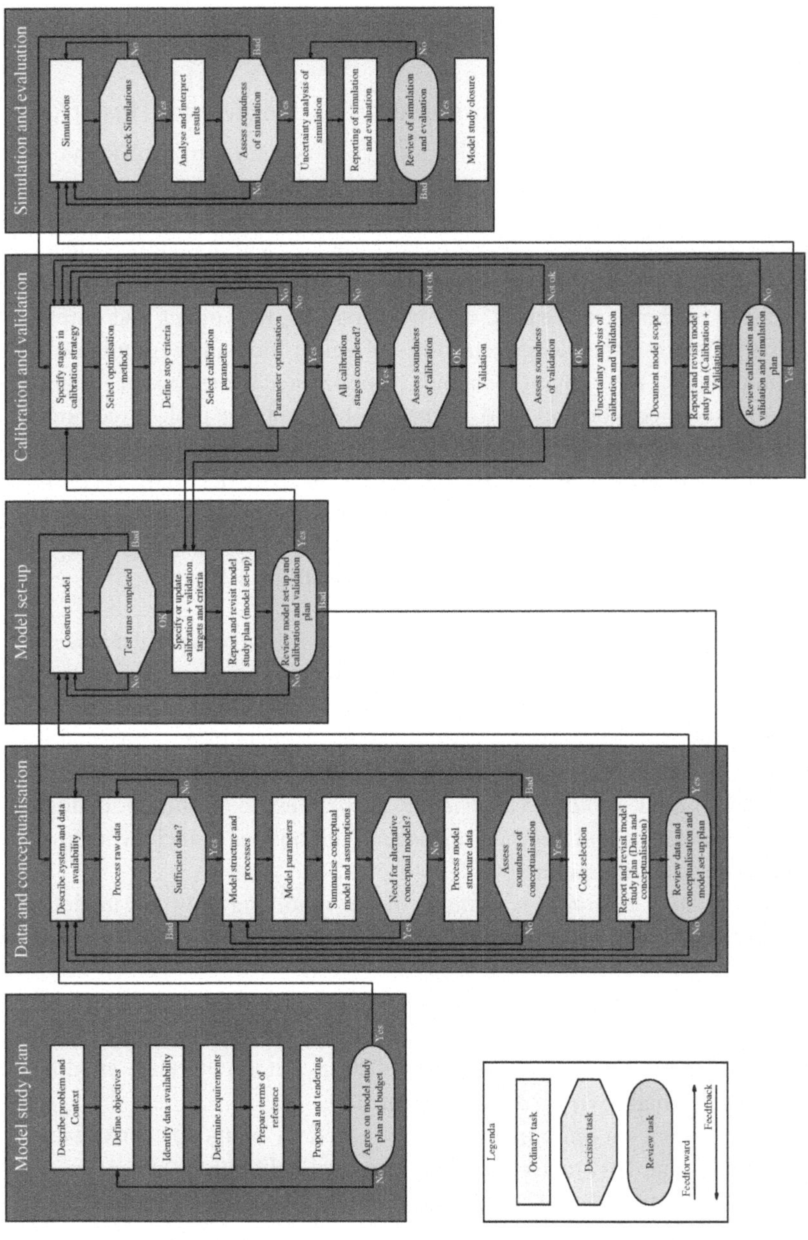

Figure 4.3. Five steps and 48 tasks of the modelling process in HarmoniQuA's modelling Knowledge Base. This representation of the modelling process is simplified in its redo-loops

MoST supports multi-domain studies and working in teams, which members play different roles (water managers, modellers, auditors, stakeholders and members of the public). It contains an interactive glossary that is accessible via hyperlinked terms. The key functionality of MoST is to

- *Guide*, to ensure a model has been properly applied. This is based on the modelling Knowledge Base.
- *Monitor*, to record decisions, methods and data used in the modelling work and in this way enable transparency and reproducibility of the modelling process.
- *Report*, to provide suitable reports of what has been done for managers/ clients, modellers, auditors, stakeholders and the general public.

4.3.3. Terminology and glossary

A consistent terminology is crucial to establish a common understanding between modeller, water manager and stakeholders. In addition, terminology may be controversial, especially with regard to terms such as validation and verification (Refsgaard and Henriksen, 2004). The KB contains a glossary of more than 1000 terms with careful definitions of some of the key terms listed in Table 4.2.

Table 4.2. Set of definitions of modelling terminology (from Refsgaard and Henriksen, 2004).

Term	Definition
Reality	*The natural system, understood here as the study area.*
Conceptual model	*A description of reality in terms of verbal descriptions, equations, governing relationships or 'natural laws' that purport to describe reality.*
Model code	*A mathematical formulation in the form of a computer program that is so generic that it, without program changes, can be used to establish a model with the same basic type of equations (but allowing different input variables and parameter values) for different study areas.*
Model	*A site-specific model established for a particular study area, including input data and parameter values.*
Model confirmation	*Determination of adequacy of the conceptual model to provide an acceptable level of agreement for the domain of intended application.*

(*continued*)

Table 4.2. *Continued*

Term	Definition
Code verification	*Substantiation that a model code is in some sense a true representation of a conceptual model within certain specified limits or ranges of application and corresponding ranges of accuracy.*
Model calibration	*The procedure of adjustment of parameter values of a model to reproduce the response of reality within the range of accuracy specified in the performance criteria.*
Model validation	*Substantiation that a model within its domain of applicability possesses a satisfactory range of accuracy consistent with the intended application of the model.*
Model set-up	*Establishment of a site-specific model using a model code. This requires, among other things, the definition of boundary and initial conditions and parameter assessment from field and laboratory data.*
Simulation	*Use of a validated model to gain insight into reality and obtain predictions that can be used by water managers.*
Performance criteria	*Level of acceptable agreement between model and reality. The performance criteria apply both for model calibration and model validation.*
Domain of applicability (of conceptual model)	*Prescribed conditions for which the conceptual model has been tested, i.e. compared with reality to the extent possible and judged suitable for use* (*by model confirmation*).
Domain of applicability (of model code)	*Prescribed conditions for which the model code has been tested, i.e. compared with analytical solutions, other model codes or similar to the extent possible and judged suitable for use* (*by code verification*).
Domain of applicability (of model)	*Prescribed conditions for which the site-specific model has been tested, i.e. compared with reality to the extent possible and judged suitable for use* (*by model validation*).

4.4. Modelling guidance

4.4.1. Decomposition

The modelling process has been decomposed into five steps. Each step includes several tasks (Figure 4.3). Each task has an internal structure i.e. name, definition, explanation, interrelations with other tasks, activities, activity related methods, references, sensitivity/pitfalls, task inputs and outputs. The KB contains knowledge specific to seven domains (groundwater, precipitation-runoff, river hydrodynamics, flood forecasting, water quality, ecology, and socio-economics), and forms the heart of the tool. Furthermore, the KB contains guidance for the different actors in the modelling process (water manager, modeller, auditor/reviewer, stakeholders and general public). Definitions of the MoST terminology are given in Appendix B.1.

4.4.2. Steps and tasks

The contents of the five modelling steps are:

- *Step1* (*Model Study Plan*). This step aims to agree on a Model Study Plan comprising answers to the questions: Why is modelling required for this particular model study? What is the overall modelling approach and which work should be carried out? Who will do the modelling work? Who should do the technical reviews? Which stakeholders/public should be involved and to what degree? What are the resources available for the project? The water manager needs to describe the problem and its context as well as the available data. A very important task is then to analyse and determine the various requirements of the modelling study in terms of the expected accuracy of modelling results. The acceptable level of accuracy will vary from case to case and must be seen in a socio-economic context. It should, therefore, be defined through a dialogue between the modeller, water manager and stakeholders/public. In this respect an analysis of the key sources of uncertainty is crucial in order to focus the study on the elements that produce most information of relevance to the problem at hand.
- *Step 2* (*Data and Conceptualisation*). In this step the modeller should gather all the relevant knowledge about the study basin and develop an overview of the processes and their interactions in order to conceptualise how the system should be modelled in sufficient detail to meet the requirements specified in the Model Study Plan. Consideration must be given to the spatial and temporal detail required of a model, to the system dynamics, to the boundary conditions and to how the model parameters can be determined from the available data. The need to model certain processes in alternative ways or to differing levels of detail in order to enable assessments of model structure uncertainty should be evaluated. The availability of existing computer codes that can address the model requirements should also be addressed.

- *Step 3 (Model Set-up).* Model Set-up implies transforming the conceptual model into a site-specific model that can be run in the selected model code. A major task in Model Set-up is the processing of data in order to prepare the input files necessary for executing the model. Usually, the model is run within a Graphical User Interface (GUI) where many tasks have been automated. The GUI speeds up the generation of input files, but it does not guarantee that the input files are error free. The modeller performs this work.
- *Step 4 (Calibration and Validation).* This step is concerned with the process of analysing the model that was constructed during the previous step, first by calibrating the model, and then by validating its performance against independent field data. Finally, the reliability of model simulations for the intended domain of applicability is assessed through uncertainty analyses. The results are described so that the scope of model use and its associated limitations are documented and made explicit. The modeller performs this work.
- *Step 5 (Simulation and Evaluation).* In this step the modeller uses the calibrated and validated model to make simulations to meet the objectives and requirements of the model study. Depending on the objectives of the study, these simulations may result in specific results that can be used in subsequent decision making (e.g. for planning or design purposes) or to improve understanding (e.g. of the hydrological/ecological regime of the study area). It is important to carry out suitable uncertainty assessments of the model predictions in order to arrive at a robust decision. As with the other steps, the quality of the results needs to be assessed through internal and external reviews.

Each of the last four steps is concluded with a reporting task followed by a review task. The review tasks include dialogues between water manager, modeller, reviewer and, often, stakeholders/public. The protocol includes many possible redo-loops (Figure 4.3).

Short definitions of the five steps and the 48 tasks are given in Appendix B.2.

4.4.3. Examples of description of tasks

The KB comprises guidance for nine modelling domains. Seven of the domains are traditional disciplines:

- Groundwater (GW)
- Precipitation-runoff (PR)

- River hydrodynamics (HD)
- Flood forecasting (FF)
- Water quality (WQ)
- Ecology/biota (BI)
- Socio-economics (SE)

The two additional KB domains are:

- Generic (GE). GE can be used as a general modelling guidance. GE is the backbone of the KB with all the guidance that is applicable to several domains. For most of the activities the GE guidance also applies to the seven domains. The other domains are different where it is required to add domain-specific knowledge, such as the description of which data to collect.
- Multi-domain (MD) which is used in integrated modelling, where several domains are included. For example the GW + PR + HD + MD domains will typically be required in case of catchment modelling focussing on groundwater/surface water interaction.

To illustrate the guidance contained in the KB the content for GE of two tasks will be outlined here. The text of the KB for the GE of these two tasks is shown in Appendix B.3 and Appendix B.4 with respect to the activities and the sensitivities and pitfalls.

Task 1.4 Document Requirements (Appendix B.3)
The task Document Requirements is number four in Step 1 (Model Study Plan). This is at the initial, framing part the modelling study, and therefore the responsibilities to carry out this task lies with the water manager with interaction as required with stakeholders and general public. The output of this task will essentially be the substance of the Terms of References for the subsequent modelling study, and therefore the modeller who is going to carry out actual modelling has no formal role in this task. If the water manager does not have sufficient modelling expertise to provide sufficiently precise specifications of the requirements he/she may need to have support from a consultant to draw up these specifications.

The activities included in this task are:

- Quality and Relevance
- Overall Quality
- Model Analyses
- Performance Criteria

- Multi-domain Couplings (only in case of integrated modelling)
- Stakeholder Expertise
- Resources
- Communication and Reporting
- Other Requirements
- Managing Model Study
- Stakeholder Opinion on Requirements
- Answers to Stakeholder Opinions and Final Stakeholder Involvement Plan
- Assess Uncertainty

The activities may be seen as a check list contributing from various perspectives to scoping the modelling study. It is also noted that they involves consultation of stakeholders on the requirements. Finally it is noted that the last activity deals with assessment of uncertainty. The reason for this is to draw attention to uncertainty at this early stage to ensure that the modelling study is designed so that it addresses the issues, where there is expected to be most uncertainty and where it is possible to reduce the uncertainty.

The sensitivities and pitfalls point at some of the difficulties that may be expected such as establishment of adequate performance criteria for the modelling.

Task 4.8 Validation (Appendix B.4)

The task Validation is one of the last tasks in Step 4 (Calibration and Validation). This task is by nature very different from the task Determine Requirements. Whereas Determine Requirements deals with scoping and involves decision makers and stakeholders the task Validation is much more technical and is carried out by the modeller.

The activities included in this task are:

- Design validation and test procedure
- Confirm performance measures
- Process validation data
- Carry out validation test

The guidance text here also differs from the Determine Requirements. The activity descriptions are less and more brief, but the technical guidance is here primarily provided in terms of a number of applicable methods, some of which may be chosen:

- Aggregated Performance Indicators
- Standard Performance Measures

- Qualitative Performance Measures
- Visualisation and Tabulation
- Contouring
- Mass Balance
- Correlation and Regression
- Differential Split-sample Test
- Aggregation
- Surface Fitting
- Infilling (temporal)
- Zonation
- Expert Evaluation of Model Performance
- Event Validity
- Extreme-condition Test
- Comparative Assessment
- Statistical Validation
- Split-sample Test
- Differential Split-sample Test
- Proxy Basin Test
- Proxy Basin Differential Split-sample Test

The sensitivities and pitfalls also here focus on some of the difficulties to be encountered, such as a proper and balanced presentation of the validation status acknowledging that often model validation is not possible, not even conditional validation.

4.4.4. Methods

The KB contains 90 methods and for each activity advice on applicable methods is given. The names of methods and to which tasks they are allocated are given in Appendix B.5. For many activities a large number of applicable methods are listed, but depending on the circumstances, only a few of them will fit in a specific case. An example of this is seen for the task Validation (Appendix B.5) where many of the listed methods are mutually exclusive. The modeller or the water manager can use the list of methods as inspiration and guidance but at the end need to compose the suite of methods in accordance with the approach they chose in their specific case.

Thus, the methods are not meant as compulsory or prescriptive, but rather as options which can support the modelling activities. Each of the methods is described by their approach and references are provided to examples and to further reading and supporting tools (textbooks, papers, websites).

4.5. Modelling Support Tool, MoST

4.5.1. Introduction to MoST

Based on Section 4.2 and especially Appendix A, the Modelling Support Tool can now be described in more detail. It can work in different modes with components to set-up and edit projects (Section 4.5.3), getting guidance (Section 4.5.4), monitoring what teams actually do in a project (Section 4.5.5) and reporting what has to be done (Section 4.5.6). Detailed descriptions of the tool and its usefulness for model-based water management are given by Scholten *et al.* (2007).

The relation between the Modelling Support Tool and the other components of the system are depicted in Figure 4.2 with emphasis on how the components interact with each other and the role of team members and domain experts in this system.

4.5.2. Starting MoST

Depending how the tool is started, users will enter in the guideline mode (if no project is opened) or in the project mode (by double clicking on a modelling journal file). In the former case the project mode component tab and the reporting mode component tab will not be visible, as will be in the latter case.

There are several ways to start working in a project

(1) For a new project:
   a. *New Online Project*: a new project has to be defined, which will be stored on a central server to allow distributed teams to co-operate.
   b. *New Local Project*: a new project has to be defined, which will be stored on a hard disk; this approach is typical for single person projects.
   c. *New Project from Template*[10]:
      (i) *Local*: a new local project, but, if a template is used, a part of the definition is included in the template (relevant subproject(s), relevant tasks, (optional) users, (optional) authorizations of users).
      (ii) *Online*: a new online project, but, if a template is used, a part of the definition is included in the template (relevant subproject(s), relevant tasks, (optional) users, (optional) authorizations of users).

---

[10]Project *templates* are empty project journals containing the relevant structure of a project, i.e.: relevant subproject(s), relevant tasks, (optional) users, (optional) authorizations of users.

(2) For an existing project
   a. *Local project*: there are 2 options:
      (i) Double click on the modelling journal file (*filename*.mpj);
      (ii) Start MoST and use (in the menu) *Open Local Project*;
   b. *Online project*: Start MoST and use (in the menu) *Open Online Project*.

The choice for a *local* or for an *online* project is not permanent, as one can switch between local and online, but there are consequences. When working in a local project, only one user can work in a project. Imagine, that a multi-user online project is transformed to a local project (menu: *Project/Export online Project to Local*) from that moment on, only one user can work in the project. If others continue to work in the online version, there will be a synchronization problem, as the online version is permanently synchronized, when some of its users change the content of the modelling journal. The independently edited local version cannot automatically be synchronized with the online one. Starting with a single member team in a local project is less tricky, as no others can participate unless on the same local version on the same computer. After transforming a local project to an online one (menu: *Project/Convert Local Project to Online*), synchronization is guaranteed, as long as it stays online.

Starting a project from a template is straightforward. Templates can be found on the hard disk in a directory called *templates*, which is a subdirectory of the directory where MoST has been installed. So far several templates are available for specific types of multidisciplinary model-based water management projects[11]. Other templates can be generated from (successful) projects (menu: *File/Save the Project as Template*), allowing to include all users and their authorizations (efficient in routine type of projects) or to exclude team member information.

4.5.3. Setting-up modelling projects

Projects have to be defined in the project initialization phase. In a *window with tabs*[12] the project initiator has to specify the project to start. After giving the project a name, the following *tabs* have to be used for setting up a project:

- **Subprojects**: selecting one or more *domains* from a list of available *domains* and adding this set to the list of *subprojects* and replacing the default *subproject* name by a meaningful one defines a *subproject*.

[11]See http://harmoniqua.wau.nl/training/help/templates/available_templates.htm.
[12]In MS Windows terminology it is called a *tab control*.

- **Tasks**: For each *subproject* one can deselect tasks from the full list or, if used a template project, select and deselect *tasks*. No *tasks* can be defined here. One can also choose between random access of the *tasks* to do or enforce strictly following the order of *tasks* given by the guidelines.
- **Users**: Users[13] are added by providing username, password, first name and last name, selecting one or more roles per user and adding the user to the user list.
- **Authorization**: For each user the project initiator has to specify authorization. First one has to decide whether to disable or enable login account and if the user is a project administrator. The project initiator is by default administrator, unless this authorization is deselected. Furthermore per user and per subproject the following privileges: *write*, *decision making* (allowed to take decision in decision tasks of that subproject) and *reading* of modelling journal content per role.
- **Scoreboard**: Project administrators are allowed to add or remove questions in scoreboards. A list of default scoreboard question is provided. Managers[14] (i.e. the clients of the project) are allowed to edit weights of scoreboard questions and during projects auditors have to fill scoreboards as part of their evaluation activities.

Projects can have more than one project administrator. Administrators have several privileges, all related to smooth the progress of the project and to assist team members in their daily practice. Most of these privileges concern project settings (i.e. project specifications specified at project initialization).

4.5.4. Guiding modelling projects

The guideline component provides three views on the KB: the *tree view*, the *flowchart view* and the *task view*. These views are the major panels on the screen in the guideline component (Figure 4.4 and Figure 4.5).

The **tree view** resembles standard MS-Windows trees[15] like in the MS-Windows Explorer and enables to browse through the steps and tasks in a similar way as Windows allows browsing through directories and files. The '+' sign will allow to expand a *step* into *tasks* and '−'will collapse a set of *tasks* into

---

[13]*Users* are *team members*.

[14]In MoST, the implementation of the Modelling Support Tool for HarmoniQuA, called 'Water manager'.

[15]A *tree view* is a graphical diagram used to display the hierarchal structure of items, such as directories and files on a disk.

the *step* these tasks belong to (see Figure 4.4 and Figure 4.5). Four symbols indicate the nature of the concepts:

(1) Red ovals for *steps*;
(2) Yellow rectangles for (*ordinary*) *tasks*;
(3) Blue diamonds for *decision tasks*;
(4) Green hexagons for *review tasks*.

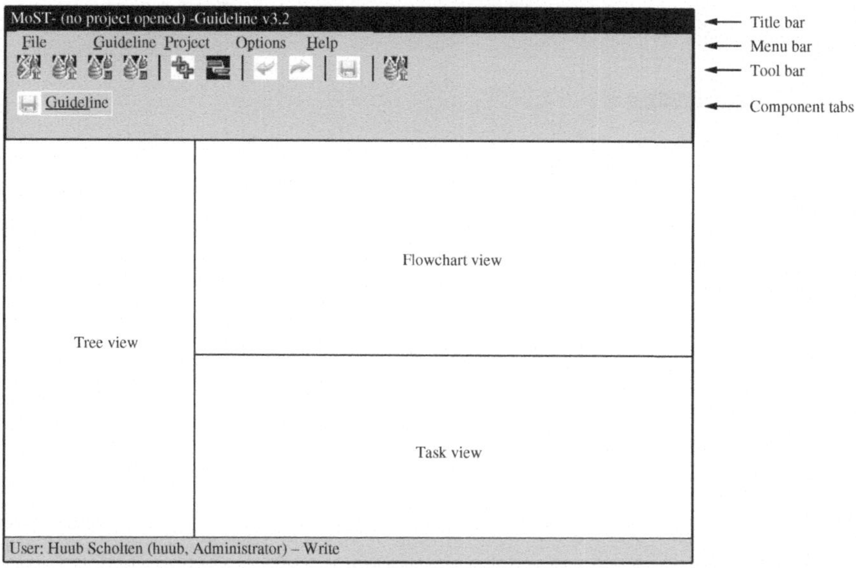

Figure 4.4. Schematized layout of MoST GUI in case no project has been opened with menu bar, tool bar and guideline component tab in the top and the tree view, the flowchart view and the task view in the central part of the screen

The **flowchart view** provides another type of browsing through the network of *steps* and *tasks*, but gives also a structured view with the order of *tasks* and of *redo-loops*. Arrows indicate the order and have tool tips to see their origin *task* or destination *task*.

The task view: presents a textual description of the chosen concept (*guideline*, *step* or *task*). It shows the content of task. If a task has been selected, several tabs can show specific parts of its content (*Task description*, i.e. definition and explanation of the task; *Activities information*, i.e. list of activities and description of each activity; *Methods information*, i.e. list of available methods and their brief descriptions; *Pitfalls and sensitivities*, i.e. an overview of the risky elements in the task; *Other information*. i.e. relations to other tasks, software aspects, references, inputs and outputs). The descriptions of tasks in this view contain discipline related terminology, which is explained in the glossary. This glossary can be accessed through the hyperlinks in the task view or by the menu (*help/glossary...*).

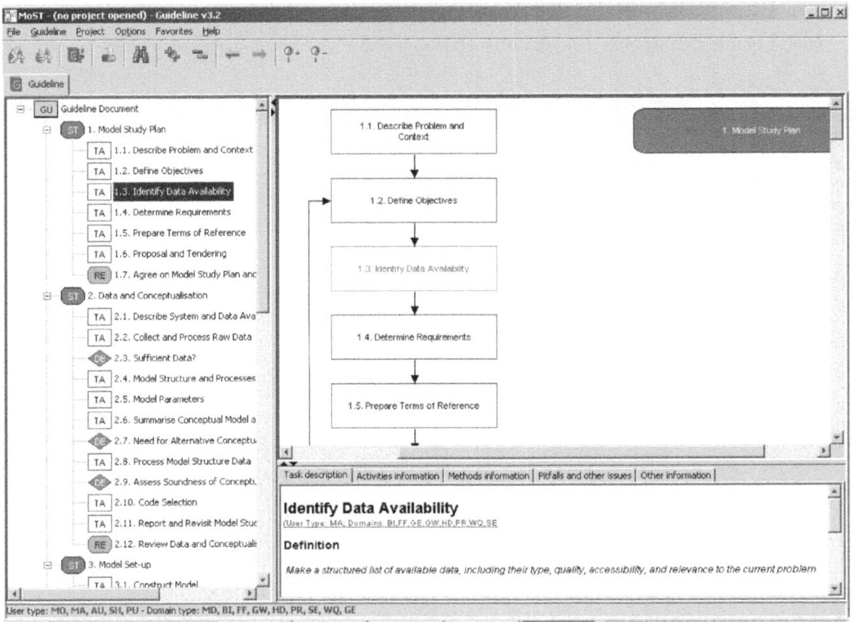

Figure 4.5. Screen dump of guideline component. In the tree view task 1.3 has been chosen and the guideline component synchronizes the other views by setting the pointer to the same activity in these views

One of the major aims of the overall system is customizing the guidelines (and what team members have to do) to their roles, the domains / disciplines involved, the complexity of the case and its purpose. Therefore the guideline component of the tool allows its users to filter for these customization properties[16]. Filtering will automatically update the *tree view*, the *flowchart view* and the *task view*.

The menu item *guideline* enables also several functionalities, including selecting a guideline version (from local computer), downloading guidelines

[16]At present *job complexity* and *application purpose*, although implemented in the tool, have been switched off, because these options were not used in modelling test cases of HarmoniQuA.

(from central server), printing the flowchart and printing the text of the active task[17]. The menu item on guidelines also allows finding terms, expanding and collapsing the tree and zooming-in or –out in the flowchart, adapting fonts and colours.

The guideline component can also be used outside the context of running a modelling project.

4.5.5. Monitor modelling projects

Keeping records is essential in the project execution phase, as it enables doing audits and allows reconstruction of projects. This monitoring component of MoST is complex in its user interface and in its functionality.

The project execution component provides three views of which two on the KB: the *tree view* and the *task view*. The third view, the activity view, presents what actually is done in the task and related characteristics at a very detailed level (*activity level*) or at a more generic level (*task level*). These views are the major panels on the screen in the project execution component (Figure 4.6 and Figure 4.7).

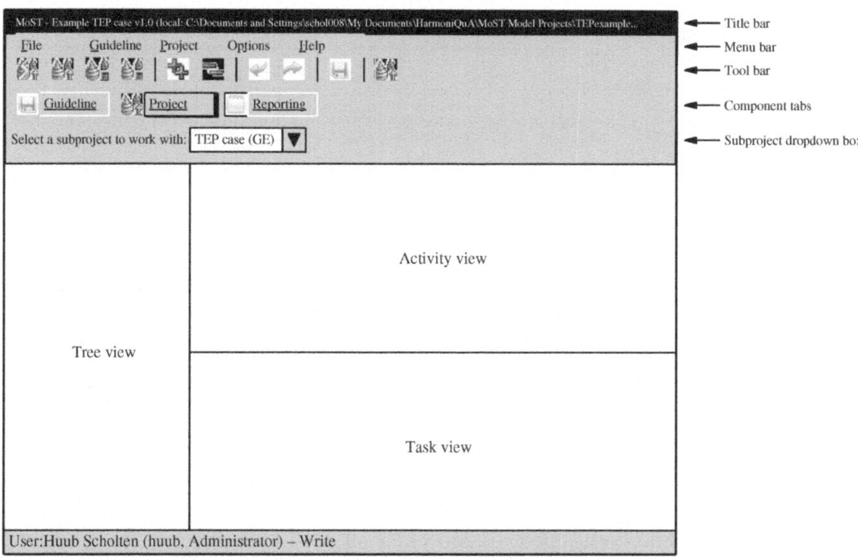

Figure 4.6. Schematized layout of MoST GUI in case a project (TEP case (GE)) has been opened with *menu bar*, *tool bar*, *component tabs* and *subproject dropdown box* in the top and the *tree view*, *task view* and *activity view* in the central part of the screen

[17]To print the full guideline it can be downloaded for printing from www.harmoniqua.org/tools.

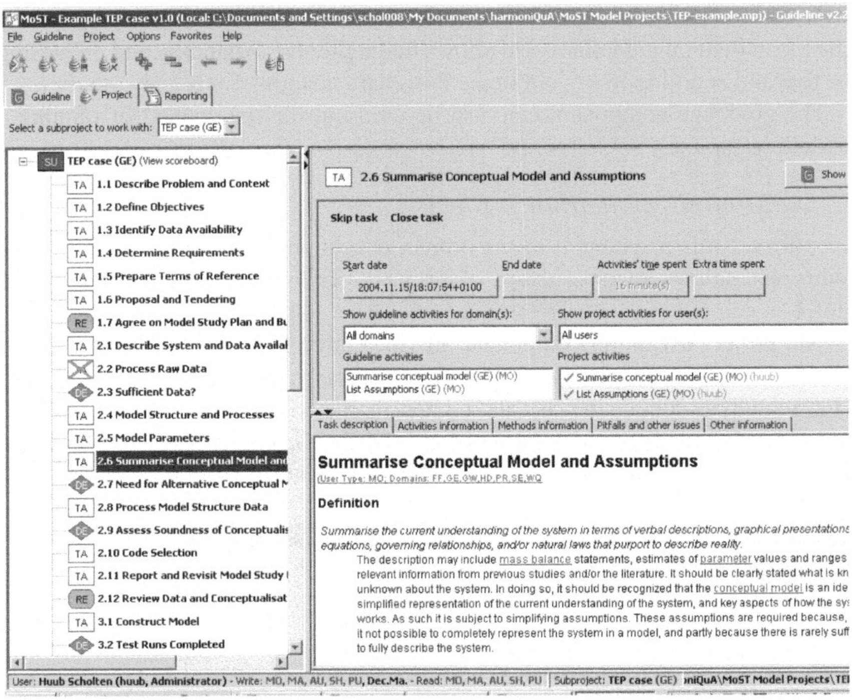

Figure 4.7. Screen dump of MoST GUI, see Figure 4.6 for explanation

The **tree view** (left in Figure 4.6 and Figure 4.7) displays all the tasks in the project and the status of these tasks. This view can be used to browse through the tasks and also to inspect the status of each task in the subproject. Different icons are used to show this status. A transparent rectangle indicates a task not yet started. A yellow rectangle indicates a task that has already started, but not yet finished. If it is finished, a green checkmark is displayed. Skipped task are shown with a red cross.

The **task view** (lower right in Figure 4.6 and Figure 4.7) is similar to the task view in the guideline component of MoST (Section 4.5.4). It provides guidance on the task that is selected in the tree view, with tabs to select the sections on the guidance for that task.

The **activity view** (upper right panel in Figure 4.6 and Figure 4.7) is the main panel to record what team members do. What team members do be recorded in a modelling journal (e.g. in case of modelling a modelling journal). After selecting a *task* in the *tree view*, there are three possibilities:

(1) **Open at Task level**: all activities of the task at hand are handled as a single activity;
   a. Menu options:
      (i) **Skip task**: see below;
      (ii) **Close task**: when completed a task can be closed, prohibiting that users continue to work on that task; only project administrators can re-open it;
      (iii) **Skip activity**: similar to skip task, see below;
      (iv) **Finish activity**: when the combined *activities* of the *task* at hand are finished, a small window pops-up showing the time between start and end of the combined *activities* and allowing to fill-in time spent;
      (v) **Manage time**: a small window pops-up showing the time between start and end of the combined *activities* and allowing to fill-in time spent;
      (vi) **Work at activity level**: switch from *task level* to *activity level*;
   b. Provide *detail, actions and outcomes* of the combined *activities* of that *task*;
   c. Select *methods* used;
   d. Allow to attach and retrieve files to the modelling journal, accessible from the task at hand;
   e. In case of a *decision task* and *review task*, decide the following *task* to do (i.e. going back to redo a previous *task* or continue with the next one).

(2) **Open at activity level**: all *activities* of the *task* at hand are handled as separate *activities*.
   a. Menu options:
      (i) **Work at task level**: switch from *activity level* to *task level*;
      (ii) **Skip task**: see below;
      (iii) **Close task**: when completed a *task* can be closed, prohibiting that users continue to work on *activities* of that *task* or start new *activities* in that *task*; only project *administrators* can re-open it;
   b. A complex panel:
      (i) Shows time management issues;
      (ii) Allows to filter *activities* per domain;
      (iii) Allows to filter *activities* per *user type* (role);
      (iv) Allows to select an *activity* to start or redo; here several menu options allow you to
         - **Skip activity**: similar to skip task, see below;
         - **Finish activity**: when an activity is finished, a small window pops-up showing the time between start and end of the combined activities and allowing to fill-in time spent;

- **Manage time**: a small window pops-up showing the time between start and end of the combined activities and allowing to fill-in time spent;

Further more one work on the activity by:

- Provide *detail, actions and outcomes* of that *activity*;
- Select *methods* used;
- Allow to attach and retrieve files to the *modelling journal*, accessible from the *task* at hand;
- In case of a decision task and review task, decide the following task to do (i.e. going back to redo a previous *task* or continue with the next one).

(3) **Skip Task**:
   a. Menu options:
      (i) **Perform task**: still perform task;
      (ii) **Work at activity level**: switch from *task level* to *activity level*;
   b. Provide a reason why to skip the task.

Browsing through the *tasks* should be done in the *tree view*, but – depending on if the prescribed order in the guidelines is enforced, *tasks* have to be performed in the right order.

The menu item *project* facilitates many functions, including:

- Convert local projects to an online project;
- Export an online project to local;
- Save projects as project templates, which include all project characteristics (domains, tasks, etc., but no users);
- Access to the scoreboard (project administrators can edit the default questions, managers can change the weights per step and per question and others can see the questions and results, if authorized);
- Change settings on whether or not confirmations are needed for all sorts of user interactions;
- Logon as different user;
- Change password;
- Collapse the tree into tasks or expand it;
- Select small medium or large tree icons;
- Change project settings, which allows project administrator to adapt all project settings, which are specified during project set-up, e.g. to add new team members, edit authorizations, etc.

4.5.6. *Reporting modelling projects*

During a modelling project and after its completion there is a strong need for fancy reports of the project. All reports will be generated from the content of a modelling journal. Two aspects of reports are of interest here: its *intended use* and its *audience*. The use may vary from informing other team members, management checks on progress, to the final report for the client of the modelling study done in the project. Audiences are also diversified, including scientists, managers, professional engineers, (lay) stakeholders and interested members of the general public.

Selecting what information has to be included in reports for specific use and audience is the major functionality of the report component. The selection mechanism of the reporting component is powerful, but also rather complex (see Figure 4.8 and Figure 4.9).

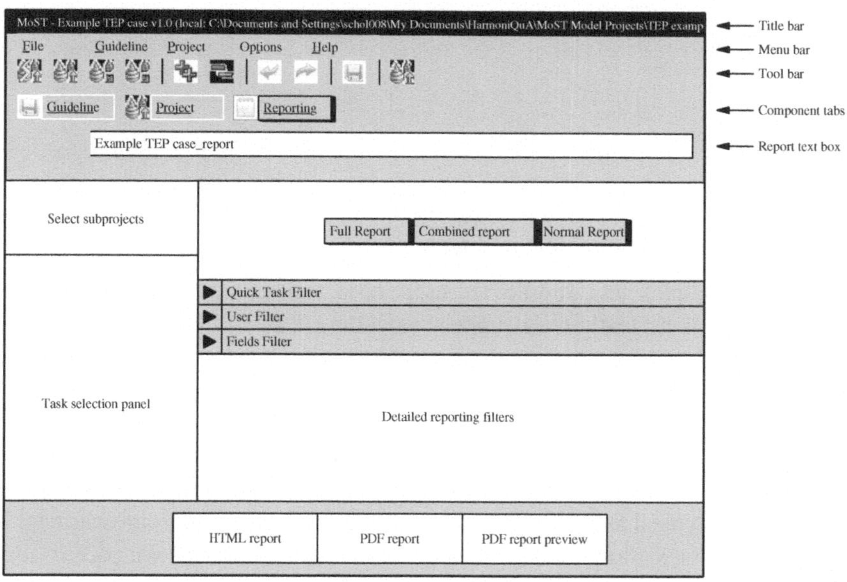

Figure 4.8. Schematized layout of MoST GUI of the report component in case a project (TEP case (GE)) has been opened with menu bar, tool bar, component tabs and report text box. Centrally selection options and in the bottom three options to generate reports

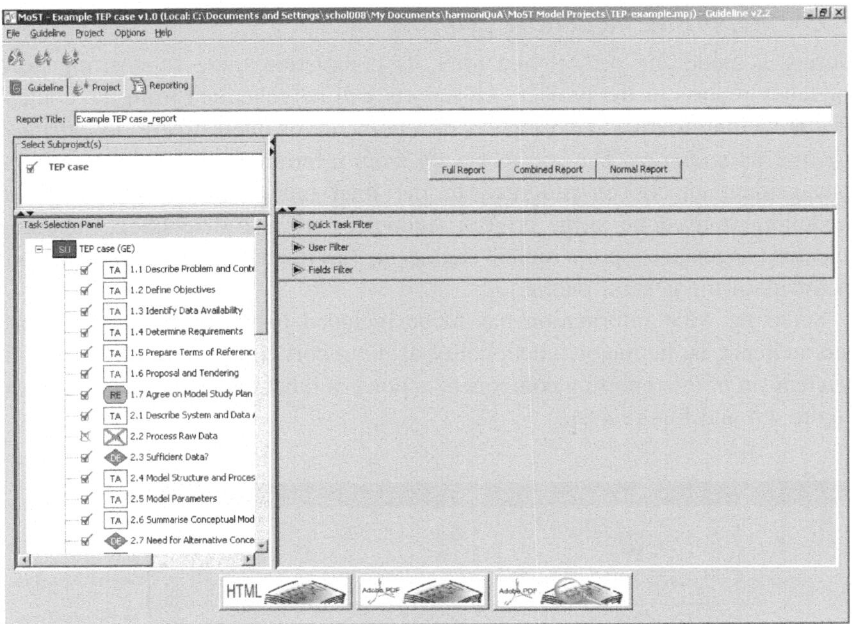

Figure 4.9. Screen dump of MoST GUI, see Figure 4.8 for explanation

4.5.7. *Training material and help*

MoST is a complex tool and novice users need some training before they can benefit from all its features to support modelling teams in their daily routine. Therefore, comprehensive training material has been developed for students and professional modellers in water management[18]. This training material consists of written material, presentations, exercises that encourage using MoST in a test case project and many screen-recording movies on MoST, its KB, a case study and some background information. The screen recordings are Macromedia Flash® movies with a typical length of one to two minutes, accompanied by an instructor's voice. The movies allow trainees to work individually, at their own pace. Using the movies also allows the instructor both to explain in words and to show by demonstration. The movies are the core part of the training material and aim at helping users to work with MoST and will act as a sort of animated help facility.

In addition to the training material a *help function* has been implemented with the summarized information of the training material.

[18]The training material is accessible from www.HarmoniQuA.org/training.

5. REFERENCES

ASTM (1992). Standard practice for evaluating mathematical models for the environmental fate of chemicals, American Society for Testing and Materials, Standard E978-92. www.astm.org.

ASTM (1994). Standard guide for application of a ground-water flow model to a site-specific problem, American Society for Testing and Materials, Standard D5447-93. www.astm.org.

ASTM (1996). Standard guide for Conceptualization and Characterization of Ground-Water Systems, American Society for Testing and Materials, Standard D5979-96. www.astm.org.

BDMF (2000). Protocols for Water and Environmental Modeling. Bay-Delta Modeling Forum. Ad hoc Modeling Protocols Committee.

CEN (2004). Compass – European standardisation in a nutshell, 6 pp. http://www.cenorm.be/cenorm/aboutus/information/otherpublications/compassen1.pdf.

CEN (2005). CEN Workshop Agreements. http://www.cen.eu/boss/production/production+processes+-+index/cen+workshop+agreements/cen+workshop+agreements. asp.

Crosby, P. (1979). Quality is free: the art of making quality certain, McGraw-Hill Book Company, New York.

De Groof, A., Dries, V., Grøn, C. and Ruwiel, E. (2005), The quality thing: How do we do it, and how do you? Presentations in Special Session 27, Consoil, Bordeaux, 3–5 October 2005.

EC (2004a). Common Implementation Strategy for the Water Framework Directive (2000/60/EC) Guidance Document No 11, Planning Processes. European Communities. Available on http://www.waterframeworkdirective.wdd.moa.gov.cy/docs/GuidanceDocuments/Guidancedoc11PlanningProcess.pdf (accessed 26 November 2007).

EC (2004b). Common Implementation Strategy for the Water Framework Directive (2000/60/EC) Guidance Document No 3, Analysis of Pressures and Impacts, IMPRESS. Available on http://www.waterframeworkdirective.wdd.moa.gov.cy/docs/GuidanceDocuments/Guidancedoc3IMPRESS.pdf (accessed 26 November 2007).

German Geological Society, (2001). Hydrogeological Models. Guidelines for Clients, Consultants and Regulatory Officers, 36 pp.

Hansen, J.B. and Grøn, C. (2006). Sampling of soil and waste, Report to the Danish Environmental Protection Agency (DHI Water & Environment (unpublished – in Danish).

Henriksen, H.J., Troldborg, L. Nyegaard, P. Sonnenborg, T.O. Refsgaard J.C. and Madsen, B. (2003). Methodology for construction, calibration and validation of a national hydrological model for Denmark. *Journal of Hydrology* **280**, 52–71.

Humphrey, W.S. (1989). Managing the software process, Addison-Wesley, Reading etc., 494 pp.

IH (1999). Flood Estimation Handbook. 5 Volumes, Center for Ecology and Hydrology, Wallingford.

Jakeman, A.J., Letcher, R.A. and Norton, J.P. (2006). Ten iterative steps in development and evaluation of environmental models. *Environmental Modelling & Software* **21**, 602–614.

Kassahun, A. and Scholten, H. (2006). A knowledge base system for multidisciplinary model-based water management, Summit on Environmental Modelling and Software, 3rd Biennial meeting of the International Environmental Modelling and Software Society, Burlington, Vermont, USA, July 9–12, 2006.

Klemes, V., (1986). Operational testing of hydrological simulation models. *Hydrological Sciences Journal* **31**, 13–24.

Middlemis, H. (2000). Murray-Darling Basin Commission. Groundwater flow modelling guideline., Aquaterra Consulting Pty Ltd., South Perth. Western Australia, Project no. 125. www.harmoniqua.org/public/Reports/existing%20guidelines/MBDC_Groundwater_Modeling.pdf.

NRC, (1990). Ground Water Models: Scientific and Regulatory Applications. National Research Council, National Academy Press, Washington, DC.

Packman, J.C. and Old, G.H. (2005). European Quality Assurance Standards in River Basin Modelling Practice, Centre for Ecology and Hydrology, Wallingford, HarmoniQuA-report D-WP4-6, 18 pp. http://harmoniqua.wau.nl/public/Reports/Standards/EQS-final.pdf.

Packman, J.C. (2002). Quality Assurance in the UK. In: J.C. Refsgaard (Ed.), State-of-the-Art Report on Quality Assurance in modelling related to river basin management, Geological Survey of Denmark and Greenland, Copenhagen, ISBN, Chapter 17, www.harmoniqua.org.

Pahl-Wostl, C. (2002). Towards sustainability in the water sector – the importance of human actors and processes of social learning. *Aquatic Sciences* **64**, 394–411.

Pascual, P., Stiber, N. and Sunderland, E. (2003). Draft guidance on the development, evaluation, and application of regulatory environmental models, US – EPA, Council for Regulatory Environmental Modeling, Washington D.C., 60 pp.

Refsgaard, J.C. and Henriksen, H.J. 2004. Modelling guidelines – terminology and guiding principles. *Advances in Water Resources* **27**, 71–82.

Refsgaard, J.C., Henriksen, H.J. Harrar, B. Scholten H. and Kassahun, A. (2005). Quality assurance in model based water management – review of existing practice and outline of new approaches. *Environmental Modelling & Software* **20**, 1201–1215.

Refsgaard, J.C., Van der Sluijs, J.P. Højberg, A.L. and Vanrolleghem, P.A. (2007). Uncertainty in the environmental modelling process – a framework and guidance. *Environmental Modelling & Software* **22**, 1543–1556.

Refsgaard, J.C., van der Sluijs, J.P. Brown J.D. and Van der Keur, P. (2006). A framework for dealing with uncertainty due to model structure error. *Advances in Water Resources* **29**, 1586–1597.

Reilly, T.E. and Harbaugh, A.W. (2004). Guidelines for evaluating ground-water flow models. U.S. Geological Survey Scientific Investigations Report 2004-5038, 30 pp.

Scholten, H. and Udink ten Cate, A.J. (1995). Testing simulation modelling quality, Mathematical modelling and simulation in agriculture and bio-industries, vol. I, IMACS/IFAC, Brussels, Belgium, ISBN, pp. II.A.3.1-6.

Scholten, H. and Udink ten Cate, A.J. (1999). Quality assessment of the simulation modeling process. *Computers and Electronics in Agriculture* **22**, 199–208.

Scholten, H. and Beulens, A.J.M. (2005). Testing ontological support for multi-disciplinary model-based problem-solving for water management. In: M. Makowski (Editor), Knowledge Creation and Integration for Solving Complex Problems. The 19th International Workshop on Complex Systems Modeling (CSM) jointly with the 6th International Symposium on Knowledge and Systems Sciences (KSS), August 29–31, 2005, IIASA, Laxenburg, Austria, ISBN, pp. 51–55.

Scholten, H. and Van der Tol, M.W.M. (1998). Quantitative validation of deterministic models: when is a model acceptable? In: M.S. Obaidat, F. Davoli, D. DeMarinis (Editors), The proceedings of the Summer Computer Simulation Conference, vol., SCS, The Society for Computer Simulation International, San Diego, CA, USA, ISBN, pp. 404–409.

Scholten, H. and Groot, S. (2002). Dutch guidelines. In: J. Refsgaard (Editor), State-of-the-art report on Quality Assurance in modelling related to river basin management, vol. HarmoniQuA-report D-WP1-1, www.HarmoniQuA.org., GEUS, Copenhagen, ISBN, pp. ch.12, 1-10 www.harmoniqua.org.

Scholten, H., (2006). Final HarmoniQuA Report 2002 – 2005, Wageningen University, Wageningen, HarmoniQuA-report D-final, 88 pp.

Scholten, H., Kassahun A. and Refsgaard, J.C. 2006, Managing multidisciplinary model based water management projects. In: P. Gourbesville, J. Cunge, V. Guinot, S.-Y. Liong (Editors), 7th International Conference on HydroInformatics, Nice, France, 4–8 September 2006, Research Publishing, Volume 3, pp. 2231–2238.

Scholten, H., Kassahun, A. Refsgaard, J.C. Kargas, T. Gavardinas, C. and Beulens, A.J.M. (2007). A methodology to support multidisciplinary model-based water management. *Environmental Modelling & Software* **22**, 743–759.

Sonnenborg, T.O. and Henriksen, H.J. (2005). Handbook in groundwater modelling. Geological survey of Denmark and Greenland. Report 2005/80, 313 pp (In Danish).

Van Asselt, M.B.A. and Rotmans, J. (2002) Uncertainty in integrated assessment modelling – From positivism to pluralism. Climatic Change, **54**(1–2), 75–105.

Van Waveren, R.H., Groot, S. Scholten, H. Van Geer, F. Wösten, H. Koeze, R. and Noort, J.J. (1999). Good Modelling Practice Handbook, STOWA/RWS-RIZA, Utrecht/ Lelystad, in Dutch, ISBN 90-5773-056-1, 149 pp. (http://harmoniqua.wau.nl/public/ Reports/Existing%20Guidelines/GMP111.pdf).

6. APPENDICES

APPENDIX A FUNCTIONAL REQUIREMENTS AND DESIGN

A.1. Modelling KB and associated tools

The proposed quality assurance consists of two major parts: a modelling Knowledge Base (modelling KB) and a Modelling Support Tool MoST. For the development and maintenance of the modelling KB additional tools are required. This section of Appendix A will present and discuss the design of the modelling KB and associated tools and the next section will do the same for the Modelling Support Tool MoST. Both sections start with a summary of functional requirements and subsequently each requirement will be discussed, especially how this requirement is handled in the design and implementation of the modelling KB and MoST.

The modelling KB has been designed and implemented in an ontological way. This means that the structure of the Knowledge Base is an ontology and its content is implemented as instances of this ontology. An ontology is a concise and precise, formal specification shared by its users (whether human or machine) and providing sufficient vocabulary that a piece of knowledge can be formalized for its intended purpose. The modelling KB content is realized by decomposing the modelling process, according to the ontological structure of the Knowledge Base.

Functional requirements to the modelling KB and associated tools are summarized in Table A.1 and how these requirements are met in the design and implementation of the modelling KB will be discussed subsequently.

Table A.1. Functional requirements to the modelling KB and associated tools (based on Scholten and Beulens, 2005, Scholten, 2006; Scholten *et al.*, 2007). Discussions of the requirements in the text will refer to the numbers in this table

Requirement
1. The KB should make modelling knowledge *explicit*.
2. A proper *granularity* (i.e. level of detailing) of the knowledge in the KB should be chosen.
3. KB should be *flexible* (i.e. easy to change its structure).
4. *Maintenance* of the KB should be easy.
5. *Updating* of the knowledge in the KB should be easy.
6. The KB should be designed in way that its ontological format does not hinder moving to other formats.
7. The tool to develop the ontology should provide all basic functionality that can be expected from such a tool, including handling of names of concepts and relations, allocating identification numbers to all concepts and relations, basic terminology.
8. Inexperienced domain knowledge specialists (i.e. expert modellers for water management) should be helped in defining, uploading and editing of their knowledge in the model KB, preferably with a web based tool that act as front-end to the selected ontology editor.
9. Adding and editing the KB should be *protected* by an adequate authorization system.
10. The KB should be '*open*', allowing all interested persons to comment on the KB
11. A glossary with model-based water management terminology should be a part of the modelling KB.
12. Some authority or board should *control the content* of the KB.

(*continued*)

Table A.1. *Continued*

Requirement
13. The part of the KB that has to be *'understood' by computers* (the ontological structure) should be as formal as possible.
14. The part of the Knowledge Base that has to be *understandable for persons* should be less formal and more textual.
15. The KB should be *consistent*.
16. The KB should be *complete*.
17. The knowledge should be *shared* by a substantial fraction of the professional community.
18. The KB and the support tool should reflect the work of (distributed) teams working in *multidisciplinary* model-based projects.
19. The KB should enable different *roles* in the team.
20. The KB should enable to execute different *domains/disciplines* synchronously or asynchronously.
21. The KB should enable different types of *project complexity*.
22. The KB should be suited for *processes in general* including, but not restricted to various types of simulation and applying mathematical models.
23. The KB should be structured in a way that enables *reuse* of parts of the Knowledge Base, especially the ontological structure.
24. The developed KB and the KB Editor should be *platform* (operating system) *independent*.
25. The KB tools should be designed in an *object orientated* way.
26. The basic architecture has to be *client-server* with a thin server and a big client.
27. The *server* should contain all *shared information* (i.e. KB with guidance and the work done by the team).
28. *Adding and editing* knowledge (i.e. guidance stored in the KB) should be carried out on the *client side* and stored on the *server*.

Making knowledge explicit is an intrinsic quality of Knowledge Bases in general and applicable to the modelling KB, discussed here (*requirement 1*). To realise the modelling KB, the modelling process has been decomposed into *steps*, steps into *tasks* and tasks into *activities* (and associated *methods*). This resulted in an appropriate granularity (i.e. level of detailing) of the modelling KB (*requirement 2*). Choosing for an ontological approach in developing the modelling KB dealt with various requirements (*requirement 3*, *requirement 4* and *requirement 5*). When the ontological structure is changed the ontological development tool has to protect the integrity and consistency of the KB. A proper ontology development tool enables the KB to be flexible. Here the free, open source ontology editor and Knowledge Base framework, Protégé[19], has been used to set-up the ontological KB with modelling guidelines. Protégé has been extended by building a plug-in for XML-export, according to a predefined XML-format, interpretable for the modelling support tool that has to co-operate with the KB. Protégé supports all kinds of (export) formats, which allows switching to other ontology editors (*requirement 6*). Furthermore, Protégé provides all functionality that may be expected from such a tool (*requirement 7*).

To fill the KB a (web based) KB-Editor has been developed, which acts as front-end between domain experts, unskilled in knowledge engineering, and the Knowledge Base implemented in Protégé. Domain experts, responsible for parts of the modelling knowledge for that part, use the KB-editor to upload (modelling) knowledge to the Knowledge Base or to edit it (*requirement 8*). The KB Editor handles also authorization issues (*requirement 9*) and assures that users can work concurrently in the system by locking knowledge elements under revision to overcome authors entering conflicting updates (Kassahun and Scholten, 2006). The modelling KB is accessible to all interested, but without editing authorization only comments can be given by all registered users to the content of the modelling KB (*requirement 10*). The KB-Editor is at present extended to support better other types of collaborative research processes than only modelling. Furthermore, the KB-editor enables to develop glossaries related to the content of a process KB, including the modelling KB. A glossary for model-based water management is part of the Knowledge Base (*requirement 11*).

To guarantee a consistent and accepted modelling KB and to maintain and improve MoST, a Post HarmoniQuA Collaboration Framework has been established with expert partners from the HarmoniQuA project, but interested parties can join[20] (*requirement 12*).

The ontological approach realizes that those parts that have to be understood by computers and software is as formal as possible (*requirement 13*) and the part that has to be understandable by humans is in instances of the ontology, formulated by expert modellers, is more informal and should be clear to human users (*requirement 14*).

The consistency and completeness of the modelling KB (*requirement 15* and *requirement 16*) are hard to evaluate, although the Knowledge Base has been tested in 21 test case studies and has been assed in many other ways, which is reported elsewhere (Scholten and Beulens, 2005; Scholten, 2006).

Although the need for a modelling KB, encouraging good modelling practices, is widely shared within the professional community of those involved

---

[19]See for information on Protégé: http://protege.stanford.edu/.
[20]See http://harmoniqua.wau.nl/public/organisation.htm.

in model-based water management, it is not very well appreciated nor used (*requirement 17*). Modellers feel it as a straitjacket that forces them to work according to strict guidelines and regard it as just *overhead* instead of *help*. But, if they used it anyway, they experience the ease of use and the benefits of making explicit what they actually did.

The modelling KB has been developed by professional modellers for water management (with support of knowledge engineers and software engineers) and reflects therefore the complexity nowadays often found in multidisciplinary model-based water management projects, in which team members cooperate and contribute their disciplinary expertise to accomplish the goals of the project. In this way *requirement 18* is met, as the modelling KB facilitates mutual understanding between team members with different disciplinary backgrounds.

The modelling KB distinguishes that team member play different roles (user types: (*water*) *manager*, *modeller*, *auditor*, *stakeholder* and *public*), allocates different pieces of a modelling project to each of them and provides dedicated guidance filtered per role from the Knowledge Base (*requirement 19*). Furthermore, disciplinary specialists can work within their own domain, e.g. *hydrodynamics*, *groundwater* (quantity and quality), *surface water quality*, *precipitation-runoff*, *flood forecasting ecology* and *socio-economics*. This handles *requirement 20*.

Modelling projects range from simple routine jobs to complex integrated modelling approaches or research oriented projects. Each requires its own subset of the full guidelines in the modelling KB (*requirement 21*). This is achieved by providing templates for various types of model complexity.

Although the modelling KB has been designed and implemented for water management, its design and especially its ontological structure allows to extend it for modelling in other water management domains and even for modelling with simulation models for other application domains, other types of modelling than just simulation and, finally, for other processes (*requirement 22*). This has been realised by organising the ontological structure in layers ranging from very generic to very detailed, which allows that more generic parts of the ontological Knowledge Base can be reused (*requirement 23*).

The KB-editor is a web based tool and the modelling KB is exported from Protégé in XML-format, which makes them operating platform independent (*requirement 24*). Furthermore, the KB-editor is designed according to object oriented software engineering principles (*requirement 25*).

The whole modelling support system has been set-up in a client-server architecture, discussed in Section 4.2 (*requirement 26*) with the ontological structure and the modelling KB content on some central server (*requirement 27*) and the KB-editor as web based tool at the client side (*requirement 28*).

A.2. Modelling Support Tool, MoST

Functional requirements to the Modelling Support Tool, MoST, are summarized in Table A.2 and how these requirements are met in the design and implementation of MoST will be discussed subsequently.

The Modelling Support Tool, MoST aims at supporting the full life-cycle for multidisciplinary teams in model-based problem-solving for water management. The support consists of the following features: *providing guidance* from the modelling KB, *monitoring* (and storing in modelling journals what multi-disciplinary teams do in a model project plus their results) and *reporting* (the content of a modelling journal). In this way *requirement 1* is met.

MoST understands and can handle concepts like user type, i.e. a role of a team member (*requirement 2*) and supports the following roles: (*water*) *manager*, *modeller*, *auditor*, *stakeholder* and *public*. Furthermore MoST allows subprojects operating in one or more application domains (*requirement 3*), at present including *hydrodynamics*, *groundwater* (quantity and quality), *surface water quality*, *precipitation-runoff*, *flood forecasting ecology* and *socio-economics*.

MoST's Graphical User Interface (GUI) gives access to its functionality (*requirement 4*). Most can run in different modes: setting up projects, providing guidance, monitoring what teams do and helping to report projects (see Section 4.5).

MoST should help with the complex task to *set up modelling projects* (*requirement 5*), including naming of the project and indicating version, defining subprojects and associated domains/disciplines, selecting relevant tasks per subproject, defining team members, specifying roles per team member, authorizing them per subproject and role and appointing project administrator(s) and editing the scoreboard template questions.

In the *guidance mode*, MoST provides three views on the KB (*requirement 6*): (1) *tree view* (resembles standard windows trees like and allow easy browsing through *steps* and *tasks*), (2) *flowchart view* (allows browsing through the network of *steps* and *tasks*, but provides also a structured view with the order of *tasks* and redo-loops) and (3) *task view* (is a detailed, textual description of a task). Tasks and activities are filtered per user and per application domain.

The *project mode* of MoST enables to set up projects and to monitor what team members actually do in a modelling project (*requirement 7*). Within a task, team members are guided on (1) which activities to perform, (2) how to perform these activities and (3) which methods are available. Furthermore they are encouraged to keep a log in the modelling journal that is shared by (authorized) team members.

Table A.2. Functional requirements to the Modelling Support Tool, MoST (based on Scholten and Beulens, 2005; Scholten, 2006; Scholten *et al.*, 2007)

Requirement
1. MoST should support *multidisciplinary* teams during the whole project lifecycle.
2. MoST should support that member of multidisciplinary teams have different user types, i.e. they can play different *roles*.
3. MoST should support that subprojects can handle parts of a modelling project within one or more water *management domains*. The following water management domains are included: *hydrodynamics*, *groundwater* (quantity and quality), *surface water quality*, *precipitation-runoff*, *flood forecasting ecology* and *socio-economics.*
4. A *GUI menu* should give access to all (most) functionality.
5. MoST should enable to *set-up projects.*
6. MoST should *present modelling guidelines* from the KB in various ways in order to support its use efficiently and effectively.
7. MoST should *monitor* of what team members do.
8. MoST should help making *reports.*
9. MoST should make use of a *glossary* that provides terminology per application domain.
10. The GUI of the tool should be easy to use without a steep *learning curve*, but it should also include most options wished by professionals that have to work with it on a daily base.
11. The Modelling Support Tool should be *portable* to other operating systems (platforms).
12. The components of the tool (providing guidance, project initialization, project execution, reporting) should be *reliable*, *high potential*, *easy to maintain/ update* and *fast.*
13. The Modelling Support Tool should *automatically* use changes in the KB.
14. The Modelling Support Tool should be designed in such way that it facilitates adapting it for *other languages*.
15. The Modelling Support Tool should be used by a team working *distributed* in a network setting (LAN or Internet).
16. The components of the system should be distributed in a *client-server architecture.*
17. A single *central server* should be used for all versions of the KB in order to facilitate maintenance and upgrading.
18. Projects should run on *project servers.*
19. Computers with connection to the central server and to the project server should run a client application for executing project work per team member.
20. A server application should *synchronize* contributions of team members in the modelling journal.

The *report mode* of MoST helps to make all kind of reports for various audiences and purposes (*requirement 8*).

MoST generates hyperlinks to terms in the task and activity descriptions, which connect these terms to the modelling KB glossary (*requirement 9*).

In the design of MoST much effort is spent to keep the tool as simple as possible, but nevertheless it appeared difficult for novice users and therefore extensive multimedia training material has been developed to flatten the steep learning curve (*requirement 10*).

MS Windows™ is the dominant operating platform, but a substantial part of modellers uses other platforms, usually UNIX (or its variants). The Modelling Support Tool and the server application are developed in Java. This guarantees platform independency (*requirement 11*).

Reliability is an essential functionality in teamwork support. Reliability includes the following aspects: (1) what people did should be carefully recorded and filed, (2) high potentials are needed for extensive use by professionals, (3) easy maintenance and updates are obvious for the Knowledge Base, (4) the tool should be adaptable to new insights and (5) the size of the KB requires that the tool should be able to present the relevant parts of the KB fast (*requirement 12*).

MoST works with a local copy of the modelling KB (in XML-format). Changes in the modelling KB can be used by MoST, as soon as a copy of the new version is downloaded from the central server (*requirement 13*).

MoST (until version 3.1.5) is not designed to be used with a GUI in other languages than English. Its successor ProST (Process Support Tool) is prepared for that. All tool related terminology is stored in a single file. Terms translated in another language can replace the English equivalents (*requirement 14*).

A typical project setting, in which MoST can help managing project progress and quality, is as follows. Each team member has her own PC (desktop or notebook), connected to the internet or a LAN (*requirement 15*). PCs of team members and a project server are organised in a client-server setting (*requirement 16*). All team members can download the latest version of the modelling KB (or another version) from a central server (*requirement 17*). Team members share a modelling journal, which is stored on some project server on Internet or LAN (*requirement 18*). MoST runs on the PCs of team members and acts as client application (*requirement 19*).An application running on the server handles information exchange between client application (i.e. MoST on team members' PCs) and server application on the project server. The server application guarantees synchronization of the modelling journal (*requirement 20*).

APPENDIX B DETAILS OF HARMONIQUA'S MODELLING GUIDELINES

B.1. Terminology

Table B.1. Terminology used in HarmoniQuA's modelling KB and MoST

Term	Definition
Activity	Each task in HarmoniQuA is decomposed into a number of activities comprising descriptions of things to be done within the specific tasks.
Auditor	A person that is conducting some kind of review of a modelling study. The review may be more or less comprehensive depending on the requirements of the particular case. The auditor is typically appointed by the water manager to support the water manager to match the modelling capability of the modeller. The auditor may belong to the water manager's organisation or to an independent organisation. The task of the auditor should be distinguished from internal reviews, which the modeller's organisation often carries out as part of its internal quality assurance procedure. The reviews performed by the auditor are done over and above any internal reviews that the modeller may carry out.
Biota domain	The biota (ecological) modelling domain comprises water quality processes in all surface water elements such as: • River systems including flood plain, lakes and reservoirs • Estuaries and coastal waters The biota (ecological) domain in HarmoniQuA corresponds to the ecological aspects in 'river' 'lake' 'transitional water' and 'coastal water', as these terms are defined in the Water Framework Directive
Client	Synonym of Water Manager
Consultant	Synonym of Modeller
Domain	In HarmoniQuA the description of the modelling process is decomposed into seven different domain: • groundwater (including solute transport and geochemistry) • precipitation runoff (including non-point pollution) • hydrodynamics (flows in pipes, rivers, estuaries, coastal water and associated sediment transport and morphology) • flood forecasting

(*continued*)

Table B.1. *Continued*

Term	Definition
	• surface water quality • biota (ecology) • socio-economics In addition there is a generic domain composed of general guidelines. This is available for the MoST user as 'others'
Flood forecasting domain	The flood forecasting modelling domain comprises all modelling activities related to operational, real-time forecasting of floods in rivers and coastal waters. The flood forecasting domain in HarmoniQuA is applicable to the 'river', 'transitional water' and 'coastal water', as these terms are defined in the Water Framework Directive.
Generic	The general guidelines that covering all general aspects of modelling. The generic domain serves two functions: • It is the backbone of the Knowledge Base for the seven specific domains implying that for many tasks and activities the text in the respective domains comes directly from the generic domain • A MoST user which do not find that his modelling study fits well to any of the seven specific domains may use the generic domains by selecting 'Others' when customising his project.
Ground-water domain	The groundwater modelling domain comprises modelling activities related to groundwater, i.e.: • Groundwater flow (heads and flows) • Solute transport (advection, diffusion and hydrodynamic dispersion) • Geochemistry (ad/absorption, ion exchange, complexation, degradation, etc.) The groundwater domain in HarmoniQuA corresponds to the term 'groundwater' as this term is defined in the Water Framework Directive.
Hydrodynamic domain	The hydrodynamic modelling domain comprises flows of water, sediment transport and morphological aspects in all surface water elements such as: • Water supply systems • Sewer drainage systems • River systems including flood plain, lakes and reservoirs • Estuaries and coastal waters

(*continued*)

Table B.1. *Continued*

Term	Definition
	The hydrodynamic domain in HarmoniQuA corresponds to the water flow and sediment aspects in 'river' 'lake' 'transitional water' and 'coastal water' as these terms are defined in the Water Framework Directive.
Method	A description of a methodology for carrying out an activity, possibly with links to supporting tools.
Modeller	A specialist who undertakes the technical modelling activity. A modeller should have expert knowledge of both the issue and the sound model modelling approaches. The modeller should present and interpret the model outputs to the water manager and to the stakeholders. The modeller is the counterpart of the water manager. If the modeller and the water manager belong to different organisations their roles will typically be denoted consultant and client, respectively.
Multi-domain	When a modelling project in MoST is composed of more than one domain it is denoted as a multi-domain project
Precipitation-runoff domain	The precipitation-runoff domain comprises modelling activities related to precipitation-runoff processes, i.e.: • Flow processes (precipitation, evapotranspiration, runoff etc., typically at catchment scale) • Non-point pollution (e.g. nitrate, pesticides originating from agricultural practise) The precipitation-runoff domain in HarmoniQuA corresponds more or less to the term 'surface water' related to 'river basin' as these terms are defined in the Water Framework Directive.
Public	The public is composed of partners that are not directly involved (as modeller, manager or auditor) in a modelling study but that have a legitimate interest in the modelling results. The public may typically be either interest groups/stakeholders or the general public.
Socio-economics domain:	The socio-economic modelling domain comprises all modelling activities related to human interventions and evaluations. The model elements include the driving socio-economic factors and impacts of water use, the costs and benefits of water management, the socio-economic optimisation procedures, and the water policies.

(*continued*)

Table B.1. *Continued*

Term	Definition
Stakeholder	The interested parties, i.e. those with a stake in the water management issue, either their interest is in exploiting or protecting the resource. Stakeholders include the following different groups: (i) competent water resource authority; (ii) interest groups; and (iii) general public. The water manager typically belongs to one of the first two groups.
Step	HarmoniQuA decomposes the modelling process into five major steps and 48 tasks. The five steps are: (1) Model Study Plan (2) Data and Conceptualisation (3) Model Set-up (4) Calibration and Validation (5) Simulation and Evaluation
Surface water quality domain	The surface water quality modelling domain comprises water quality processes in all surface water elements such as: • River systems including flood plain, lakes and reservoirs • Estuaries and coastal waters The surface water quality domain in HarmoniQuA corresponds to the water quality aspects in 'river' 'lake' 'transitional water' and 'coastal water', as these terms are defined in the Water Framework Directive.
Task	HarmoniQuA decomposes the modelling process into five major steps and 48 tasks. The five steps are: (1) Model Study Plan (2) Data and Conceptualisation (3) Model Set-up (4) Calibration and Validation (5) Simulation and Evaluation
Water manager	A water manager is the person or organisation responsible for the management or protection of the water resource, and thus of the modelling study and its outcome (the problem owner). The water manager is the counterpart of the modeller. If the water manager and the modeller belong to different organisations their roles will typically be denoted client and consultant, respectively.

B.2. *Steps* and *tasks* of HarmoniQuA's modelling guidelines

Table B.2. List of the five steps and the 48 tasks of HarmoniQuA's modelling guidelines. Steps are left aligned and **bold**. Tasks are right aligned and SMALL CAPS. All task definitions are from HarmoniQuA's modelling guidelines, version 2.3

Concepts	Meaning
Model Study Plan	This step aims at setting-up a project by performing the following tasks: • Describe Problem and Context • Define Objectives • Identify Data Availability • Determine Requirements • Prepare Terms of reference • Proposal and Tendering • Agree on Model Study Plan and Budget
DESCRIBE PROBLEM AND CONTEXT	*Task* definition of the *step Model Study Plan*: A clear, precise (not necessarily quantitative) specification of the known problem details, the context of the study and the type of calculations to be made.
DEFINE OBJECTIVES	*Task* definition of the *step Model Study Plan*: Specification of the goals to be achieved.
IDENTIFY DATA AVAILABILITY	*Task* definition of the *step Model Study Plan*: Make a structured list of data, including its relevance, availability and quality for the problem at hand.
DETERMINE REQUIREMENTS	*Task* definition of the *step Model Study Plan*: Specify requirements to the quality of the model study.
PREPARE TERMS OF REFERENCE	*Task* definition of the *step Model Study Plan*: Terms of Reference is the specifications for the modelling job to be carried out.
PROPOSAL AND TENDERING	*Task* definition of the *step Model Study Plan*: Tendering is the process of selecting a consultant for a modelling job on the basis of technical and financial proposal(s).
AGREE ON MODEL STUDY PLAN AND BUDGET	*Task* definition of the *step Model Study Plan*: The water manager and the modeller reach an agreement on the technical and financial conditions of the modelling job.

(*continued*)

Table B.2. *Continued*

Concepts	Meaning
Data and Conceptualization	This *step* aims at • An adequate description of the system • Correct handling and evaluation of data • An adequate description • Evaluation of conceptual models • Selecting (existing) model code by performing the following *tasks*: • Describe System and Data Availability • Process Raw Data • Sufficient data? • Model Structure and processes • Model Parameters • Summarize Conceptual Model and Assumptions • Need for Alternative Conceptual Models? • Process Model Structure Data • Assess Soundness of Conceptualization • Code Selection • Report and Revisit Model Study Plan (Data and Conceptualization • Review Data and Conceptualization and Model Set-up Plan
DESCRIBE SYSTEM AND DATA AVAILABILITY	*Task* definition of the *step Data and Conceptualization*: Identify the processes to be represented, and the data needed, to setup and validate a model of sufficient complexity to meet the objectives defined in the Model Study Plan.
PROCESS RAW DATA	*Task* definition of the *step Data and Conceptualization*: Collection, pre-processing and evaluation of raw data.
SUFFICIENT DATA?	*Task* definition of the *step Data and Conceptualization*: Determine if there is sufficient qualified data to fulfil the project objectives as defined in the Model Study Plan. Qualified data refers to data that has been collected, evaluated and deemed applicable for use in the model study.

(*continued*)

Table B.2. *Continued*

Concepts	Meaning
MODEL STRUCTURE AND PROCESSES	*Task* definition of the *step Data and Conceptualization*: Delineation of the physical extent, dimensionality, internal framework, boundaries and processes of the model domain, and a description of how they will be represented mathematically.
MODEL PARAMETERS	*Task* definition of the *step Data and Conceptualization*: Framing of parametric representation or parameterization of system characteristics.
SUMMARIZE CONCEPTUAL MODEL AND ASSUMPTIONS	*Task* definition of the *step Data and Conceptualization*: Summarize the current understanding of the system in terms of verbal descriptions, graphical presentations, equations, governing relationships, and/or natural laws that purport to describe reality
NEED FOR ALTERNATIVE CONCEPTUAL MODELS?	*Task* definition of the *step Data and Conceptualization*: Review conceptual model to determine whether there is sufficient uncertainty over the model structure and processes to warrant the development and evaluation of alternative formulations.
PROCESS MODEL STRUCTURE DATA	*Task* definition of the *step Data and Conceptualization*: Transform and/or organize the qualified data into a database for use in constructing the model.
ASSESS SOUNDNESS OF CONCEPTUALIZATION	*Task* definition of the *step Data and Conceptualization*: An evaluation of the credibility and suitability of the conceptual model based upon professional judgment.
CODE SELECTION	*Task* definition of the *step Data and Conceptualization*: Decide on appropriate modelling software that is consistent with the conceptual model and the study objectives.
REPORT AND REVISIT MODEL STUDY PLAN (DATA AND CONCEPTUALIZATION)	*Task* definition of the *step Data and Conceptualization*: Report on step Data and Conceptualisation, to cover the database and metadata, and also the conceptual model and its soundness. Draft associated chapters of the Project Report. Revise the Model Study Plan with respect to the tasks in the next modelling step.
REVIEW DATA AND CONCEPTUALIZATION AND MODEL SET-UP PLAN	*Task* definition of the *step Data and Conceptualization*: Review the work carried out during the modelling step Data and Conceptualisation.

(*continued*)

Table B.2. *Continued*

Concepts	Meaning
Model Set-up	This *step* aims at setting-up the model to be used by performing the following *tasks*: • Construct Model • Test Runs Completed • Specify and Update Calibration and Validation Targets and Criteria • Report and Revisit Model Study Plan (model Set-up) • Review Model Set-up and Calibration and Validation Plan
CONSTRUCT MODEL	*Task* definition of the *step Model Set-up*: Construct model is the process of transforming a conceptual model into a model set-up for the study area using the selected model code, where both set-up data and simulations parameters have been defined.
TEST RUNS COMPLETED	*Task* definition of the *step Model Set-up*: A test run is carried out in order to identify any problems with the constructed model (set-up data and computational control parameters) and/or execution of the selected code.
SPECIFY AND UPDATE CALIBRATION AND VALIDATION TARGETS AND CRITERIA	*Task* definition of the *step Model Set-up*: Calibration and validation accuracy targets should be proposed as measures for the acceptance criteria prior to undertaking model calibration.
REPORT AND REVISIT MODEL STUDY PLAN (MODEL SET-UP)	*Task* definition of the *step Model Set-up*: Report the Model set-up and write the associated chapter of the Project Report and revising the Model Study Plan with respect to the tasks in the next modelling step.
REVIEW MODEL SET-UP AND CALIBRATION AND VALIDATION PLAN	*Task* definition of the *step Model Set-up*: Review the work carried out during the modelling step Model set-up.
Calibration and Validation	This *step* aims at calibrating and validating the model to be used by performing the following *tasks*: • Specify Stages in Calibration Strategy • Select Optimization Method • Define Stop Criteria • Select Calibration parameters

(*continued*)

Table B.2. *Continued*

Concepts	Meaning
	• Parameter Optimization • All Calibration Stages Completed? • Asses Soundness of Calibration • Validation • Asses Soundness of Validation • Uncertainty Analysis of Calibration and Validation • Scope of Applicability • Report and Revisit Model Study Plan (Calibration and Validation) • Review Calibration and Validation and Simulation Plan
SPECIFY STAGES IN CALIBRATION STRATEGY	*Task* definition of the *step Calibration and Validation*: Subdivision of the entire calibration process into sub-steps.
SELECT OPTIMIZATION METHOD	*Task* definition of the *step Calibration and Validation*: Select a parameter optimization method for model calibration
DEFINE STOP CRITERIA	*Task* definition of the *step Calibration and Validation*: Define one or more criteria to determine when to stop calibration.
SELECT CALIBRATION PARAMETERS	*Task* definition of the *step Calibration and Validation*: Determine which parameters (and other model input, e.g. decision variables) will be used for calibration and which will be kept constant for all calibration model runs.
PARAMETER OPTIMIZATION	*Task* definition of the *step Calibration and Validation*: Execute parameter optimization.
ALL CALIBRATION STAGES COMPLETED?	*Task* definition of the *step Calibration and Validation*: To evaluate whether all steps in calibration strategy have been finalized satisfactorily.
ASSES SOUNDNESS OF CALIBRATION	*Task* definition of the *step Calibration and Validation*: The assessment of the soundness of the calibration is an evaluation of the credibility of the calibrated model based upon professional judgment.
VALIDATION	*Task* definition of the *step Calibration and Validation*: Substantiation that a model within its domain of applicability possesses a satisfactory range of accuracy consistent with the intended application of the model.

(*continued*)

Table B.2. *Continued*

Concepts	Meaning
ASSES SOUNDNESS OF VALIDATION	*Task* definition of the *step Calibration and Validation*: The assessment of the soundness of the validation is an evaluation of the credibility of the calibrated model based upon professional judgment.
UNCERTAINTY ANALYSIS OF CALIBRATION AND VALIDATION	*Task* definition of the *step Calibration and Validation*: Quantification of uncertainty in the calibrated and validated model due to incomplete knowledge of model parameters, input data, boundary conditions and conceptual model. In an uncertainty analysis the combined effects of these uncertainties are taken into account.
SCOPE OF APPLICABILITY	*Task* definition of the *step Calibration and Validation*: This task describes the circumstances or conditions under which the model has documented predictive capabilities.
REPORT AND REVISIT MODEL STUDY PLAN (CALIBRATION AND VALIDATION)	*Task* definition of the *step Calibration and Validation*: Reporting the model calibration and validation and writing the associated chapters of the Project Report and revising the Model Study Plan with respect to the tasks in the next modelling step.
REVIEW CALIBRATION AND VALIDATION AND SIMULATION PLAN	*Task* definition of the *step Calibration and Validation*: Review of the tasks carried out within the modelling step Calibration and validation.
Simulation and Evaluation	This *step* aims at defining and running the scenarios that will be used to solve the problem at hand by performing the following *tasks*: • Set-up Scenario • Simulations • Check Simulations • Analyze and Interpret results • Assess Soundness of Simulation • Uncertainty Analysis of Simulation • All Scenarios Completed • Reporting of Simulation and Evaluation • Review of Simulation and Evaluation • Need for Post Audit • Model Study Closure

(*continued*)

Table B.2. *Continued*

Concepts	Meaning
SET-UP SCENARIO	*Task* definition of the *step Simulation and Evaluation*: Adjust validated model to match the scenario, and prepare any time-series data needed to run the model.
SIMULATIONS	*Task* definition of the *step Simulation and Evaluation*: Use of a validated model to quantify the response of the system to possible future events. Real-time runs are aimed at producing model output to estimate future states of the system within the lead time of the forecast. Simulations are made by running the model with a set of validated parameters and imposing expected future stresses and/or management scenarios.
CHECK SIMULATIONS	*Task* definition of the *step Simulation and Evaluation*: Screening the results of the simulations for possible errors. The check of simulations is an inspection of the model results with the objective of exposing extreme or incorrect model output.
ANALYZE AND INTERPRET RESULTS	*Task* definition of the *step Simulation and Evaluation*: Critical examination of the results of the simulations.
ASSESS SOUNDNESS OF SIMULATION	*Task* definition of the *step Simulation and Evaluation*: A qualitative assessment of the model simulations that takes into account quantitative measures and practical experience.
UNCERTAINTY ANALYSIS OF SIMULATION	*Task* definition of the *step Simulation and Evaluation*: Assessment of uncertainty in model results due to incomplete knowledge of model parameters, input data, boundary conditions and conceptual model. Furthermore the uncertainty originating from the decision context (external factors) may be included.
ALL SCENARIOS COMPLETED	*Task* definition of the *step Simulation and Evaluation*: Decide whether another scenario needs to be analysed. If not, compare results from those already analysed.
REPORTING OF SIMULATION AND EVALUATION	*Task* definition of the *step Simulation and Evaluation*: Reporting of the results from the modelling step Simulation and Evaluation.
REVIEW OF SIMULATION AND EVALUATION	*Task* definition of the *step Simulation and Evaluation*: Review of the tasks carried out within the modelling step Simulation and Evaluation.

(*continued*)

Table B.2. *Continued*

Concepts	Meaning
NEED FOR POST AUDIT	*Task* definition of the *step Simulation and Evaluation*: A post audit (or a post project appraisal) is a review of the modelling study carried out after some time when new information is available.
MODEL STUDY CLOSURE	*Task definition* of the *step Simulation and Evaluation*: Formal completion of the modelling job.

B.3. Details of the *task* 'Determine requirements' (*Step 1*)

Table B.3. Description of activities and sensitivities & pitfalls for the task 'Determine Requirements' under Step 1. The description is taken for the generic domain

Name of Task: Determine Requirements

Definition of task:

Specify quality standards required of the model, and define the project management structure and the resources required.

Name of activity: Quality and Relevance

User Responsible for performing activity: Water Manager

Description of activity:
If a model is to support decision making, requirements should be specified regarding the quality and relevance of its output. These requirements include:

- Which model outputs are essential?
- Which outputs have the highest priority?

Name of activity: Overall Quality

User responsible for performing activity: Water Manager

Description of activity:
It is essential that the overall requirements of the model study are clearly defined, or the model study may be useless. The requirements should be quantified in terms of output uncertainty, preferably after calibrating and validating the model – though this may be a difficult task for models that are multi-variate and dynamic in space and time. Uncertainty may be quantified as say an 'error' averaged over time and space or by some other measure of overall model quality. The criteria should be decided at this stage of the model study so that all parties can contribute to prioritising the study resources, and later evaluate the overall success of the work.

However, numerical criteria of model quality do not always capture the full essence of how good or poor a model is. A visual appreciation of overall model quality and of where errors occur (e.g. at high or low flows, in extreme or average conditions, in quick or slow response rates, in winter or summer conditions, etc) can be as or even more important. Similarly, although formal calibration and validation are vital for getting the believable best out of a model, they are usually based on numerical criteria, and may also require considerable input and observational data that may not exist.

Accepting greater uncertainty, a raw (uncalibrated and unvalidated) model could be used. However, some basic level of calibration and validation should still be possible, transferring parameters found in previous studies, and checking model performance against less objective measures of system behaviour. For example, in the PR and HD domains: model peak flows may be compared with field observations of channel capacities, or with trash lines and level marks at bridges and channel structures; model 'times to peak' and flood extents may be compared with information gained from onlookers or newspaper reports. In the GW domain, model water tables may be compared with reported spring lines or extremes in well levels. Linking the WQ, BI and HD domains, modelled and observed areas of sedimentation and habitat types may be compared. Such expert but less formalised model assessments may be compared with the Assess Soundness tasks described throughout the MoST flowchart.

Depending on the overall model objectives, other circumstances in which a raw model might be used include: trying to develop a theoretical understanding of a problem; assessing or broadly ranking a number of strategic options; or even defining what additional data are most needed to develop and validate the model, and where they should be collected.

Name of activity: Model Analyses

User responsible for performing activity: Water Manager

Description of activity:
Define the model analyses that are needed and their purpose. Only an initial assessment is required, and this may be updated later, based on recommendations from the auditor and the modeller, in the decision tasks at the end of the five model steps.

The need to model steady state and dynamic conditions (as individual events or continuous time series) should be considered. Different approaches may suit different domains, with processes occurring over a range of time scales, and data available at a mix of time steps (from under an hour to maybe three months or more). If dynamic effects are a concern, event models are generally more suitable where time-series data is scarce or has many gaps, where the problem relates to events (e.g. floods), or where model complexity would require excessive computer resources. Also, event models may omit processes that have little short term impact, and calibration can focus on conditions of most concern. By contrast, continuous simulation can involve running over long periods of little interest, and calibration may favour average rather than critical conditions. However, continuous models are usually better at representing the linkages between domains and processes that respond at different characteristic rates, and outputs do not depend on arbitrary choices of the initial and boundary conditions to use for specific events.

The requirements should also define whether the analyses should include:

- Sensitivity analysis, to enable the selection of parameters and other 'factors' to be used in calibration.
- Calibration, to reduce differences between model output and system observations.
- Validation, to check whether the model is generic or specific to the calibration data.

Of these analyses, calibration has the highest priority and validation the highest value. Calibration can be a very complex procedure involving not just sensitivity analysis to select the parameters to adjust, but 'decoupled' calibration of model components (domains/processes/spatial units), a range of manual/automatic optimisation methods, and various performance criteria. Relevant guidance can be found under the task Specify or Update Calibration and Validation Targets and Criteria in Step 3 (Model Set-up), and under the first five tasks in Step 4 (Calibration and Validation).

Validation is of a more generic nature and increases the credibility of the model. The system response data is partitioned, often about a selected date, with the earlier events or time-series used for calibration and the later events or time-series used for validation. Results can be affected by changes to the system (or inputs) during the data record, but also the data could be partitioned in a way that tests the model response to such changes. Alternatively, if data are available at several locations, some locations could be excluded from calibration, but used to validate the model's spatial structure. Guidance on validation methods can be found under the task Specify or Update Calibration and Validation Targets and Criteria in Step 3 (Model Set-up), and under the tasks Validation in Step 4 (Calibration and Validation).

It should also be recognised that scenario analysis may involve making changes to the validated model. Such changes should properly be validated by post project appraisal.

Name of activity: Performance Criteria

User responsible for performing activity: Water Manager

Description of activity:
Performance criteria used to assess model quality may include subjective measures (based on past experience and an appreciation of model performance and parameter values), but they should also include numerical scores of how well the model performance matches observed system behaviour. Such 'goodness of fit' criteria (including the mean error and mean squared error between modelled and observed time series of system outputs) can become 'objective functions' to

optimise in model calibration. Many numerical criteria could be used, each prioritising the goodness of fit to different aspects of system behaviour (water levels, high flows, low flows, annual peaks, forecast accuracy at different lead times, pollutant flushes, annual loads, biodiversity scores, etc.). Not all these criteria are equally useful or can be easily interpreted. The choice of performance criteria is a subjective decision, particularly in multi-objective studies, where the balance of priorities between achieving the various objectives must also be considered.

To improve communication between the water manager and the modeller, the numerical criteria to be used and their method of calculation should be specified. Similarly, a precise definition of the test sample should also be given by the client. This will prevent the modeller only choosing those events where the model performs adequately. However, model performance should not be assessed by numerical criteria alone, and guidance on the role of qualitative criteria is given in the tasks All Calibration Stages Completed and Assess Soundness for calibration and validation.

If the water manager wants model performance to be assessed comparatively to a benchmark model (i.e. a reference model) then this should be specified. This may be a simple model (just derived from data) or more sophisticated. The objective is to provide a reference that should be exceeded to ensure good model credibility. For example, the mean observation may be used as a reference, comparing the mean square error between observed and modelled values with the mean square difference between observed values and their mean (see Nash-Sutcliffe). For real-time forecasting, a more useful reference is the current value (i.e. no change during the forecast lead time), comparing the mean square forecast error with the mean square difference between observed values separated by the forecast lead time (see Persistence Criterion).

The requirements defined here may be reviewed during the project. Additional guidance is given under task Specify or Update Calibration and Validation Targets and Criteria in Step 3 (Model Set-up).

Name of activity: Multi-domain Couplings

User responsible for performing activity: Water Manager

Description of activity:

Describe the interactions between the different domains. In case of interactions the coupled modelling can be carried out in two fundamentally different ways: either (i) as fully coupled models, where feed back between model domains are ensured, or (ii) as uncoupled (or chained) models, where there is no feedback. Fully coupled models are usually troublesome to operate, because the modelling system becomes much more complex, both with respect to code and numerical stability. On the other hand lack of important feedback may some times result in significantly poorer quality of results. It is therefore particularly important to describe the importance of possible feedbacks between domains. This information should be included in the Terms of References. By using appropriate coupling techniques, such as OpenMI, separated models can be linked, including feedback loops. The main advantage of such an approach is that stand-alone models can be coupled, so that the development of a single complex model is not needed to cover several different domains.

Applicable Method: OpenMI

Name of activity: Stakeholder Expertise

User responsible for performing activity: Water Manager

Description of activity:
Assess and describe the requirements of stakeholders in terms of domain expertise.
To address a specific problem with a model, stakeholders must have expertise in appropriate domains. If the required expertise cannot be provided by the project participants, then there must be a plan of how to acquire it.

Name of activity: Resources

User responsible for performing activity: Water Manager

Description of activity:
An estimate must be made of the resources required to complete the project. There may be external forces or other reasons which restrict the amount of time/ manpower available and these must be taken into account during the planning of the project. There must be a clear relationship between the size of the project (time, manpower, computing equipment and budget) and its level of ambition. An indication of the global time required may be expressed as follows (the percentages below vary from case to case):

- Step 1: Model Study Plan – 10%
- Step 2: Data and Conceptualisation – 20%
- Step 3: Model Set-up – 20%
- Step 4: Calibration and Validation – 30%
- Step 5: Simulation and Evaluation – 20%

Feedback loops mean that some steps and tasks are performed more than once. The time expenditure given here refers to the total time to be spent. Occasionally, there may be a ready-made model which can be deployed directly and, in this case, Steps 1, 2 and 3 will take far less time.

Name of activity: Communication and Reporting

User responsible for performing activity: Water Manager

Description of activity:
The requirements for Communication and Reporting should be clearly defined. In the HarmoniQuA flowchart, each step ends with a decision task:

- Step 1: Agree on Model Study Plan and Budget
- Step 2: Review Conceptualisation and Model Set-up Plan
- Step 3: Review Model Set-up and Calibration and Validation Plan
- Step 4: Review Calibration and Validation and Simulation Plan
- Step 5: Review of Simulation

These decision tasks facilitate regular discussions between the water manager and modeller, based on the modeller's report on the work done in that step and draft updates to the Model Study Plan for the next step. This should diminish the risk of different perceptions on the frequency and form of communication and reporting. However, the water manager and modeller should be explicit on other communication and reporting requirements by specifying their focus, direction and frequency.

The level of detail required in communication and reporting will depend on the job complexity, and for relatively straight-forward studies, reporting and review tasks could be handled by less formal progress reports and project management meetings.

Name of activity: Other Requirements

User responsible for performing activity: Water Manager

Description of activity:
Discuss and list the additional requirements that should be included in the Terms of Reference. These may include:

- Use of the results from other models and the requirements which must be set for that purpose.
- Supply of results from current modelling project for use by other models.
- Scientific reporting.
- The hypotheses on which the model is based.
- The quality of the field data.
- Formulation of responsibilities for the final completion of the model study.
- How to deal with results originating from two or more different approaches (with various models and/or model programmes).
- Transfer of model results and/or model implementation to the client.
- Transfer of uncertainty information to the client.
- Translating model results into policy advice.
- Provision of digital files of model results.
- Archiving of results and model.

- Service in maintenance/warranty period.
- Possible post audit (post project appraisal) as a follow up to the present modelling study.

Name of activity: Managing Model Study

User responsible for performing activity: Water Manager

Description of activity:
Decide and document whether model study will be undertaken in-house or put out for tender. If put out for tender then it should be considered (decided under the task Proposal and Tendering) whether there will be a call for a single tender or multiple tenders and whether tenders will involve pre-qualifications or be open. In cases of domains and/or modelling projects that are relatively specialised, single or limited tenders are often advantageous.

Name of activity: Stakeholder Opinion on Requirements

User responsible for performing activity: Water Manager + Stakeholders + General Public

Description of activity:
Document stakeholder opinions on requirements.
HarmoniQuA addresses the minimum requirements, of the WFD, for stakeholder and general public participation (i.e. information provision and consultation) by allowing them to provide input at the start of the project (Model Study Plan) and to comment on work as it progresses during review tasks.

When requirements are determined, stakeholders and the general public shall be allowed to provide comments and opinions on the Model Study Plan which includes the Stakeholder Involvement Plan (SIP) developed by the manager and discussed with the modelling team.

According to the WFD it is the responsibility of managers to provide information to stakeholders and the general public about management timetables, issues and the participants, which are considered to be the foundation for all further participation activities. Furthermore, the WFD states, that the manager is responsible for proper consultation which should include encouraging written and oral responses.

Stakeholders and the general public should provide feedback on needs for adjustment of preferred interaction and communication tools (according to SIP) that are to be used for the subsequent step until next stakeholder feedback in task 2.12 Review Data and Conceptualisation and Model Set-up Plan (se method Interaction and communication ABC).

Comments and opinions should be structured according to the SIP (see Activity Prepare Stakeholder Involvement Plan – SIP).

Applicable Method: Interaction and Communication ABC

Name of activity: Answers to Stakeholder Opinions and Final SIP

User responsible for performing activity: Water Manager

Description of activity:
Evaluate and respond to stakeholder opinions on requirements. Responses should be structured according to the SIP (see activity Stakeholder Involvement Plan). The final and adjusted SIP should be included in the Terms of Reference, discussed with the modeller and subsequently included in the Model Study Plan.

During this process where adjustments to the SIP are carried out, Interaction and Communication Tools and actor regrouping may be necessary (see methods Interaction and Communication ABC and Actor grouping).

Applicable Method: Interaction and Communication ABC

Name of activity: Assess Uncertainty

User responsible for performing activity: Water Manager + Stakeholders + General Public

Description of activity:
Document the requirements for assessing uncertainty. The water manager should assess the uncertainty involved in the modelling study in consultation with the stakeholders/public. This activity will also require some modelling expertise. If the water manager does not have sufficient modelling expertise in his own organisation he will need to get support from external experts for this activity.

Uncertainty should be assessed in a broad context and it must not be limited to statistical uncertainty that is commonly addressed in comprehensive modelling studies. Thus the uncertainty should also include types of uncertainty that cannot be characterised by probabilities, but may be characterised in terms of scenarios or in qualitative terms. Furthermore, areas and processes where knowledge is known to be incomplete should be included.

The purpose of this activity at this early stage of the modelling study is to draw attention to uncertainty to ensure that the modelling study is designed so that it addresses the issues where there is expected to be most uncertainty and where it is possible to reduce the uncertainty.

The activity should use the Uncertainty Matrix or a similar qualitative method as a dialogue platform between the various parties.

Applicable Method: Uncertainty Matrix

Sensitivities and pitfalls:

- *Performance Criteria*: Establishment of performance criteria specifying the required accuracy of the model simulations is a very difficult task. The level of acceptable risk or uncertainty in a decision situation cannot be generalised. It will vary from country to country and from river basin to river basin according to the site specific socio-economic conditions. Or in other words insituations where much is at stake either economically and/or politically more accurate modelling simulations may be required. Usually, higher performance requirements will require better data and more thorough modelling studies and will therefore often be more costly. The trade-off between accuracy requirements and costs involved need to be made by the water manager after consultation with other relevant parties.
- *Importance of clearly defined Requirements*: This is a difficult part of the model study. Many disputes between the modeller and the client are caused by a lack of clear appointments on many of the activities of this task. For jobs of a more basic nature standard formulation can be used, but even then one should be explicit.
- *Public Participation*: Public participation can be weakened by a lack of resources (time, money, staff), a lack of rules of participation, a lack of in-depth involvement of authorities, a lack of hands-on in use of MoST for the stakeholders and general public and a lack of professional supervision of the process

---

B.4. Details of the *task* 'Validation' (*Step 4*)

Table B.4. Description of activities and sensitivities & pitfalls for the task 'Validation' under Step 4. The description is taken for the generic domain

---

Name of Task: Validation

Definition of task:

Substantiation that a model within its domain of applicability possesses a satisfactory range of accuracy consistent with its intended application.

Thus, validation is a test to determine whether a calibrated model can be used as a predictive tool, by demonstrating that the calibrated model provides an adequate representation of the physical system. Model simulations are, to the extent possible, compared with independent data to determine whether the model is able to reproduce the past with the required accuracy and whether the model is suitable for use as a predictive tool.

Name of activity: Design validation test procedure

User Responsible for performing activity: Modeller

Description of activity:
If a model is to support decision making, requirements should be specified regarding the quality and relevance of its output. These requirements include:

- Which model outputs are essential?
- Which outputs have the highest priority?

Applicable Methods:

- Proxy Basin Test
- Differential Split-sample Test
- Expert Evaluation of Model Performance
- Proxy Basin Differential Split-sample Test
- Comparative Assessment
- Split-sample Test

Name of activity: Confirm performance measures

User Responsible for performing activity: Modeller

Description of activity:
Review performance measures decided in the Task Specify or Update Calibration and Validation Targets and Criteria.

Applicable Methods:

- Standard Performance Measures
- Aggregated Performance Indicators
- Visualisation and Tabulation
- Event Validity
- Extreme-condition Test
- Statistical Validation
- Qualitative Performance Measures

Name of activity: Process validation data

User Responsible for performing activity: Modeller

Description of activity:
Process any additional time-series data necessary to conduct the planned validation strategy, extending the inventory of data prepared in Task Collect and Process Raw Data (Activity Evaluate pre-processed data) to cover any new sub periods of time-series data.

Applicable Methods:

- Visualisation and Tabulation
- Contouring

- Mass Balance
- Correlation and Regression
- Differential Split-sample Test
- Aggregation
- Surface Fitting
- Infilling (temporal)
- Zonation

Name of activity: Carry out validation test

User Responsible for performing activity: Modeller

Description of activity:
The tests are conducted by running the model, and the objective function values and other performance measures are stored in the model run log (established in Task Test Runs Completed) and attached to the Model Study Journal. The results are analysed and conclusions are derived with respect to whether the model meets the specified accuracy criteria. On this basis the validation status can be assessed.

Applicable Methods:

- Aggregated Performance Indicators
- Visualisation and Tabulation
- Differential Split-sample Test
- Expert Evaluation of Model Performance
- Standard Performance Measures
- Proxy Basin Test
- Event Validity
- Extreme-condition Test
- Proxy Basin Differential Split-sample Test
- Comparative Assessment
- Statistical Validation
- Qualitative Performance Measures
- Split-sample Test

Sensitivities and pitfalls:

- *Conditional Validation*: A model should only be assumed valid with respect to outputs for which they have explicitly been validated. This means, for instance, that a model which is validated against catchment runoff cannot automatically be assumed valid also for simulation of soil erosion on a hillslope within the catchment, because smaller scale processes may dominate here: it will need validation against hillslope soil erosion data.
- *Use of Model if Validation fails or not possible*: It must be realised that the validation tests proposed in this chapter are so demanding that many applications today would fail to meet them. This does not imply that these modelling studies are not useful, only that their output should be recognised to be somewhat more uncertain than is often stated and that they should not make use of the term ‘validated model.

B.5. *Methods* described in HarmoniQuA's modelling guidelines

Table B.5. List of the 90 methods described in HarmoniQuA's modelling guidelines with an indication of which task each method is included as an optional support

	Method	Used in Tasks
1	Stakeholder Summary Table	Describe Problem and Context
		Review Calibration and Validation and Simulation Plan
2	Interaction and Communication ABC	Determine Requirements
		Describe Problem and Context
		Review Data and Conceptualisation and Model Set-up Plan
		Review Model Set-up and Calibration and Validation Plan
		Review Calibration and Validation and Simulation Plan
		Review of Simulation and Evaluation
3	Cost-effectiveness analysis (CEA)	Describe System and Data Availability
		Describe Problem and Context
		Summarise Conceptual Model and Assumptions
		Model Structure and Processes
4	Cost-benefit analysis (CBA)	Describe System and Data Availability
		Describe Problem and Context
		Summarise Conceptual Model and Assumptions
		Model Structure and Processes
5	Multi-criteria analysis (MCA)	Describe System and Data Availability
		Describe Problem and Context
		Summarise Conceptual Model and Assumptions
		Model Structure and Processes

(*continued*)

Table B.5. *Continued*

	Method	Used in Tasks
6	Actors grouping	Determine Requirements
		Describe Problem and Context
7	Economic optimisation models	Describe System and Data Availability
		Describe Problem and Context
		Summarise Conceptual Model and Assumptions
		Model Structure and Processes
		Code Selection
		Construct Model
8	Metadata table	Identify Data Availability
9	Uncertainty Matrix	Determine Requirements
10	OpenMI	Determine Requirements
		Model Structure and Processes
		Code Selection
		Construct Model
11	Invited Tender without Pre-qualification	Proposal and Tendering
12	Invited Tender with Pre-qualification	Proposal and Tendering
13	Open Tender	Proposal and Tendering
14	Proposal Evaluation	Proposal and Tendering
15	Sole Source	Proposal and Tendering
16	Cost assessment methodologies	Describe System and Data Availability
		Summarise Conceptual Model and Assumptions
17	Economic benefit assessment methodologies	Describe System and Data Availability
		Summarise Conceptual Model and Assumptions
		Model Structure and Processes

(*continued*)

Table B.5. *Continued*

	Method	Used in Tasks
18	Visualisation and Tabulation	All Calibration Stages Completed?
		Specify or Update Calibration and Validation Targets and Criteria
		Collect and Process Raw Data
		Set-up Scenarios
		Check Simulations
		Assess Soundness of Simulation
		Model Parameters
		Test Runs Completed
		Specify Stages in Calibration Strategy
		Parameter Estimation
		Validation
		Construct Model
19	Site visits and Photographs	Collect and Process Raw Data
20	Aggregation	Collect and Process Raw Data
		Set-up Scenarios
		Check Simulations
		Assess Soundness of Simulation
		Test Runs Completed
		Specify Stages in Calibration Strategy
		Validation
21	Correlation and Regression	Collect and Process Raw Data
		Set-up Scenarios
		Model Parameters
		Specify Stages in Calibration Strategy
		Validation

(*continued*)

Table B.5. *Continued*

	Method	Used in Tasks
22	Infilling (temporal)	Collect and Process Raw Data
		Set-up Scenarios
		Specify Stages in Calibration Strategy
		Validation
23	Spatial Interpolation and Integration	Collect and Process Raw Data
		Set-up Scenarios
		Model Parameters
		Test Runs Completed
		Specify Stages in Calibration Strategy
		Validation
		Construct Model
24	Zonation	Collect and Process Raw Data
		Set-up Scenarios
		Model Parameters
		Specify Stages in Calibration Strategy
		Validation
		Construct Model
25	Contouring	Collect and Process Raw Data
		Set-up Scenarios
		Model Parameters
		Specify Stages in Calibration Strategy
		Validation
		Construct Model
26	Surface Fitting	Collect and Process Raw Data
		Set-up Scenarios
		Model Parameters
		Test Runs Completed
		Specify Stages in Calibration Strategy
		Validation
		Construct Model

(*continued*)

Table B.5. *Continued*

	Method	Used in Tasks
27	Double Mass Curves	Collect and Process Raw Data
28	Rating curves	Collect and Process Raw Data
29	Scatter plots	Collect and Process Raw Data
30	Pumping tests and slug tests	Collect and Process Raw Data
31	Mass Balance	Collect and Process Raw Data
		Set-up Scenarios
		Check Simulations
		Assess Soundness of Simulation
		Specify Stages in Calibration Strategy
		Validation
32	Cost assessment methodologies	Model Structure and Processes
33	Data Assimilation, Updating (general)	Set-up Scenarios
		Model Structure and Processes
34	Modifications of input	Set-up Scenarios
		Model Structure and Processes
35	Update of Parameters	Set-up Scenarios
		Model Structure and Processes
36	State Variable Updating	Set-up Scenarios
		Model Structure and Processes
37	Error Prediction	Set-up Scenarios
		Model Structure and Processes
38	BMW Toolbox	Code Selection
39	Verification of Code	Code Selection
40	Register of Ecological Models	Code Selection
41	Model Appraisal	Review Data and Conceptualisation and Model Set-up Plan
		Review Model Set-up and Calibration and Validation Plan
		Review Calibration and Validation and Simulation Plan
		Review of Simulation and Evaluation

(continued)

Table B.5. *Continued*

	Method	Used in Tasks
42	Peer Review	Review Data and Conceptualisation and Model Set-up Plan
		Review Model Set-up and Calibration and Validation Plan
		Review Calibration and Validation and Simulation Plan
		Review of Simulation and Evaluation
43	Checklist for Reviews/Audits/ Appraisals	Review Data and Conceptualisation and Model Set-up Plan
		Review Model Set-up and Calibration and Validation Plan
		Review Calibration and Validation and Simulation Plan
		Review of Simulation and Evaluation
44	Model Audit	Review Data and Conceptualisation and Model Set-up Plan
		Review Model Set-up and Calibration and Validation Plan
		Review Calibration and Validation and Simulation Plan
		Review of Simulation and Evaluation
45	Potential Evapotranspiration Estimation	Construct Model
46	Finite Difference	Construct Model
47	Finite Element	Construct Model
48	Thiessen Polygon Method	Construct Model
49	Global behaviour	Test Runs Completed
50	Robustness	Test Runs Completed
51	Mass balance	Test Runs Completed
52	Stability Test	Test Runs Completed
53	Standard Performance Measures	All Calibration Stages Completed?
		Specify or Update Calibration and Validation Targets and Criteria
		Parameter Estimation
		Validation

(*continued*)

Table B.5. *Continued*

	Method	Used in Tasks
54	Aggregated Performance Indicators	All Calibration Stages Completed?
		Specify or Update Calibration and Validation Targets and Criteria
		Validation
55	Qualitative Performance Measures	All Calibration Stages Completed?
		Specify or Update Calibration and Validation Targets and Criteria
		Validation
56	Event Validity	Specify or Update Calibration and Validation Targets and Criteria
		Validation
57	Statistical Validation	Specify or Update Calibration and Validation Targets and Criteria
		Validation
58	Extreme-condition Test	Specify or Update Calibration and Validation Targets and Criteria
		Validation
59	Automatic Optimisation	Select Calibration Method
		Parameter Estimation
60	Pure Random Search (GE,GW,WQ,BI,SE)	Select Calibration Method
61	Controlled Random Search	Select Calibration Method
62	Multistart and Clustering Methods	Select Calibration Method
63	Manual Optimisation	Select Calibration Method
		Parameter Estimation
64	Pure Random Search (FF,PR)	Select Calibration Method
65	Genetic Algorithms	Select Calibration Method
66	Combined Methods	Select Calibration Method
67	Parameter Ranges in Sensitivity Analysis	Select Calibration Parameters
68	Response Surface Method	Select Calibration Parameters
69	Regionalized Sensitivity Analysis	Select Calibration Parameters
70	Resampling Techniques	Select Calibration Parameters

(continued)

Table B.5. *Continued*

	Method	Used in Tasks
71	Monte Carlo Simulation	Uncertainty Analysis of Simulation
		Uncertainty Analysis of Calibration and Validation
		Select Calibration Parameters
72	Sensitivity Analysis	Uncertainty Analysis of Simulation
		Uncertainty Analysis of Calibration and Validation
		Select Calibration Parameters
73	Internal Validity	Assess Soundness of Calibration
74	Traces	Assess Soundness of Calibration
75	Split-sample Test	Validation
76	Differential Split-sample Test	Validation
77	Proxy Basin Test	Validation
78	Proxy Basin Differential Split-sample Test	Validation
79	Comparative Assessment	Validation
80	Expert Evaluation of Model Performance	Validation
81	Regression Techniques	Uncertainty Analysis of Simulation
		Uncertainty Analysis of Calibration and Validation
82	Sensitivity Analysis (PR)	Uncertainty Analysis of Simulation
		Uncertainty Analysis of Calibration and Validation
83	NUSAP	Uncertainty Analysis of Simulation
		Uncertainty Analysis of Calibration and Validation
84	Expert Elicitation	Uncertainty Analysis of Simulation
		Uncertainty Analysis of Calibration and Validation
85	Data Uncertainty (HarmoniRiB)	Uncertainty Analysis of Simulation
		Uncertainty Analysis of Calibration and Validation

(continued)

Table B.5. *Continued*

	Method	Used in Tasks
86	Error Propagation Equation	Uncertainty Analysis of Simulation
		Uncertainty Analysis of Calibration and Validation
87	State Space Techniques	Uncertainty Analysis of Simulation
88	Analytical Solutions of Stochastic PDEs	Uncertainty Analysis of Simulation
89	Scenario Analysis	Uncertainty Analysis of Simulation
90	Ensemble Prediction	Uncertainty Analysis of Simulation

Guidance Report I.2

Model calibration and validation in model-based water management

Henrik Madsen, Ann van Griensven and Anker Lajer Højbjerg

Abstract

Any model-based river basin management project requires proper model calibration and validation. The overall objective of model calibration is to select model parameters so that the model as closely as possible simulates the behavior of the considered system. A calibration protocol is outlined that includes the following steps: (i) Choice of calibration parameters, (ii) Specification of calibration objectives, (iii) Estimation of model parameters, and (iv) Validation of the calibrated model.

Concepts such as non-identifiability/equifinality of overparameterised models are introduced and the use of sensitivity analysis to select parameters to be estimated is presented. Often leading to multiple objectives to be maximized, the estimation problem can be formulated in a general multi-objective framework, considering multi-variable measurements, multi-site measurements and multi-response modes. The process of parameter estimation is either done manually or automatically, the latter gaining increasing attention. Manual methods take full advantage of the modeler's experience and knowledge. However, experiences are not easily transferred, potentially leading to different calibration results by different modelers. Conversely, automatic procedures are objective and require considerably less labor. However, they require numerical performance measures that may not be that easy to formulate. A section is devoted to the optimization algorithms, distinguishing local and global search methods and discussing the trade-off between effectiveness of the algorithm and the required computational load.

Model validation is essential to a river basin modeling project in order to document the predictive capabilities and credibility of the calibrated model. A hierarchical validation test scheme with increasing power is presented that distinguishes between model predictions performed on the same river basin as used for parameter estimation or on a different river basin, and whether predictions are done under the same conditions as present in the calibration period or under changed conditions. Finally, uncertainty aspects of model calibration are highlighted, with special attention devoted to the issue of non-uniqueness of the parameter estimates.

The whole is illustrated with four calibration examples, for a lumped, conceptual rainfall-runoff model, for a groundwater model, for a distributed integrated model and for a water quality model. Finally, a number of available optimization algorithm packages are described.

Keywords

Calibration, equifinality, identifiability, multi-objective optimization, parameter estimation, uncertainty, validation

 Modelling Aspects of Water Framework Directive Implementation Volume 1. By Peter A. Vanrolleghem. ISBN: 9781843392231. Published by IWA Publishing, London, UK.

ABOUT THIS GUIDELINE

In the quality assurance (QA) guideline for model based water management developed as part of the HarmoniQua project (Refsgaard *et al.*, 2005, see also the first part of this volume), the modelling process is divided into 5 major tasks:

(1) Formulation of model study plan
(2) Collection of data, conceptualisation and choice of model code
(3) Model setup
(4) Model calibration and validation
(5) Model simulations and evaluation of results

This document provides a guideline related to Task 4 on model calibration and validation. Although model calibration is considered as an isolated task in this document, it has important links to the other modelling tasks. If the calibration does not provide results of a sufficient accuracy according to some pre-specified criteria, the entire model setup phase in Task 3 should be reviewed. If the final model applications for which the model development was intended provide unsatisfactory results in Task 5, the entire calibration process should be reconsidered.

The objective of this document is to provide a guideline for calibration of models related to different modelling aspects in river basin management. As such, the aim of this guideline is to cover calibration aspects related to both water resources and water quality models, including groundwater, surface water and integrated modelling. Calibration and validation of economic and socio-economic modelling tools are not considered in this guideline.

A review of existing QA guidelines in different modelling domains is given in Refsgaard *et al.* (2005). Hill *et al.* (2004) provide a review of selected groundwater modelling guidelines. Guidelines with specific focus on model calibration include Hill (1998) and Sonnenborg and Henriksen (2005) for groundwater modelling, and Sorooshian and Gupta (1995) who describe model calibration mainly related to conceptual rainfall-runoff modelling.

Sensitivity and uncertainty analysis is an integrated part of model calibration. In the Harmoni-CA project separate guidelines have been prepared that cover these topics (see part 4 of this volume). Thus, in the present guideline sensitivity and uncertainty analysis will only briefly be discussed in relation to specific parts of the model calibration process.

This document is mainly targeted to the modeller to provide guidelines in the model calibration and validation phase of a river basin management project.

An overview of the different elements of the calibration process is provided, and recommendations are given for use of different analysis methods depending on the model application being considered. The document includes four application examples that exemplify the different steps in the model calibration process for different river basin management models. It is not intended to give a detailed treatment of all aspects of model calibration and validation. The guideline provides an extensive list of references to existing literature where more detailed information can be found.

The guideline includes an executive summary and a summary of each main chapter, which provide an overview of the contents of the guideline with emphasis on aspects that are particularly relevant to water managers to evaluate the model calibration and validation phase of a river basin management project.

1. EXECUTIVE SUMMARY

Model calibration and validation is a prerequisite in any model-based river basin management project. To obtain a credible river basin model that can be used for subsequent simulations, model parameters should be calibrated against measurements of different model responses being available (river runoff, groundwater levels, water quality parameters etc.). The calibration protocol is shown in the Figure 2.2.

The first step in model calibration is the selection of the parameters to be included in the calibration. Some river basin models, such as lumped, conceptual rainfall-runoff models, include relatively few parameters to be estimated and normally all parameters are included in the calibration without further consideration. More complex distributed models include a huge number of model parameters and model structures to be analysed, and in this case a rigorous model parameterisation and conceptualisation process is required in order to reduce the number of calibration parameters, and hence make the calibration manageable. In the model parameterisation available field data (e.g. geological, soil type, vegetation, and land use maps) are utilised to describe the spatial variations of a model parameter using few representative parameters. To screen the importance of the parameters defined in the model parameterisation and identify those that should be further refined via calibration sensitivity analysis can be applied. At this stage sensitivity analysis can also be applied to investigate the sensitivity of model responses that are of particular interest in the modelling study. If a parameter is insensitive with respect to the model responses for which calibration data are available (and hence cannot be well-determined by calibration), but sensitive for the model prediction considered, the model credibility for that particular modelling aspect is questionable.

The second step in model calibration is selection of data and formulation of calibration objectives. Data requirements for model calibration depend on the type of river basin model considered. For lumped, conceptual rainfall-runoff models the only available source of calibration data is the runoff at the catchment outlet. For calibration of distributed models both catchment integrated and distributed data should be included in the model calibration. To better constrain the model calibration it is important to include data of different types (e.g. both groundwater and runoff data for an integrated surface water-groundwater model), and define performance measures of different model response modes (e.g. overall water balance, high flow and low flow). The calibration data should reflect the natural variability of the variable being considered, e.g. include both dry and wet periods. To evaluate the model skills numerical performance measures are calculated. These measures are used together with visual judgement in manual calibration and as explicit objective functions to be optimised in automatic calibration. Scaling errors are of particular concern when comparing observed point measurements with simulated values that represent an average value within a computational element. In this case one may deliberately accept a bias in the model simulations to achieve a good correspondence of the dynamical model behaviour. Weights can be assigned to the observations used in the calibration to reflect the uncertainty of the observed data or to put emphasis on particular modelling aspects, thereby tailoring the model calibration to the specific model application being considered.

The third step in model calibration is the parameter estimation process itself. This can be done either manually or by using automated procedures. Manual procedures have traditionally been used in calibration of river basin models. These methods take full advantage of the experience and knowledge gained by modellers. However, experiences are not easily transferred, and hence different modellers may obtain very different model calibrations. Another major drawback of the manual approach is the huge labour requirements. Automatic calibration procedures eliminate the major drawbacks of the manual approach. It provides an objective model calibration and requires considerably less human resources. The main drawback of automatic calibration, however, is its dependency on the formulation of the numerical performance measures (objective functions) to be optimised. By using different numerical performance measures in the optimisation process that measure different characteristics of the modelled system it is possible to closely resemble the measures used by modellers in evaluating the goodness of the calibrated model. Automated procedures may also be explicitly tailored to specific model codes and implemented in so-called expert systems. Several studies based on different river basin models have demonstrated the benefits and recommended the use of automatic calibration, and in recent years the use of these techniques has advanced considerably.

Numerical optimisation procedures are used in model calibration to optimise the model parameters with respect to a defined objective function that measures the model fit. Optimisation algorithms are divided into two classes; local and global procedures. Local optimisation procedures evolve a single solution from an initial guess using information about the objective function value and, for some procedures, information about the gradient of the objective function. Local search procedures work well for objective functions with one well-defined optimum, but when several local optima exist they are less effective. Global optimisation procedures are especially designed to locate the global optimum of a complex objective function with multiple local optima, but are, in general, computationally more expensive than local search procedures. For a particular application the trade-off between the effectiveness of the algorithm and the computational requirements is important to consider when choosing between different algorithms and tuning algorithmic parameters. For calibration of conceptual rainfall-runoff models local search procedures have been found to be less effective, and global procedures have been widely applied. In groundwater modelling local gradient-based procedures have traditionally been applied. However, for groundwater models that include surface water exchange processes local procedures may be less effective. For calibration of distributed and integrated river basin models global optimisation procedures have been found to be effective. Optimisation algorithms are directly included in some river basin modelling software, and generic optimisation software exist that can be easily linked to any modelling software.

Model calibration is, in general, better constrained (more well-defined model parameter estimates are obtained) when multiple sources of information are used. The optimisation problem can be formulated in a general multi-objective framework, considering multi-variable measurements, multi-site measurements and multi-response modes. Mathematically, the solution to a multi-objective calibration problem is the set of Pareto solutions determined by the trade-off between the different objectives. For a Pareto solution none of the objective functions can be improved without violating one or more of the others. Pareto optimisation allows choosing preferred solutions according to the specific model application being considered, thus trading-off the different objectives using other type of information. In practical applications estimation of the entire Pareto front may be computationally infeasible, and one may only be interested in part of the solutions. In this case weights can be applied at the calibration stage to reflect the relative priorities given to certain objectives.

Model validation is essential to a river basin modelling project in order to document the predictive capabilities and credibility of the calibrated model. Validation must be carried out on independent data that have not been used in the calibration, and must assess the expected accuracy of the model predictions with respect to the model application being considered. Multi-site and multi-variable validation is necessary to document the capability of the model for prediction of the spatial distribution of different variables. Although properly validated with respect to some model responses, care should be taken assessing the confidence of model predictions of variables for which validation has not been performed. A hierarchical validation test scheme with increasing power is presented that distinguishes between model predictions performed on the same river basin as used for parameter estimation or on a different river basin, and whether predictions are done under the same conditions as present in the calibration period or under changed conditions. If the model cannot be properly validated it is necessary to go back to the initial stages of the model calibration protocol and reconsider the model parameterisation and choice of calibration parameters.

Uncertainty is an unavoidable and inherent element in river basin modelling. The main uncertainty sources can be grouped into (i) perceptual model uncertainty, (ii) data uncertainty, (iii) parameter uncertainty, and (iv) model structural uncertainty. The disagreement between observations and model predictions is a combined effect of these different error sources, and if not appropriately accounted for in the model calibration can lead to biased model parameter estimates. One way to explicitly deal with uncertainties in model calibration is to apply weights that reflect the relative uncertainties of the data used for calibration. Parameter non-uniqueness may be a problem in model calibration. Non-uniqueness in terms of equifinality refers to a situation where different parameter sets may provide virtually similar model responses with respect to a given performance measure. This basically means that the available information used in the model calibration is insufficient to distinguish between different parameter sets. The problem of non-uniqueness can be reduced by properly constraining the parameter estimation using different types of calibration data. The effect of the different uncertainties on the model predictions should be quantified as part of the modelling. However, how to properly include all the different uncertainty sources is a complex and yet unsolved problem. Classical statistical non-linear regression techniques and Monte-Carlo based procedures can be used to quantify the effect of some of these uncertainties on the model predictions.

2. CALIBRATION FRAMEWORK

Summary

Model calibration and validation is a prerequisite in any model-based river basin management project. To obtain a credible river basin model that can be used for subsequent simulations, model parameters should be calibrated against measurements of different model responses being available (river runoff, groundwater levels, water quality parameters etc.). A calibration protocol is outlined that includes the following steps: (i) Choice of calibration parameters, (ii) Specification of calibration objectives, (iii) Estimation of model parameters, and (iv) Validation of the calibrated model. These different components are described in detail in the following.

Application of numerical modelling in river basin management generally requires models to be customised to the specific characteristics and conditions of the river basin being considered. This includes tuning or estimation of model parameters. Some model parameters may be estimated from observable physiographic, climatic and physical characteristics of the river basin, but for most parameters often only a range of likely values can be given. For example, lumped, conceptual types of river basin models include parameters that are associated with aggregate hydrological process descriptions, which cannot be directly related to observable characteristics of the river basin. Parameters in physically-based models that solve the governing partial differential equations of the processes being modelled should, in principle, be assessable from observable characteristics. However, due to experimental constraints and scaling problems (i.e. the difference between the measurement scale, the model scale, and the scale at which the governing process descriptions are derived) this is not possible. Thus, to obtain a credible river basin model, parameters should be estimated using observed data of the different responses being modelled.

The process of parameter estimation where model responses are compared to observed data is illustrated in Figure 2.1. For a given set of parameters, the simulation model is run for a certain period of time, referred to as the calibration period. Model outputs are compared with the available observations, and based on this evaluation parameters are adjusted. The simulation model is then run with this new set of parameters. The process is repeated until a satisfactory agreement between model simulations and observations has been achieved.

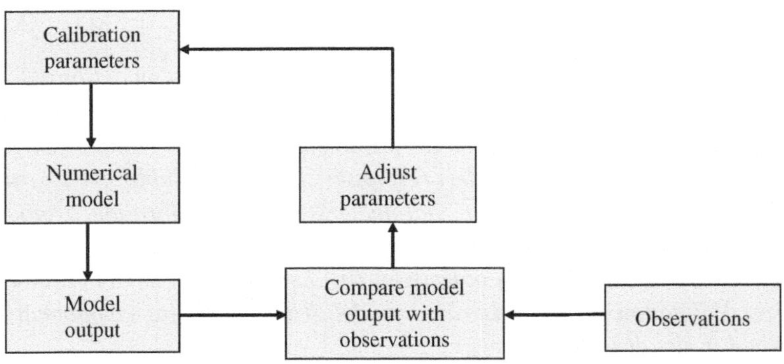

Figure 2.1. Parameter estimation process.

The parameter estimation process as illustrated in Figure 2.1 is the core of model calibration. The overall objective of model calibration is to select model parameters so that the model as closely as possible simulates the behaviour of the considered system, and hence provides a credible model that can be used for subsequent simulation studies.

A proper calibration protocol includes the following steps:

(1) *Choice of calibration parameters*. Based on the model setup key model parameters are selected for calibration. Some river basin models include relatively few parameters to be estimated and normally all parameters are included in the calibration without further consideration. Other, more complex models include a huge amount of free

model parameters. In this case a rigorous model parameterisation and conceptualisation process is required in order to reduce the number of calibration parameters, and hence make the calibration manageable.

(2) *Specification of calibration objectives.* Observations of modelled responses (e.g. river runoff, groundwater levels and water quality data) are a prerequisite for model calibration. Based on the available data, calibration objectives (or calibration criteria) are specified. These objectives are used in the parameter estimation process to evaluate the fit of the simulated responses to the observed data.

(3) *Parameter estimation.* When the calibration parameters and the calibration objectives have been defined, the parameter estimation illustrated in Figure 2.1 is carried out. For this step, different parameter adjustment methodologies can be applied. These include manual calibration, automated procedures based on numerical optimisation algorithms, and a combination of the two.

(4) *Model validation.* An evaluation of the credibility of the calibrated model is important for the subsequent model applications. In this regard the model must be validated against independent data that have not been used in the parameter estimation process. Proper validation tests should be applied to investigate and document the expected performance of the model and the accuracy of modelled responses for subsequent applications.

The calibration protocol is illustrated in Figure 2.2. The different steps in the protocol are further elaborated below. It should be emphasised that calibration is an iterative process. If a calibration results in poorly defined model parameters, the model parameterisation should be reconsidered and a simpler parameterisation should be defined that includes fewer calibration parameters. On the other hand, if the variability reflected in the observations cannot be sufficiently resolved with the present parameterisation, a more elaborate parameterisation should be considered. This feedback loop is shown in Figure 2.2.

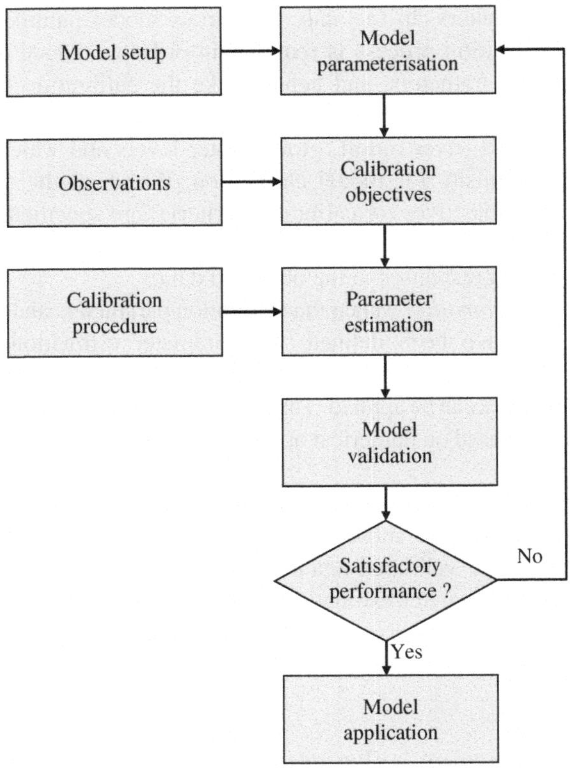

Figure 2.2. Model calibration protocol

3. MODEL PARAMETERISATION AND CHOICE OF CALIBRATION PARAMETERS

Summary

The first step in model calibration is the selection of the parameters to be included in the calibration. Some river basin models, such as lumped, conceptual rainfall-runoff models, include relatively few parameters to be estimated and normally all parameters are included in the calibration without further consideration. More complex distributed models include a huge number of model parameters and model structures to be analysed, and in this case a rigorous model parameterisation and conceptualisation process is required in order to reduce the number of calibration parameters, and hence make the calibration manageable. In the model parameterisation available field data (e.g. geological, soil type, vegetation, and land use maps) are utilised to describe the spatial variations of a model parameter using few representative parameters. Different parameterisation procedures that have been applied in river basin modelling are described. To screen the importance of the parameters defined in the model parameterisation and identify those that should be further refined via calibration sensitivity analysis can be applied. At this stage sensitivity analysis can also be applied to investigate the sensitivity of model responses that are of particular interest in the modelling study. If a parameter is insensitive with respect to the model responses for which calibration data are available (and hence cannot be well-determined by calibration), but sensitive for the model prediction considered, the model credibility for that particular modelling aspect is questionable.

3.1. Principle of parameter parsimony

The first step in model calibration is the selection of the parameters to be included in the calibration (see Figure 2.2). One of the key principles in modelling, and in model calibration in particular, is the principle of parameter parsimony. That is, the model should be kept as simple as possible to describe the processes in the system to a sufficient degree of detail in accordance with the available observations. Too many free parameters may turn model calibration into a pure curve fitting exercise that leads to an over-parameterised model, which provide good results in the calibration period but has poor predictive skills. On the other hand, if too few parameters are included in the calibration, the model may not be able to account for certain processes and characteristics, which leads to biased model predictions.

Some river basin models such as lumped, conceptual rainfall-runoff models are intrinsically parsimonious. They typically include 5–10 model parameters, which are all normally included in the calibration. More complex, distributed river basin models include a huge number of model structures and model parameters to be analysed. In this case a rigorous model parameterisation is crucial to define a parsimonious model. This is even more important when automatic procedures are applied for parameter estimation in order to obtain a well-posed and manageable calibration problem.

Following the principle of parsimony, Hill (1998) recommends starting with a simple model with few key parameters that represent the main features of the system and then add complexity. However, one may also start with a complex model parameterisation and then include explicitly the principle of parsimony in the parameter estimation process. This can be done by including a regularisation term in the objective function to be optimised (e.g. Doherty, 2003). Such a regularisation scheme basically trade-offs the value of additional model complexity against the improvement in model performance supported by the available data.

3.2. Parameterisation methodologies

Model parameterisation or model conceptualisation generally refers to the definition of functions with few representative parameters to reflect the general pattern (normally spatial pattern) of a model parameter. From the available field data such as geological descriptions from well-logs and geophysical measurements, pumping test analyses, soil type maps, soil analyses (texture, density, retention curves), and vegetation maps the most significant variations of the parameters can be defined. Four approaches that are widely applied in model parameterisation are: (i) classification, (ii) zonation, (iii) pilot points, and (iv) multipliers.

In the parameter classification approach a conceptual model is defined with parameter classes according to the available field data. For instance, based on the geological information a number of representative geological classes are defined. Each computational element in the river basin model is then assigned a geological class, and the hydraulic parameters that define this class are then subject to calibration. Similarly, hydraulic parameters for the unsaturated zone can be defined in soil classes obtained from soil maps, evapotranspiration parameters can be defined in vegetation and crop classes from land use maps etc.. The main advantage of the classification approach is that the number of free parameters can be reduced significantly, while still keeping a sufficient spatial variability of the parameter. The major challenge is to define an appropriate number of classes that adequately trade-offs the within-class and between-class variability. That is, if too many classes are defined, the between-class variability is small and parameters may not be uniquely defined. On the other hand, if too few classes are defined, the within-class variability is large and the modelling skills are reduced. Examples of application of the classification approach can be found in Refsgaard (1997) and Blasone *et al.* (2007).

In the parameter zonation approach the modelling domain is divided into a number of sub-regions or zones where model parameters are assumed constant (or, in general, assumed to vary in a predefined manner). The zones are interpreted from the available field data. The zonation approach is adequate for representing abrupt changes in parameter values such as faults in hydrogeological models and land use changes, but cannot represent gradual changes in parameter values. The main disadvantage is related to the delineation of the zones which may be difficult from the available information. If the zones are not well-defined, parameter estimation may be biased. The delineation of the zones may be included as an integrated part of the calibration to be estimated along with the parameter values. The zonation approach has been widely applied in groundwater modelling for parameterisation of hydraulic conductivities of the saturated zone. Examples include D'agnese *et al.* (1999) and Barlebo *et al.* (2004).

In the pilot point approach parameters to be calibrated are specified at a number of computational elements in the model domain. Parameter values in the entire domain are then calculated from the parameter values at the pilot points using spatial interpolation. In groundwater modelling kriging is often applied for the interpolation. The advantage of the pilot point approach is that it provides a continuous pattern of the parameter values, which may be more appropriate for description of spatial patterns than the classification and zonation methods. However, an important drawback of the method is that the appropriate number and locations of the pilot points may be difficult to determine prior to calibration. Often a trial-end-error procedure is required to obtain the optimal number and locations of the pilot points. LaVenue and Pickens (1992) presented a methodology where parameter estimation at the pilot points is combined with an optimal selection of the locations. Alternatively, one can apply a large number of pilot points and then use regularisation techniques in the optimisation to obtain the optimal locations (Doherty, 2003; Doherty and Skahill, 2006; Moore and Doherty, 2006).

In the parameter multiplier approach a few selected parameters are calibrated, while other parameters are adjusted simultaneously according to predefined ratios. Thus, in this case a spatial pattern of a parameter can be defined prior to the calibration, and during the parameter estimation process the spatial distribution of the parameter is then modified as a whole. The advantage of this approach is that the number of free parameters can be reduced significantly. However, the predefined spatial pattern of the parameter is crucial for the success of this method, which may be difficult to obtain from the available information. Examples of application of the multiplier approach can be found in Eckhardt and Arnold (2001) and Frances *et al.* (2006). The multiplier approach is also often used in combination with the other parameterisation methods. For instance, in groundwater modelling it is common to fix the anisotropy ratio between the vertical and horizontal hydraulic conductivity.

3.3. Sensitivity analysis

In the model parameterisation process, sensitivity analysis can be applied to investigate the importance of the different parameters to the model responses. Such a sensitivity analysis can be used to screen the importance of the parameters defined in the parameterisation process and identify those that should be further refined via calibration.

A number of different sensitivity analysis techniques exist. The most commonly applied sensitivity measure is based on locally evaluated sensitivities around a nominal parameters set. They are evaluated by perturbing one parameter at a time and calculating the resulting effect on the model response. Denote by R a particular model response, the sensitivity of a parameter θ_i with respect to that response can be calculated as:

$$S_i = \frac{R(\theta_1, \theta_2, ..., \theta_i + \Delta\theta_i, ..., \theta_p) - R(\theta_1, \theta_2, ..., \theta_i, ..., \theta_p)}{\Delta\theta_i} \quad (1)$$

where $\Delta\theta_i$ is the perturbation of the parameter. Instead of using a forward finite difference approximation as done in Eq. (1), a backward or a central difference approximation can be applied.

In order to compare the local sensitivity coefficients between parameters of different scales of magnitude, scaled sensitivity values are calculated. These are obtained by multiplying the sensitivity coefficient in Eq. (1) with a value representing the scale of the parameter, typically the nominal parameter value itself or a user-defined parameter range. The latter approach has the advantage that sensitivity coefficients are scaled according to the prior knowledge of the parameter. Similarly to scaling with respect to the parameter value, the sensitivity coefficient can be scaled with respect to the model response R in order to compare parameter sensitivities for different model responses (e.g. Brun *et al.*, 2001).

The scaled sensitivities provide a ranking of the parameters with respect to the importance of the parameters for the considered model responses. Larger absolute values indicate more sensitive parameters. As a rule of thumb, parameters are said to be insensitive if their scaled sensitivity value is less than about 0.01–0.02 times the maximum scaled sensitivity value (in terms of absolute values). However, care should be taken using this as a strict threshold measure. Hill (1998) argues that parameters that have absolute scaled sensitivities that are less than about 0.01 times the largest absolute scaled sensitivity cannot be uniquely determined by calibration.

The model response R may be a particular simulated model state variable or some aggregate measure. If R is related to the predicted responses corresponding to the observations used in the calibration, the scaled sensitivity coefficients are used to evaluate which parameters should be included in the calibration. Sensitive parameters can be well-determined by calibration, whereas insensitive parameters cannot. If R is related to a model prediction that is not measured, the scaled sensitivity coefficients are used to evaluate which parameters are important for that particular model prediction. The interpretation of parameter sensitivity measures with respect to measured and unmeasured model predictions is shown in Figure 3.1. This matrix indicates the problem of using a model for prediction of variables that have not been included in the calibration. If the parameter is insensitive with respect to the model responses for which calibration data are available, but sensitive for the model prediction considered, the model credibility for that particular modelling aspect is questionable.

		Sensitivity with respect to unmeasured model prediction	
		High	Low
Sensitivity with respect to measured model prediction	High	• Well-determined by calibration • Important for model predictions	• Well-determined by calibration • Not important for model predictions
	Low	• Cannot be determined by calibration • Important for model predictions	• Cannot be determined by calibration • Not important for model predictions

Figure 3.1. Interpretation of sensitivities for measured and unmeasured model predictions

It is important to emphasise that the sensitivity coefficients given by Eq. (1) provide the sensitivity of the model parameters around a specified parameter set, and hence give information about the importance of the parameters only at that location in parameter space. If the simulation model is highly non-linear in its parameter-output interaction, sensitivity measures may vary considerably in the parameter space. Thus, parameters that are insensitive for certain parameter sets may be highly sensitive for other parameter sets, and vice versa. In addition, the sensitivity measures do not properly account for parameter correlations, implying that parameters that seem to be insensitive may have important correlations with other parameters that are important for the model predictions.

It is recommended to evaluate sensitivities around different parameter sets to obtain more robust conclusions regarding parameter sensitivity. A more formal methodology for evaluating sensitivities for different parameter sets is the Morris design (Morris, 1991). The method estimates the sensitivity of a parameter by computing a number of local sensitivities cf. Eq. (1) at different points in the parameter space. The parameter sets are selected in such a way that each parameter is varied over its feasible range. The average of the local sensitivities is then a measure of the global sensitivity. In addition, the variability of the local sensitivities provides a measure of the interaction between the parameters and the existence of non-linear effects.

An overview of sensitivity analysis methods is given by Saltelli *et al.* (2000). Guidelines for use of sensitivity analysis in river basin modelling can be found in the Harmoni-CA guideline Model Sensitivity Analysis (see part 3 of this volume).

3.4. Use of prior information

Inclusion of prior information about parameter values in the parameter estimation process is normally given in terms of a feasible parameter space by defining lower and upper bounds on each parameter. These bounds can be chosen according to physical characteristics and other information of the river basin being considered and modelling experiences. For some parameters the bounds are determined by mathematical constrains in the model, e.g. a runoff coefficient that determines the division of net rainfall into direct runoff and infiltration has a feasible value between 0 and 1. Most numerical optimisation procedures explicitly use prior information to restrict the search to the defined feasible parameter space, thus defining a constrained optimisation problem.

By explicitly using lower and upper bounds on the parameters in the parameter estimation process, infeasible or unrealistic optimised parameter values are avoided. However, one should be careful constraining the optimisation problem. Unrealistic optimised parameter values (as a result of an unconstrained optimisation) may indicate a fundamental model error. For that reason Hill (1998) recommends not using parameter constraints in the optimisation. If an optimised parameter in constrained optimisation approaches one of the bounds, it is recommended to relax the bound and repeat the optimisation process in order to evaluate if the parameter estimation tends to be biased due to data or model structural errors.

Instead of specifying bounds on the parameters, prior information can be included in the optimisation using a penalty function. In this case the objective function to be optimised includes a term that measures the deviations of the parameters from prior parameter values. The optimisation then trade-offs the improvement in goodness-of-fit of the model response compared to the calibration data and the deviations from the initial parameter values. A deviation from the initial parameter values is weighted against the corresponding increase in goodness-of-fit. The relative weighting between the parameter penalty term and the goodness-of-fit is specified by the user and determines how much emphasis should be given to prior information in the parameter estimation.

4. CALIBRATION DATA AND CHOICE OF CALIBRATION OBJECTIVES

Summary

The second step in model calibration is selection of data and formulation of calibration objectives. Data requirements for model calibration depend on the type of river basin model considered. For lumped, conceptual rainfall-runoff models the only available source of calibration data is the runoff at the catchment outlet. For calibration of distributed models both catchment integrated and distributed data should be included in the model calibration. To better constrain the model calibration it is important to include data of different types (e.g. both groundwater and runoff data for an integrated surface water-groundwater model), and define performance measures of different model response modes (e.g. overall water balance, high flow and low flow). The calibration data should reflect the natural variability of the variable being considered, e.g. include both dry and wet periods. To evaluate the model skills numerical performance measures are calculated. These measures are used together with visual judgement in manual calibration and as explicit objective functions to be optimised in automatic calibration. A number of different and supplementary performance measures are presented. Scaling errors are of particular concern when comparing observed point measurements with simulated values that represent an average value within a computational element. In this case one may deliberately accept a bias in the model simulations to achieve a good correspondence of the dynamical model behaviour. Weights can be assigned to the observations used in the calibration to reflect the uncertainty of the observed data or to put emphasis on particular modelling aspects, thereby tailoring the model calibration to the specific model application being considered.

4.1. Data requirements

Selection of data and formulation of calibration objectives to be used for the parameter estimation refers to the second step in the calibration protocol cf. Figure 2.2. Basically, according to statistical estimation theory, the number of observations used for calibration should be larger than the number of parameters in order to obtain unique parameter estimates. In practice, however, due to the limited information content of the available data sources (each new data point does not provide independent, non-commensurable information), the number of observations should be much larger than the number of parameters to obtain unique parameter estimates. Limitations in information content of the data can result in parameter insensitivity, and hence parameters that are important for a particular modelling aspect can not be determined by calibration (see the section Sensitivity analysis). In addition, this may lead to an ill-posed optimisation problem. Thus, the proper choice of calibration data and formulation of objective functions are crucial for the success of the model calibration.

Data requirements for model calibration depends on the type of river basin model considered. For lumped types of rainfall-runoff models, the only available data source for calibration is the runoff at the catchment outlet. However, although only one source of information is available, it is important to define different measures that compare model responses with observations for different modelling aspects (e.g. low flow and high flow situations) in order to properly constrain the model calibration. In the calibration of a lumped, conceptual rainfall-runoff model Madsen (2000) used four different measures that focus on different modelling aspects: (i) water balance error, overall agreement of the hydrograph dynamics, (iii) agreement with respect to high flows, and (iv) agreement with respect to low flows. Boyle *et al.* (2000) partitioned the hydrograph into three components for calculating performance measures used in the calibration: (i) driven by rainfall, (ii) non-driven, quick response, and (iii) non-driven, slow response.

For calibration of distributed models it is important to evaluate the distributed behaviour of the model, rather than just the catchment-integrated behaviour reflected in the runoff at the catchment outlet. Thus, in this case calibration data should include different types of data distributed within the model domain. Multi-variable and multi-site data are important to obtain a credible model for distributed predictions of different state variables. To better constrain the model calibration it is, in general, always better to include (i) new data of a different type rather than more data of the same type, and (ii) new data at a different location rather than more data at existing locations.

For example, in groundwater modelling several studies have shown (e.g. Barlebo *et al.*, 1998) that hydraulic head data alone cannot sufficiently constrain the calibration problem. By including also some kind of flux data (e.g. flow measurements, tracer or concentration data) in the calibration the problem becomes better constrained with more well-defined parameter estimates.

Besides the diversity of response measures, data types, and measurement locations, the information content also depends on the quality of the data. Higher quality data with small data errors better constrain model calibration than data that have large errors. In addition, the variability of data with respect to different responses is important. That is, the calibration data should reflect the natural variability of the variable being considered, e.g. include both dry and wet periods. Thus, to include extreme parts of the modelled regime in the calibration a relatively long record may be required. Yapo *et al.* (1996) studied the sensitivity to calibration data for calibration of a lumped, conceptual rainfall-runoff model. For their application they showed that in the order of 8 years of data were required to obtain calibrations that were virtually insensitive to the specific calibration data selected. The appropriate length of the calibration period, however, is highly domain dependent.

From the available data sources, some data are used for the parameter estimation, while other data are kept for model validation. Data requirements for model validation are discussed in the section Model validation.

4.2. Numerical performance measures

When evaluating the modelling skills, numerical performance measures are used. These measures serve different purposes. First, at different stages in the modelling process performance criteria are used as an important element in the analysis of the model credibility, i.e. to evaluate if the model is sufficiently accurate for the purpose of the modelling being considered. Secondly, numerical performance measures are used in the calibration to evaluate the improvement in model performance as the parameter estimation process proceeds. Numerical measures are used in manual calibration together with visual judgement to evaluate the goodness-of-fit. In automatic calibration numerical performance measures are used explicitly as objective functions to be optimised.

Commonly applied numerical performance measures based on comparing time series of simulated values and observed equivalents are given in Table 4.1. The mean error (*ME*), root mean squared error (*RMSE*), and standard deviation of residuals (*STD*) are related via the equation

$$RMSE^2 = ME^2 + STD^2 \tag{2}$$

Table 4.1. Continuous time series statistics. OBS_i is the observed value and SIM_i is the simulated equivalent. $\overline{OBS}$ and $\overline{SIM}$ are the average of the observed and simulated values, respectively. The statistics are evaluated over the period with observations $i = 1,\ldots,N$

Performance measure	*Equation*
Mean error	$ME = \frac{1}{N}\sum_{i=1}^{N}(OBS_i - SIM_i) = \overline{OBS} - \overline{SIM}$
Mean absolute error	$MAE = \frac{1}{N}\sum_{i=1}^{N}\lvert OBS_i - SIM_i\rvert$
Root mean squared error	$RMSE = \sqrt{\frac{1}{N}\sum_{i=1}^{N}(OBS_i - SIM_i)^2}$
Standard deviation of residuals	$STD = \sqrt{\frac{1}{N}\sum_{i=1}^{N}(OBS_i - SIM_i - (\overline{OBS} - \overline{SIM}))^2}$
Coefficient of determination	$R^2 = \frac{\left[\sum_{i=1}^{N}(OBS_i - \overline{OBS})(SIM_i - \overline{SIM})\right]^2}{\sum_{i=1}^{N}(OBS_i - \overline{OBS})^2 \sum_{i=1}^{N}(SIM_i - \overline{SIM})^2}$
Coefficient of efficiency (Nash-Sutcliffe coefficient)	$E = 1 - \frac{\sum_{i=1}^{N}(OBS_i - SIM_i)^2}{\sum_{i=1}^{N}(OBS_i - \overline{OBS})^2}$
Index of agreement	$d = 1 - \frac{\sum_{i=1}^{N}(OBS_i - SIM_i)^2}{\sum_{i=1}^{N}\left(\lvert SIM_i - \overline{OBS}\rvert + \lvert OBS_i - \overline{OBS}\rvert\right)^2}$

The *ME* is a measure of the general offset between measurements and simulations (bias), whereas *STD* is a measure of the dynamical correspondence. The *RMSE* is an aggregate measure that includes both bias and dynamical behaviour. The *RMSE* is often used as an overall measure of comparison. The splitting of the *RMSE* into *ME* and *STD* cf. Eq. (2) provides more insight into the modelling error, i.e. if it is related to a general model bias or if the model has problems replicating the dynamics. The mean absolute error (*MAE*) is an overall measure of comparison similar to *RMSE* but puts less emphasis on the largest errors as compared to *RMSE* where errors are squared.

The *ME*, *MAE*, *RMSE* and *STD* statistics are all dimensional measures with units of the variable considered. Dimensionless measures can be defined by using relative residuals in the equations, i.e. by dividing (OBS_i – SIM_i) with the observation OBS_i within the summation. If the variable considered has values that cover a large range (over several decades), the dimensional and dimensionless measures may give different interpretations on model performance. While the dimensional measures are sensitive to high extremes, the dimensionless measures are sensitive to low extremes. To put less emphasis on high extremes one may also use statistics based on logarithmic transformed values. The effect of this is virtually similar to using relative residuals.

Table 4.1 includes also three goodness-of-fit measures; the coefficient of determination R^2, the coefficient of efficiency or Nash-Sutcliffe coefficient E (Nash and Sutcliffe, 1970), and the index of agreement d (Willmont *et al.*, 1985). All these measures are dimensionless, and hence are preferable when comparing model performance related to different observed time series (i.e. different variables, locations, and periods).

The R^2 is the square of the Pearson's product-moment correlation coefficient. It ranges from 0 to 1 with larger values indicating a better fit. The R^2 measures the best linear fit between observed and simulated values. Thus, the measure is insensitive to bias and proportional differences (i.e. if $SIM_i = aOBS_i + b$, then $R^2 = 1$), and hence large R^2 values may actually be obtained for models that have serious errors. Another drawback of the R^2 measure is that it is more sensitive to outliers than values close to the mean (Legates and McCabe, 1999).

The coefficient of efficiency is widely used to evaluate model performance in hydrological modelling. It ranges from minus infinity to 1 with larger values indicating a better fit. An important special case is $E = 0$, which can be obtained if $SIM_i = \overline{OBS}$, indicating that the average of the observed values in this case is as good a predictor as the model. Thus, one would most likely require that $E > 0$ for the model to be fit (unless, for instance, a large bias is accepted). The coefficient of efficiency measures the one-to-one relationship between observed and simulated values, and hence it is sensitive to bias and proportional effects. In this respect it provides an improvement to the R^2 measure. However, E is also sensitive to high extremes as R^2. The coefficient of efficiency using logarithmic transformed values is often used to reduce the sensitivity on high extremes. It should be emphasised that E is based on the sum of squared residuals, and hence provides the same information on goodness-of-fit as the *RMSE* measure.

The index of agreement is not as widely used as E and R^2. It ranges from 0 to 1 with large values indicating a better fit. The d measure is also based on the sum of squared residuals, but standardised according to a potential error (the term in the summation in the denominator represents the largest error that each $(SIM_i - OBS_i)^2$ can reach). As is the case with E and R^2, d is also sensitive to outliers. The difference between the three measures is illustrated in Figure 4.1.

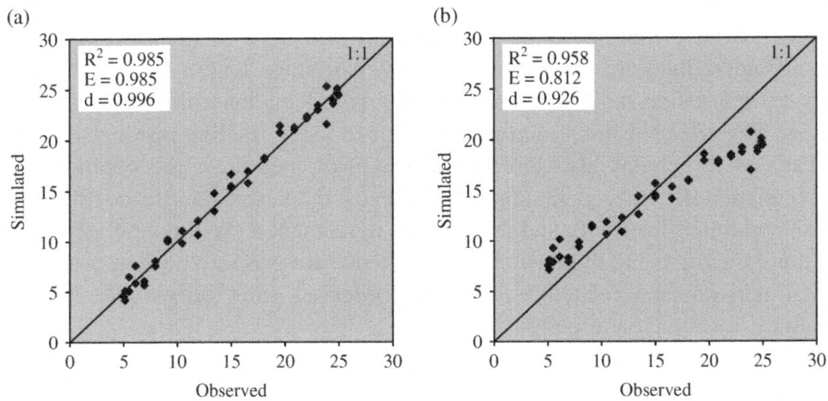

Figure 4.1. Differences of the R^2, E and d goodness-of-fit measures for two different models (a) $SIM_i = OBS_i + e_i$, and (b) $SIM_i = aOBS_i + b + e_i$, where e_i is a random error

Modified versions of E and d were presented by Legates and McCabe (1999). They introduced a general norm to be used in the equations instead of the 2nd norm. In addition they introduced a baseline adjusted measure where the observed mean in the equations is replaced by another term related to the observed time series. The persistence model efficiency (*PME*) measure proposed by Gupta *et al.* (1999) is an example of a baseline adjusted coefficient of efficiency

$$PME = 1 - \frac{\sum_{i=1}^{N} (OBS_i - SIM_i)^2}{\sum_{i=1}^{N} (OBS_i - OBS_{i-1})^2} \quad (3)$$

In this case the model prediction is compared relative to the performance of a persistence model where the model prediction at a given time step equals the observation at the previous time step. This measure is relevant when evaluating the forecast skills of the model.

To put explicit emphasis on model performance for different response modes (or variable regimes), the statistics in Table 4.1 can be defined considering only data that belong to those specific response modes. For instance, to evaluate the performance of the model with respect to high and low flow simulations, the statistics are calculated considering only data in periods where the observations are, respectively, above and below certain threshold levels.

The measures in Table 1 are used to evaluate the performance of continuous simulation models. To evaluate event-based models or if focus of a continuous simulation model is on specific events, event-based statistics are used. Some commonly applied event-based statistics (in this case focussing on high extremes) are shown in Table 4.2.

Table 4.2. Event based statistics. OBS_i is the observed value, SIM_i is the simulated equivalent, and $t(.)$ denotes time. The statistics are evaluated over an event with observations $i = 1,\ldots,n$

Performance measure	Equation
Volume error	$VE = \sum_{i=1}^{n}(OBS_i) - \sum_{i=1}^{n}(SIM_i)$
Peak error	$PE = \lvert\max\{OBS_i\} - \max\{SIM_i\}\rvert$
Time-to-peak error	$TPE = \lvert t(\max\{OBS_i\}) - t(\max\{SIM_i\})\rvert$

4.3. Scaling issues

When calculating performance statistics, observed values are compared with simulated equivalent values. In this regard, it is important that observations and simulated values refer to the same scale. For spatially aggregated values such as catchment (or sub-catchment) runoff simulated and observed values refer to the same scale and can thus be compared directly. For spatially distributed variables, however, care should be taken when comparing observed values that typically refer to the point scale with simulated values that refer to the model grid scale.

As an example of the scaling problem, comparison of simulated and observed groundwater levels is considered. The scale error in this case is primarily related to an interpolation error and an error related to the geological heterogeneity. The interpolation error depends on the applied horizontal discretisation Δx and the hydraulic gradient J. The standard deviation of the interpolation error can be expressed as (Henriksen and Sonnenborg, 2005)

$$s_{\text{int}} = 0.5J\Delta x \tag{4}$$

The error related to geological heterogeneity within the model grid is a function of the hydraulic gradient and the variance and correlation length of the hydraulic conductivity. The standard deviation of this error can be expressed as (Henriksen and Sonnenborg, 2005)

$$s_{het} = C\lambda s_{\ln K}J \tag{5}$$

where $s_{\ln K}$ and λ are, respectively, the standard deviation and the correlation length of the logarithm of the hydraulic conductivity. The parameter C depends on the properties of the aquifer. Based on modelling studies in Denmark, the standard deviation of groundwater level uncertainties related to scaling errors can be in the order of 1–2.5 m (Henriksen and Sonnenborg, 2005).

The scaling error introduces a bias when comparing point observations with simulated grid-based values. This bias is important to take into account when evaluating the model performance. Measures based on the sum of squared residuals may not be appropriate in this case. One would rather focus on the performance of the model with respect to simulating the dynamics, and thus accepting a bias. The standard deviation of residuals (*STD*) given in Table 4.1 is a measure of the dynamic model behaviour, and is preferable to the usually applied *RMSE* in the case a bias is acceptable, cf. Eq. (2). Khu *et al.* (2008) used *STD* as objective function for groundwater levels in automatic calibration of a distributed and integrated catchment model.

4.4. Weighting observations and aggregating performance measures

The performance measures given in Table 4.1 and Table 4.2 consider all measurement to be equally important in evaluating the model fit. Weights can be assigned to the observations to calculate performance measures that reflect different emphasis on the available data. Besides weighting of individual observations for calculation of the time series statistics, weighting is also relevant when combining different performance measures into aggregate, overall performance measures.

Different weighting procedures have been used in model calibration. One method is to subjectively assign weights on the observations according to the modelling purpose, such that observations in areas where more accurate simulations are wanted get higher weights than in areas that are less important. Similarly, one may assign higher weights on performance measures of certain data types than others. This method is attractive because it allows the modeller to tailor the calibration to the specific model application being considered.

Another, more objective strategy is to assign weights according to the uncertainty of the observations, or rather the uncertainty of the residual, taking all the different error sources in the modelling process into account (see discussion of model errors in the section Uncertainty sources). This weighting has a solid statistical basis and has been recommended for automatic calibration using non-linear regression methods (e.g. Hill, 1998; Hill and Tiedeman, 2003). According to the statistical theory of weighted and generalised least squares regression the weights assigned to the residuals should be proportional to the inverse of the covariance matrix of the true errors. Assuming a diagonal covariance matrix, a weighted sum of squared errors can be defined

$$SSE = \sum_{i=1}^{N} w_i (OBS_i - SIM_i)^2 = \sum_{i=1}^{N} \frac{1}{\sigma_i^2} (OBS_i - SIM_i)^2 \tag{6}$$

where the weights w_i are inversely proportional to the variance of the residual errors σ_i^2. This weighting method is intuitively sound because more certain data get higher weights in the calibration than uncertain data. In addition, the weighting provides non-dimensional residuals, thus allowing data of different types to be aggregated in an overall sum of weighted residuals that can be used as a single objective function to be optimised.

The weighting method in Eq. (6), however, has practical and theoretical limitations. First, it is in practice impossible to identify and quantify all the different error sources for estimation of the residual error variance σ_i^2. Secondly, the use of Eq. (6) is based on the premise of independent residuals. If spatial or temporal correlations exist, the information content of the observations is smaller than for independent data. If clusters of observation points in space exist, such data are typically correlated and thus will be given undue influence in the evaluation of the weighted sum in Eq. (6) as compared to observations in data sparse areas. Similarly, long time series with correlated data will be given undue weight compared to short time series with independent data. One may directly incorporate correlations between observation data by extending Eq. (6) to use a full weighting matrix on the residuals. However, it is often not possible to quantify correlations between residuals à priori, and hence some subjectivity is left for determining the weighting.

Hill (1998) argues that weighting is an important part of non-linear regression. Although it may be problematic to quantify error variances and correlations, it may in practice be less important because parameter estimation is, in general, not sensitive to moderate variations of the weighting (Hill and Tiedeman, 2003). In the calibration of a regional groundwater model using non-linear regression Leijnse *et al.* (2003) compared four different weighting schemes using both a diagonal weighting matrix cf. Eq. (6) and a full weighting matrix and found that parameter estimates were not very sensitive to the weighting applied.

Although weighting according to the uncertainties in the model residuals allows direct summation of different data types in Eq. (6), one may want to use different weights to different data types (or, in general, different performance measures). For instance, if much more observations of one data type are available for the calibration, these data may be given undue influence in the calibration if Eq. (6) is used for the weighting. In addition, one may want to assign weights to different model performances according to the modelling purpose. Often a trade-off exists between the different performance measures, i.e. it is not possible to identify a unique parameter set that minimises the model misfit for all performance measures. As an example, consider optimisation of a weighted sum of two objective functions

$$F_{agg}(\theta) = wF_1(\theta) + (1 - w)F_2(\theta) \tag{7}$$

where $0 < w < 1$ is the weight. The performance measures $F_1(.)$ and $F_2(.)$ are functions of the parameters θ. For different weight combinations the solution to the optimisation problem will provide a front of non-dominated or so-called Pareto-optimal solutions in the (F_1, F_2)-space as shown in Figure 4.2 (assuming minimisation of the two objective functions).

For different weight combinations the solution to the optimisation problem will provide a front of non-dominated or so-called Pareto-optimal solutions in the (F_1, F_2)-space as shown in Figure 4.2 (assuming minimisation of the two objective functions). A point on the Pareto front in the 2D objective function space in Figure 4.2 has the property that an improvement in one objective can only be obtained by a deterioration of the other. The weight to be used to get a specific solution of the Pareto front is a decision variable that, in general, should be chosen according to the model application being considered. For instance, if more weight should be put on objective F_1, a solution in the upper left part of the Pareto front should be chosen. On the other hand, if F_2 is more important, a solution in the lower right part is preferred. A balanced solution that puts equal emphasis on the two objectives corresponds to a solution on the central part of the Pareto front. It is, however, often difficult to à priori assign a proper weight that reflects a required relative importance of the considered objective functions.

The use of different objective functions in the parameter estimation is discussed in a general multi-objective optimisation framework in the section Multi-objective optimisation.

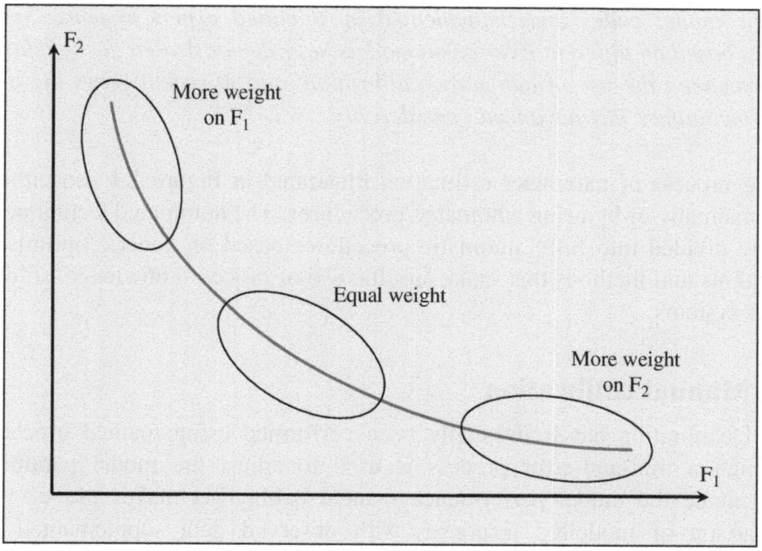

Figure 4.2. Illustration of Pareto front for optimisation of two objective functions

5. CALIBRATION METHODS

Summary

The process of parameter estimation is either done manually or by using automated procedures. Manual procedures have traditionally been used in calibration of river basin models. These methods take full advantage of the experience and knowledge gained by modellers. However, experiences are not easily transferred, and hence different modellers may obtain very different model calibrations. Another major drawback of the manual approach is the huge labour requirements. Automatic calibration procedures eliminate the major drawbacks of the manual approach. It provides an objective model calibration and requires considerably less human resources. The main draw-back of automatic calibration, however, is its dependency on the formulation of the numerical performance measures (objective functions) to be optimised.

By using different numerical performance measures in the optimisation process that measure different characteristics of the modelled system it is possible to closely resemble the measures used by modellers in evaluating the goodness of the calibrated model. Automated procedures may also be explicitly tailored to specific model codes and implemented in so-called expert systems. Several studies based on different river basin models have demonstrated the benefits and recommended the use of automatic calibration, and in recent years the use of these techniques has advanced considerably.

The process of parameter estimation illustrated in Figure 2.1 can either be done manually or by using automated procedures. The automated techniques are usually divided into fully automatic procedures based on generic optimisation algorithms and methods that make specific use of process knowledge, so-called expert systems.

5.1. Manual calibration

Model calibration has traditionally been performed using manual procedures in which a trial-and-error process is used to adjust the model parameters. To evaluate the model performance manual calibration makes use of visual comparison of modelled responses with observed data supplemented with numerical goodness-of-fit measures. To guide the parameter adjustment the modeller makes use of his/her knowledge about the characteristics of the river basin, the model structure, and the quality of the observed data. After each trial user intervention is required.

The main advantage of the manual approach is that it allows the use of several performance indicators (both subjective and objective) to be used in the calibration without the need to explicitly formulate quantitative measures to be optimised. For an experienced modeller it is possible to obtain a very good model calibration using the manual approach. However, although calibration practices are formulated in guidelines, the experiences gained by different users are not easily transferred to others, and extensive hands-on experience is required to obtain good results with manual calibration.

A major drawback of the manual approach, even with well-trained and experienced modellers, is that it is labour intensive. According to Boyle *et al.* (2000) a manual calibration of the Sacramento rainfall-runoff model (Burnash *et al.*, 1973; Burnash, 1995) may require several hundred of hours of effort. The use of graphical, interactive user interfaces has made manual calibration easier, but still it is rather time-consuming.

The subjectivity involved in manual calibration is another problem. Because there is no general objective performance measures, different modellers may obtain very different model calibrations. In addition, it is difficult to assign measures of confidence to the model predictions.

5.2. Automatic calibration

Automatic calibration procedures have been under development since the first computer-based lumped, conceptual types of rainfall-runoff models were introduced in river basin modelling in the late 1960s and early 1970s. With the rapid increase in computing power, the research and application of automatic calibration procedures for river basin models has advanced considerably. The objective of this research has been to develop calibration procedures that eliminate or reduce the major drawbacks of manual calibration.

In automatic calibration parameters are adjusted automatically according to a specified search scheme using an optimisation algorithm. The approach is fully automatic with no user intervention during the parameter estimation process itself, and hence requires considerably less human resources than manual calibration. The model performance is evaluated in terms of numerical objective functions that measure the goodness-of-fit of the modelled responses to the observations. Thus, the procedure eliminates the subjectivity of the manual approach, allowing a robust and consistent model calibration. In addition, confidence of the model predictions with the calibrated model can be explicitly stated.

The main weakness of automatic calibration is its dependency on the formulation of the numerical measures to be optimised. Typically, automatic calibration has been performed using a single objective function that measures the overall performance of the model, such as the root mean squared error (*RMSE*) function. However, a single measure of comparison is often inadequate to measure properly all the important characteristics of the model simulation, which are used by the modeller to evaluate the model performance. This aspect has caused a lot of scepticism on the use of automatic calibration, which has been claimed to be a pure curve fitting exercise that may result in physically unrealistic parameter values.

Recently, automatic optimisation procedures that allow direct use of several performance measures in the calibration have been developed and applied in river basin modelling. These methods, known as multi-criteria or multi-objective calibration methods, simultaneously optimise several user-defined objective functions that measure different aspects of the modelled responses against observations. This will result in different optimal solutions (denoted the set of non-dominated or Pareto optimal solutions), according to the trade-offs between the different objectives. Subsequently, additional criteria (including subjective measures) can then be applied to choose preferred solutions to be further analysed.

An overview of typical optimisation algorithms that have been applied in the calibration of river basin models is given in the section Optimisation algorithms.

5.3. Expert systems

As discussed above, manual and automatic calibration have different advantages and drawbacks. Calibration procedures have been developed that combine the best features from the two approaches. These methods are referred to as expert systems.

The basic philosophy behind expert systems is to automate the course of a manual calibration carried out by a well-trained and experienced modeller. The expert systems are often formulated as step-wise procedures which require some kind of user intervention between the different steps. At a certain stage during the calibration an objective measure is defined for the particular aspects being considered at that stage, and this function is then optimised using automatic procedures.

Compared to automatic calibration based on numerical optimisation procedures that basically considers the model to be calibrated as a black box, expert systems allow the inclusion of process knowledge into the calibration. However, a major drawback is that expert systems are model-specific, i.e. an expert system has to be designed for a particular river basin model.

There are different examples of application of expert systems in calibration of river basin models. Harlin (1991) and Zhang and Lindström (1997) formulated expert systems for the HBV rainfall-runoff model (Bergström and Forsman, 1973; Bergström, 1995) where groups of parameters that are related to certain processes are calibrated individually. An expert system for the MIKE 11/NAM rainfall-runoff model (Nielsen and Hansen, 1973; Havnø *et al.*, 1995) is described in the section Application examples.

5.4. Recommendations

There is a huge potential benefit of using automated calibration procedures as compared to manual calibration. Several studies have demonstrated the benefits and recommended the use of automatic calibration. Gupta *et al.* (1999) compared an automatic optimisation method and an interactive, semi-automated expert system for calibration of the Sacramento soil moisture accounting model and concluded that state-of-the-art automatic procedures perform with a level of skill resembling that of a human expert. Madsen *et al.* (2002) compared three different automated methods for calibration of the MIKE 11/NAM rainfall-runoff model. They showed that calibration based on a generic optimisation procedure compared favourably with an expert system. Poeter and Hill (1997) recommended the use of automatic parameter estimation in groundwater modelling and listed a number of different arguments in favour of automatic calibration.

Although numerous studies have demonstrated the benefit of applying automatic calibration, modellers have been hesitating to apply these techniques. In recent years, however, the development and use of automatic calibration in river basin modelling has advanced considerably. Optimisation algorithms have been implemented directly in modelling software, and generic optimisation software has been developed that can easily be linked to existing river basin models. Some of these software tools are described in the section Software.

6. OPTIMISATION ALGORITHMS

Summary

Numerical optimisation procedures are used in model calibration to optimise the model parameters with respect to a defined objective function that measures the model fit. Optimisation algorithms are divided into two classes; local and global procedures. Local optimisation procedures evolve a single solution from an initial guess using information about the objective function value and, for some procedures, information about the gradient of the objective function. Local search procedures work well for objective functions with one well-defined optimum, but when several local optima exist they are less effective. Global optimisation procedures are especially designed to locate the global optimum of a complex objective function with multiple local optima, but are, in general, computationally more expensive than local search procedures. For a particular application the trade-off between the effectiveness of the algorithm and the computational requirements is important to consider when choosing between different algorithms and tuning algorithmic parameters. For calibration of conceptual rainfall-runoff models local search procedures have been found to be less effective, and global procedures have been widely applied. In groundwater modelling local gradient-based procedures have traditionally been applied. However, for groundwater models that include surface water exchange processes local procedures may be less effective. For calibration of distributed and integrated river basin models global optimisation procedures have been found to be effective. Optimisation algorithms are directly included in some river basin modelling software, and generic optimisation software exist that can be easily linked to any modelling software.

6.1. Optimisation problem

In river basin modelling, the development and use of automatic calibration based on numerical optimisation algorithms has advanced considerably. Optimisation software has been included in modelling software, and generic optimisation software that can be easily linked to any model code has found widespread application. In this section an introduction to numerical optimisation is given, followed by a brief overview of optimisation algorithms that have been most widely applied in the calibration of river basin models.

The purpose of numerical optimisation in model calibration is to find the values of the parameters that optimises (minimises or maximises) the numerical value of a defined objective function that measures the model fit. Mathematically, this can be stated as follows (assuming minimisation of the objective function F(.))

$$\theta_{opt} : \underset{\theta}{Min}\{F(\theta)\} \tag{8}$$

where θ are the model parameters to be optimised. The optimisation problem is said to be constrained if θ is restricted to a feasible parameter space Θ. In unconstrained optimisation no restrictions on θ are imposed à priori. The optimisation problem in Eq. (8) is a single-objective problem. Formulation of numerical optimisation within a multi-objective framework is considered in the section Multi-objective optimisation.

The surface described by the objective function F(θ) is denoted a response surface. A numerical optimisation algorithm is a logical procedure that searches the parameter space to find the minimum (or maximum) of the response surface. The optimisation algorithm includes an initial guess of the solution (either in terms of a single parameter set or a population of parameter sets) and then implements a set of rules to improve the current solution. Optimisation algorithms basically differ in the way these rules are formulated. In this regard, optimisation algorithms are, in general, divided into two categories, local and global search procedures (Sorooshian and Gupta, 1995).

6.2. Local optimisation procedures

Local search methods evolve a single solution from an initial guess until no further improvement can be made. They are specifically designed to efficiently find the optimum of a response surface with one well-defined optimum (uni-modal function) since in this case the search will eventually reach the optimum irrespective of the starting point.

Local search methods can be divided into two categories, direct and gradient-based methods. Direct search methods use information only about the objective function value to guide the search. Different direct search methods have been applied to calibration of river basin models. One of the more popular methods is the simplex method of Nelder and Mead (1965), which also forms the basic search scheme of the global shuffled complex evolution algorithm presented below.

The simplex method works with a group of points (denoted a simplex) which contains (p+1) parameter sets, where p is the number of parameters to be optimised. From the simplex the worst point, i.e. the point with the largest objective function value is identified. From the rest of the points the centroid is calculated. A new point to replace the worst point θ_{worst} is then obtained using the following equation

$$\theta_{new} = \theta_{worst} + \rho C(\theta) \tag{9}$$

where $C(\theta)$ is the centroid of the best p points, and ρ is a step size. The simplex method considers different step sizes subsequently to obtain a better point θ_{new}.

Gradient-based methods use information about both the objective function value and its gradient to guide the search. Typically, gradient-based methods use the following iterative scheme

$$\theta_{k+1} = \theta_k + \rho A \nabla \theta_k \tag{10}$$

where θ_{k} is the present solution, $\nabla\theta_{\mathrm{k}}$ is the gradient of the response surface at the present solution, A is a square matrix, and ρ is a step size. The gradient determines the direction of local improvement of the objective function. Note the similarity of the gradient-based iterative scheme in Eq. (10) with that of the simplex method in Eq. (9). For the simplex method the direction of improvement is given by the centroid of the best points in the simplex.

The gradient-based methods differ in the way A and ρ are selected. Most methods are based on the Gauss-Newton method for calculation of the matrix A, such as the Marquardt-Levenberg algorithm. The Marquardt-Levenberg scheme is implemented in the generic optimisation packages PEST (Doherty, 2005) and UCODE (Poeter *et al.*, 2005). The gradient-based methods require information about the gradient of the response surface. Although a few models include explicit calculations of the gradient, most implementations of gradient-based search methods use finite difference approximations. Using for instance a forward finite difference approximation, gradients are calculated by increasing each parameter one at a time with a certain amount, run the model and compare the model output with that obtained using the present solution θ_k (similar to Eq. (1) above). Thus, to perform one iteration loop cf. Eq. (10) the method requires $(p+1)$ number of model runs.

The accuracy of the gradient approximation is important for the success of gradient-based methods. Theoretically, a parameter increment should be as small as possible to obtain a good approximation of the derivative. However, this may produce inaccurate derivatives caused by numerical round-off errors. On the contrary, a too large parameter increment may provide a bad approximation of the derivative at the present solution. Gradient-based software packages include different options for handling the accuracy of finite difference approximations. Besides forward (or backward) finite difference approximations, a central finite difference method can be used, which will provide a more accurate approximation of the derivative. However, a central difference approximation is more expensive and requires a number of model runs equal to twice the number of parameters in each iteration step. The presence of threshold processes, which is often seen in river basin models, is a particular problem for calculation of derivatives (see discussion below).

6.3. Global optimisation procedures

If the response surface is multi-modal, i.e. the objective function includes several local optima, local search procedures are less effective in locating the global optimum. In such cases convergence of local search procedures will depend on the starting point of the search. Global optimisation procedures are especially designed to locate the global optimum of a response surface with multiple optima, thus reducing the probability of being trapped in local optima compared to local search procedures.

A simple way of increasing the effectiveness of local search is to combine several local searches with different starting points, known as multi-start local search. Typically, a set of initial starting points (say r points) for the local searches are randomly selected within the feasible parameter region. Then r individual local searches are performed, and the minimum objective function obtained for these optimisations is taken as the global optimum. If the probability of a single local search to converge to a local optimum is q, the multi-start search with r individual optimisations is reduced to q^r. For instance, if the probability of convergence to a local optimum is $q = 0.8$, this probability is reduced to 0.01 when 20 local searches are performed from different starting points.

While multi-start optimisation can significantly increase the chance of converging to the global optimum, the procedure is not very efficient because individual searches are performed without any sharing of information between them. More efficient global optimisation can be obtained if information from previous searches is used when a new local search is carried out. Skahill and Doherty (2006) presented a multi-start procedure where new local searches are started from points in parameter space that are most distant from points that have been explored in the previous local searches.

Other global search procedures have been developed that simultaneously share information between individual solutions during the search. These methods typically evolve a population of solutions and are thus referred to as population-based search procedures. Evolutionary algorithms are population-based methods that have found widespread application to various optimisation problems and have also been widely applied in calibration of river basin models.

Genetic algorithms (GA) is the most well known category of evolutionary algorithms that were first introduced in calibration of a lumped, conceptual rainfall-runoff model by Wang (1991). A GA searches the parameter space using search strategies that are analogous to the natural selection process in nature (according to Darwin's theory). The parameters to be optimised are encoded as binary strings (or chromosomes). GA uses three basic operators (i) selection, (ii) cross-over (or mating), and (iii) mutation. Different selection methods exist but the most widely applied method is tournament selection. In this case members of the current population of points (chromosomes) are randomly selected according to their objective function value, so that points with a lower objective function value (better fitness) has a larger probability of being selected. Next, the selected points are mated by randomly selecting one or more crossover points where bit values between two strings are exchanged. The selection and cross-over processes are repeated until a new population of points (with a given population size) is obtained. Finally, mutation is applied by randomly selecting and exchanging a bit value of the new population with a given mutation probability.

An evolutionary algorithm that has been widely applied in calibration of river basin models is the shuffled complex evolution (SCE) algorithm (Duan *et al.*, 1992). The SCE method combines various search strategies. At the beginning of a new iteration loop, the population is divided into a given number of complexes. Each complex is then independently evolved a given number of times using the Simplex method (Nelder and Mead, 1965). Finally, the evolved complexes are combined (shuffled) to form a new population to be evolved. The complex shuffling ensures that information between complexes is shared, which improves the global effectiveness of the method. The SCE search strategy is illustrated in Figure 6.1.

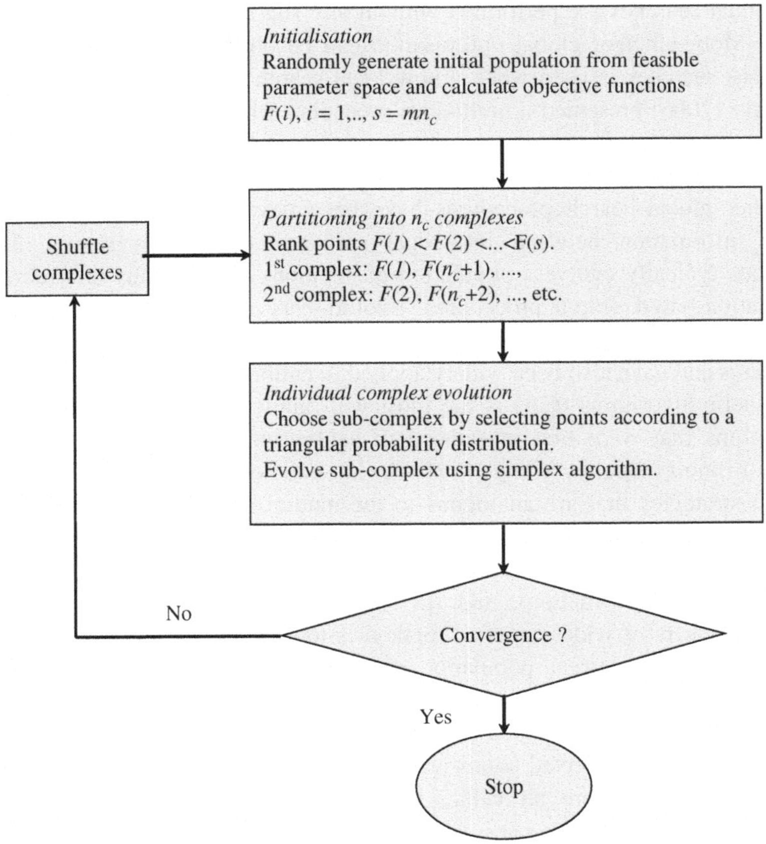

Figure 6.1. Flow chart of the SCE method. n_c is the number of complexes and m is the number of points in each complex

Although GA and SCE are based on different search strategies, they share some common features. First, they are probabilistic in nature in the sense that the initial population is selected randomly within the feasible parameter space, and in each iteration loop a random element is included in the selection of new points. A random seed can be used to initialise the random generator. The use of different random seeds may provide different solutions to the optimisation problem, hence providing a measure of the effectiveness of the algorithm for locating the global optimum.

Another common property of GA and SCE is that their performance, in general, depends on the chosen population size. The larger population is chosen, the larger probability of the algorithm to convergence to the global optimum. However, using a larger population usually requires more model runs to converge. In practice, one should choose the population size to balance this trade-off between robustness of the optimisation algorithm and required computing time. The dimensionality of the calibration problem (i.e. number of parameters to be optimised) is an important factor for determining a proper population size; that is, the more parameters are included in the optimisation the larger population size should be chosen. Recommendations for choice of population size and other algorithmic parameters are provided by Reed *et al.* (2000) for GA and by Duan *et al.* (1994) for SCE.

6.4. Recommendations

For calibration of lumped, conceptual rainfall-runoff models local search methods have been found to be less effective. Duan *et al.* (1992) conducted a comprehensive analysis of the structure of the response surface of a rainfall-runoff model and found several characteristics that highlighted the difficulties of local search methods for converging to the global optimum. The most important characteristic is that the response surface is characterised by multiple optima at different scales with several local optima located within different regions of attraction. Thus, local procedures have a relatively high probability of converging into a local optimum which may be located far from the global optimum of the response surface. In addition, the response surface is characterised by a rough structure with discontinuous derivatives due to the highly non-linear model output – parameter interaction and the existence of threshold processes. The shape of the response surface is non-convex with long curved ridges. These characteristics make gradient-based search procedures less efficient and may provide numerical problems.

Several studies have been conducted that compare different optimisation algorithms for calibration of lumped, conceptual rainfall-runoff models (e.g. Duan *et al.*, 1992; Gan and Biftu, 1996; Cooper *et al.*, 1997; Kuczera, 1997; Franchini *et al.*, 1998; Thyer *et al.*, 1999). The main conclusion from these studies is that global population-based search procedures are more effective than multi-start local search procedures, which in turn are more effective than pure local search. There are numerous examples of application of global optimisation methods for calibration of lumped rainfall-runoff models. Examples of application of SCE include Sorooshian *et al.* (1993), Gan and Biftu (1996), Yapo *et al.* (1996), Cooper *et al.* (1997), Franchini *et al.* (1998), Thyer *et al.* (1999), Madsen (2000), and Madsen *et al.* (2002). Examples of applications of GA include Wang (1991), Cooper *et al.* (1997), Franchini and Galeati (1997),

Franchini *et al.* (1998), and Cheng *et al.* (2002). In a recent study, Skahill and Doherty (2006) applied a multi-start, gradient-based optimisation procedure.

In groundwater modelling, local gradient-based search procedures have traditionally been applied. Recent application examples include Hill *et al.* (1998), D'agnese *et al.* (1999), Doherty (2003), Sonnenborg *et al.* (2003), and Moore and Doherty (2006) who all applied the Levenberg-Marquardt algorithm to calibration of a groundwater flow model. Also for calibration of combined flow and transport models, gradient-based methods have been applied, e.g. Medina and Carrera (1996), Barlebo *et al.* (1998), Barlebo *et al.* (2004). Gradient-based methods have shown to work well, in general, in groundwater model applications probably due to the smaller degree of model output-parameter non-linearities as compared to lumped, conceptual rainfall-runoff models. However, if the groundwater model includes threshold processes such as activation of drainage flow when the hydraulic head exceeds a certain value or creation of overland flow if the groundwater level exceeds the terrain level, gradient-based methods are expected to be less effective. In a recent study, Blasone *et al.* (2007), showed that a transient groundwater model that includes drainage and overland flow threshold processes has a response surface with several regions of attraction, see Figure 6.2.

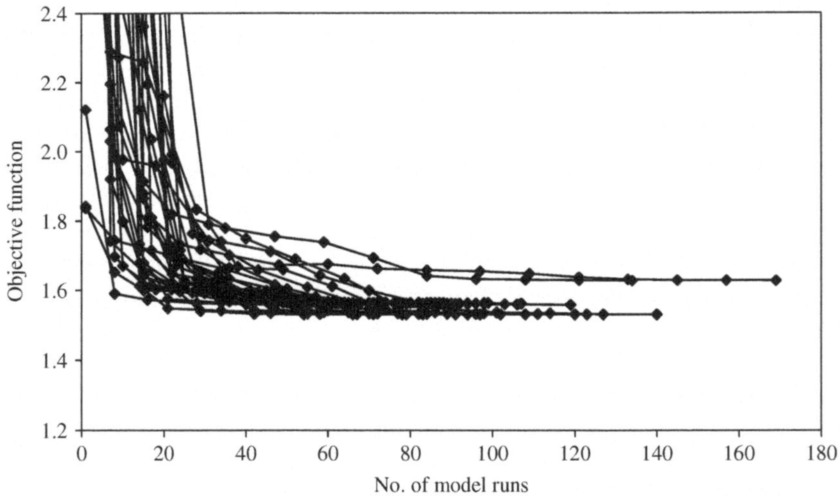

Figure 6.2. Convergence of 33 single gradient-based searches in the calibration of a MIKE SHE groundwater model (adapted from Blasone *et al.*, 2007)

To calibrate groundwater models where highly non-linear features are expected global optimisation methods are preferable. Examples of application of GA in groundwater modelling include Haidari and Ranjithan (1998), Solomatine *et al.* (1999) and Mayer and Huang (1999). Examples of application of SCE include Contractor and Jenson (2000), Agyei and Hatfield (2006) and Blasone *et al.* (2007). In Agyei and Hatfield (2006) and Blasone *et al.* (2007) SCE is combined with a Levenberg-Marquardt algorithm to obtain a computationally more efficient optimisation as compared to SCE alone.

The use of numerical optimisation procedures for calibration of distributed and integrated river basin models has not yet found that widespread application as for calibration of lumped, conceptual rainfall-runoff models and groundwater models. It is the general impression that distributed, integrated catchment models have similar features as lumped rainfall-runoff models, i.e. includes highly non-linear parameter – model output interactions, resulting in response surfaces with numerous local optima (see Figure 6.3). Thus, in general, global optimisation procedures are preferable. Examples of application of numerical optimisation in distributed catchment models include Eckhardt and Arnold (2001), Brath *et al.* (2004) and Francés *et al.* (2006) who applied the SCE algorithm using catchment runoff as calibration objective. Madsen (2003) and Blasone *et al.* (2007) also applied the SCE algorithm, but besides catchment runoff also included groundwater level measurements in the calibration.

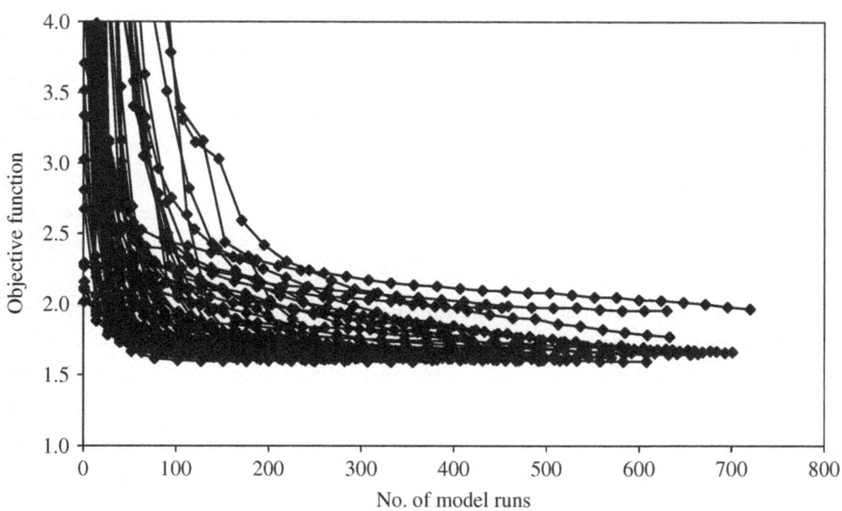

Figure 6.3. Convergence of 69 single gradient-based searches in the calibration of a MIKE SHE integrated model (adapted from Blasone *et al.*, 2007)

The use of multiple sources of information is a particular characteristic of calibration of distributed and integrated models, which becomes even more apparent when water quality processes are added. Muleta and Nicklow (2005) used a two-step calibration approach based on GA where flow parameters were first calibrated using runoff data, and subsequently sediment parameters were calibrated using sediment concentration data. A multi-step calibration approach can be followed if the model parameters that are relevant for specific processes where measurement data are available can be isolated. This is, however, seldom the case, and in this case a more effective calibration is obtained if all available information is used in a single calibration step including all parameters, thereby better constraining the model calibration (see discussion in the section Calibration data and choice of calibration objectives). In the calibration of a water quality catchment model Van Griensven and Bauwens (2003) combined flow and concentration data of different species into an aggregate measure that was optimised numerically using the SCE algorithm. When multiple sources of information are available for calibration, the numerical optimisation problem is preferably defined within a multi-objective framework as described in the section Multi-objective optimisation.

6.5. Computational requirements

An important aspect to consider when using numerical optimisation is related to the computational requirements. Optimisation requires numerous evaluations of the model with different parameter combinations. In general, local search procedures are computationally more efficient and require fewer model evaluations for convergence as compared to global, population-based procedures. However, as discussed above, global optimisation procedures are more effective for locating the global optimum as compared to local search procedures. This trade-off between efficiency and effectiveness is important to take into account when tuning a particular algorithm or when choosing between different algorithms for the calibration problem being considered.

Gradient-based local search requires $(p+1)$ number of model evaluations (p is the number of model parameters to be optimised) for each iteration loop (in the case of forward or backward finite difference approximation of the gradient). Typically, in the order of $2p$ iterations are needed for convergence (Hill, 1998), thus requiring a total of $2p(p+1)$ model evaluations. For instance, if the optimisation includes 9 parameters, in the order of 180 model evaluations are required. For the SCE algorithm the number of complexes n_c determines the trade-off between efficiency and effectiveness; that is, the larger value of n_c is chosen the higher probability of converging into the global optimum but at the expense of a larger number of model evaluations, and vice versa. Kuzcera (1997) recommended setting n_c equal to the number of model parameters to be optimised. Following the recommendations by Duan *et al.* (1994) each complex should be evolved $2p+1$ times in each iteration loop. Typically, in the order of 10 iterations are needed for SCE to converge, thus requiring in this case in the order of $10p(2p+1)$ model evaluations, corresponding to about 10 times more model evaluations than required by a local gradient-based method. Often, however, a smaller number of complexes can be used without decreasing the effectiveness of the SCE algorithm. For instance, Madsen (2003) used only 2 complexes in the application of SCE to a distributed, integrated catchment model. GA and other global optimisation algorithms require, in general, a similar number of model evaluations as SCE for convergence.

For calibration of lumped, conceptual rainfall-runoff models computational requirements are, in general, not an issue since a single model run is in the order of few seconds or less. For more complex river basin models with typical running times of several minutes or hours, optimisation may require several days, weeks or even months of computing time, and hence parameter optimisation quickly becomes prohibitive on a single computer. To make the optimisation manageable, distributed or parallel computing can be utilised. The optimisation problem is, in general, highly parallel in nature (also known as an embarrassingly parallel problem). Within an iteration loop, parallel processing of different parameter sets can be performed to speed up the optimisation considerably. For instance, for a local gradient-based method where the gradient is calculated using forward or backward finite differences $(p+1)$ model evaluations can be run in parallel. For the SCE algorithm the evolution of the individual complexes can be run in parallel. For GA all parameter sets in the population can be evaluated in parallel. Generic optimisation software includes today facilities for distributed computing (see the section Software).

7. MULTI-OBJECTIVE OPTIMISATION

Summary

Model calibration is, in general, better constrained (more well-defined model parameter estimates are obtained) when multiple sources of information are used. The optimisation problem can be formulated in a general multi-objective framework, considering multi-variable measurements, multi-site measurements and multi-response modes. Mathematically, the solution to a multi-objective calibration problem is the set of Pareto solutions determined by the trade-off between the different objectives. For a Pareto solution none of the objective functions can be improved without violating one or more of the others. Pareto optimisation allows choosing preferred solutions according to the specific model application being considered, thus trading-off the different objectives using other type of information. In practical applications estimation of the entire Pareto front may be computationally infeasible, and one may only be interested in part of the solutions. In this case weights can be applied at the calibration stage to reflect the relative priorities given to certain objectives.

7.1. Optimisation problem

When different types of information are used in the model calibration, the numerical optimisation problem is preferably defined within a multi-objective framework. In general, multi-objective calibration can be performed on the basis of:

- Multi-variable measurements, i.e. the use of different types of observed data.
- Multi-site measurements, i.e. the use of data distributed within the model domain.
- Multi-response modes, i.e. the use of objective functions that measure different model responses such as e.g. the general water balance, peak flows, and low flows.

Mathematically, the multi-objective calibration problem can be stated as follows

$$Min\{F_1(\theta), F_2(\theta), ..., F_m(\theta)\} \tag{11}$$

where $F_i(\theta)$, $i = 1, 2, \ldots, m$ are the different objective functions considered. The solution to Eq. (11) will not, in general, be a single unique parameter set but will consists of the set of Pareto (or non-dominated) solutions according to the trade-offs between the different objectives (as illustrated in Figure 4.2 for a two-objective optimisation problem).

For a Pareto solution none of the objective functions can be further minimised without an increase of some of the other objective functions, thus forming the trade-off. Each of the Pareto solutions are from a multi-objective point of view equally good. An essential component of solving the multi-objective optimisation problem in Eq. (11) is how to compare parameter sets when there are two or more calibration objectives. Available techniques can be grouped into (i) aggregation approaches, and (ii) Pareto domination approaches.

7.2. Optimisation using aggregation

In the aggregation approach individual weights are assigned to the objective functions and aggregated into one measure that is used in the optimisation. This aggregate measure is then optimised using a single-objective optimisation procedure as described above. The aggregation of the different objective functions is, however, not trivial since they reflect different information contents, different scales of magnitudes, etc. (see discussion about the weighting problem in the section Weighting observations and aggregating performance measures).

Madsen (2003) proposed a two-step aggregation method. In the first step, data of the same type within a specified area of the model domain (a sub-catchment or the entire catchment) are aggregated with respect to a certain model response mode. In this case data are weighted taken the residual uncertainties, and the spatial and temporal correlation between the residuals into account (see Eq. (6) and the discussion following the equation). The first aggregation step results in a number of objective functions that measure model performance of a certain data type, in a certain area, with respect to a certain model response. In the second aggregation step these objective functions are then aggregated into one measure using

$$F_{agg}(\theta) = \sum_{i=1}^{m} w_i g_i\big(F_i(\theta)\big) \tag{12}$$

where $F_i(.)$ are the aggregated objective functions from the step-one aggregation, w_i are the weights, and $g_i(.)$ are transformation functions assigned to each objective function. The transformation functions are applied to compensate for the differences in the magnitudes of the different objective functions so that all $g_i(.)$ have about the same influence on the aggregated objective function near the optimum.

The transformation functions are not easily determined prior to the optimisation, but can be estimated as part of the optimisation. When using a population-based search algorithm, such as GA or SCE described above an initial population of points is generated. From this initial population the transformation functions can be determined. Madsen (2000) used an Euclidian distance measure to define the transformations. Van Griensven and Bauwens (2003) applied a transformation into a 0–1 scale by assigning a probability distribution to the individual objective functions.

By performing several individual optimisation runs with different weight combinations in Eq. (12) the entire Pareto surface can be explored. Madsen (2000) used the aggregation approach together with the SCE algorithm for exploring the Pareto front using low flow and high flow objective functions in the calibration of a lumped, conceptual rainfall-runoff model. The methodology was later applied to multi-objective calibration of the MIKE SHE integrated model using groundwater levels and catchment runoff as objective functions (Madsen, 2003).

In practical applications estimation of the entire Pareto front may be computationally infeasible, and one may only be interested in part of the Pareto solutions. In this case the user can specify weights in Eq. (12) to reflect the priorities given to certain objectives, depending on the model application. The relative weighting between the objectives make sense in this case because of the transformations applied. A balanced optimum solution where the same emphasis is put on all objectives is obtained from Eq. (12) by using the same weight on all objective functions. In the calibration of the ESWAT water quality model Van Griensven and Bauwens (2003) aggregated 13 different objective functions based on runoff data and concentration data of different chemical species at different locations into one measure that was optimised using the SCE algorithm.

7.3. Optimisation using Pareto dominance

The Pareto domination approach does not rely on a single comparative measure but on whether one solution is dominated by another for the different calibration objectives considered. In the optimisation a Pareto rank is then assigned each parameter set according to the domination criterion.

By definition, the Pareto set has following properties (e.g. Gupta *et al.*, 1998):

(1) For each non-member, it is possible to find at least one member of the Pareto set for which all its objective functions are smaller.
(2) It is not possible to find any member of the Pareto set for which all its objective functions are better than those of another member.

In other words, the parameter space can be divided into “good” (Pareto optimal) and “bad” (dominated) solutions, and a member of the Pareto set is better than other members with respect to some objective functions but not all.

Based on the definition of Pareto optimality different definitions of Pareto ranking has been proposed. Goldberg (1989) used the method of “onion peeling” to define the Pareto rank. First, all the non-dominated points in the population are assigned a rank equal to 1. Next, rank 1 points are removed and the non-dominated points of the remaining population are then assigned a rank equal to 2. This process is repeated until all points are assigned a rank. Vrugt *et al.* (2003b) introduced a Pareto rank measure that better discriminates between points in the central part of the Pareto surface and at the extreme parts (corresponding to the optima of the individual objective functions).

Numerical optimisation procedures that are especially suited for multi-objective optimisation using the Pareto domination approach are evolutionary algorithms. An algorithm that has been widely used to different multi-objective optimisation problems is the elitist non-dominated sorted genetic algorithm (NSGA-II) developed by Deb *et al.* (2000). Khu and Madsen (2005) used NSGA-II for calibration of a lumped, conceptual rainfall-runoff model considering four objectives (overall water balance error, overall RMSE, RMSE of high flow events, and RMSE of low flow events).

Different multi-objective extensions of the SCE algorithm have been developed and applied to calibration of river basin models. Yapo *et al.* (1998) developed the multi-objective complex evolution (MOCOM) algorithm. Boyle *et al.* (2000) applied MOCOM for calibration of a lumped, conceptual rainfall-runoff model considering three objectives that measured different response modes of the hydrograph. Beldring (2002) used MOCOM for calibration of a rainfall-runoff model using groundwater levels and catchment runoff as objectives. In the calibration of a catchment hydrochemical model Meixner *et al.* (2002) applied MOCOM considering different combinations of 21 hydrological and chemical performance criteria.

Vrugt *et al.* (2003b) developed the multi-objective shuffled complex evolution metropolis (MOSCEM) algorithm, which combines the SCE algorithm with the Metropolis Markov chain Monte Carlo sampling procedure. They demonstrated the use of the algorithm on calibration of a lumped, conceptual rainfall-runoff model considering two response modes of the hydrograph. Schoups *et al.* (2005) applied MOSCEM for calibration of a regional groundwater model. Recently, Tang *et al.* (2006) compared MOSCEM, a variant of the NSGA-II algorithm, and a benchmark algorithm, (the strength Pareto evolutionary algorithm (SPEA2)). They applied the algorithms for calibration of a lumped, conceptual rainfall-runoff model and a distributed, integrated hydrological model.

7.4. Single vs. multi-objective optimisation

Model calibration based on Pareto optimisation is a powerful method that has several important advantages. First, it allows a comprehensive evaluation of the trade-offs between different calibration objectives, and hence highlights possible model structural deficiencies. In the case of a perfect model and ideal measurements the Pareto surface will collapse to one single point in objective function space with a unique best parameter set. Departures from this ideal solution are an indicator of model structural errors.

Secondly, Pareto optimisation offers an elaborate framework for comparison of different models or model conceptualisations by considering several performance criteria in a consistent manner. For instance, one model structure may be better in simulating one model response than another response as compared to another model structure, and vice versa. Such differences will not be highlighted if only an aggregated performance measure is used in the optimisation.

Thirdly, Pareto optimisation allows choosing single Pareto-optimal solutions according to the specific model application being considered. The modeller or decision-maker avoids having to specify preferences to any of the calibration objectives at the calibration stage but can choose a preferred solution among the Pareto optimal solutions at a later stage using information not used in the optimisation (e.g. subjective information which cannot be easily expressed in terms of an objective function).

8. MODEL VALIDATION

Summary

Model validation is essential to a river basin modelling project in order to document the predictive capabilities and credibility of the calibrated model. Validation must be carried out on independent data that have not been used in the calibration, and must assess the expected accuracy of the model predictions with respect to the model application being considered. Multi-site and multi-variable validation is necessary to document the capability of the model for prediction of the spatial distribution of different variables. Although properly validated with respect to some model responses, care should be taken assessing the confidence of model predictions of variables for which validation has not been performed. A hierarchical validation test scheme with increasing power is presented that distinguishes between model predictions performed on the same river basin as used for parameter estimation or on a different river basin, and whether predictions are done under the same conditions as present in the calibration period or under changed conditions. If the model cannot be properly validated it is necessary to go back to the initial stages of the model calibration protocol and reconsider the model parameterisation and choice of calibration parameters.

Having completed the parameter estimation process the next step in the model calibration protocol is model validation (see Figure 2.2). There are two key principles in model validation (e.g. Refsgaard and Henriksen, 2004):

- Validation must be carried out on independent data that have not been used in the model calibration
- The model should be tested to assess the expected accuracy of the model predictions when being applied to the specific problem the model is intended for.

Validation tests are essential in order to document the predictive capability and credibility of the calibrated model. If the model is to be applied for prediction of the spatial distribution of different state variables, validations using data of the different state variables at different sites in the modelling domain are required. In this regard it must be emphasised that care should be taken for assessing the confidence of the model predictions for state variables that have not been explicitly validated.

If the calibrated model does not pass the model validation test it is necessary to go back to the initial stages of the model calibration protocol and reconsider the model parameterisation and choice of calibration parameters (feedback loop in Figure 2.2).

8.1. Hierarchical test scheme

Klemes (1986) proposed a hierarchical test scheme for model validation with increasing power (the term power is used in this context as a measure of the ability to discriminate between good and bad models). The scheme distinguishes between model predictions performed on the same river basin as used for parameter estimation or on a different river basin, and whether predictions are done under the same conditions as present in the calibration period or under changed conditions. The four tests are described in the following.

8.1.1. Split-sample test

The split-sample test is used when validation is performed on the same river basin and insituations with similar conditions as in the calibration period. Typically, the available data are spilt into two portions, where one portion is used for model calibration and the other portion is used for the validation. It is important to split the data set so that both the calibration and the validation data cover the same range of variability (both temporally and spatially) with respect to the different model responses being considered.

The split-sample test is the classical validation test. For lumped, conceptual rainfall-runoff models, the available record of catchment runoff data is typically spilt into two periods. One period is used for calibration and the other period is used for validation. For validation of distributed models additional split-sample validations are required in order to validate the distributed model performance. In this case, the available spatially distributed data are split into two groups that both cover the range of spatial and temporal variability of the considered model response in the river basin. One group of distributed data are used for calibration and the other group is used for validation.

The multi-site validation test requires a large amount of spatial data to sufficiently cover the spatial variability for both calibration and validation. Alternatively, a cross-validation test can be applied. In this case, calibration is performed using data at all sites except one, which is kept for validation. The calibration and validation process is then repeated until all sites have been used for validation. The cross-validation test has less data requirements than the multi-site, split-sample test but is rather time consuming since several calibrations have to be carried out.

The most powerful split-sample test is a multi-response validation test; that is, different model responses are used for the model validation. In this regard, two different strategies may be followed using either all model responses for both calibration and validation or keeping data of particular model responses for model validation only. Both tests are more powerful than validation based only on a single model response. Of the two strategies the second seems to be the most powerful but this may not be the case (Mroczkowski *et al.*, 1997). As discussed in the section Calibration data and choice of calibration objectives, it is important to include different types of data in the calibration to properly constrain the parameter estimation. Thus, if certain model response data are kept for the validation only, parameter identification and uniqueness problems may occur in the model calibration (cf. the trade-off of sensitivity to measured and unmeasured model predictions shown in Figure 3.1), thus making such a validation test less powerful.

The different variants of the split-sample test based on multi-response data are shown in Figure 8.1.

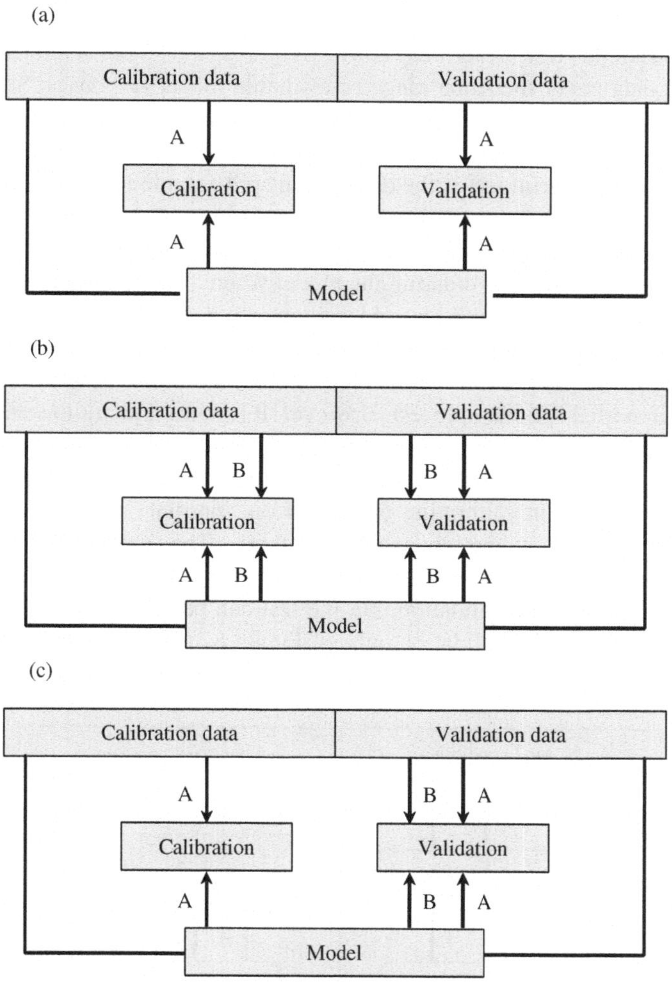

Figure 8.1. Illustration of different split-sample validation strategies using two model responses (A and B) (adapted from Mroczkowski *et al.*, 1997). (a) Single-response split-sample, (b) Multi-response split-sample, (c) Multi-response independent split-sample

8.1.2. Differential split-sample test

The split-sample test described above requires that both calibration and validation data cover the same range of available model responses. Such a test has relatively low power because only the performance of the model with respect to prediction within a similar range as experienced in the calibration data is tested. A more powerful test is the differential spilt-sample test where the model performance is tested on prediction beyond the range of calibration data. Such situations arise when validating on climatic extremes not encountered in the calibration data, or when validating the model where the hydrological regime of the river basin has experienced a sudden change, e.g. in the form of major land-use changes, changed groundwater abstraction or other man-made interventions. The ability of the model to predict future climate change should also be tested using a differential split-sample test. However, it is more difficult to select proper validation data in this case because climate change effects often vary slowly, and hence a long record is needed to test changes in the hydrological regime.

The test consists in calibrating the model on "normal" conditions and then testing the model on the changed conditions. The validation strategy is illustrated in Figure 8.2. The figure shows validation using all model response for both calibration and validation, but the test can be formulated using part of the model responses for validation only cf. Figure 8.1(c). In the validation some parameters may need to be recalibrated due to e.g. changed land-use conditions.

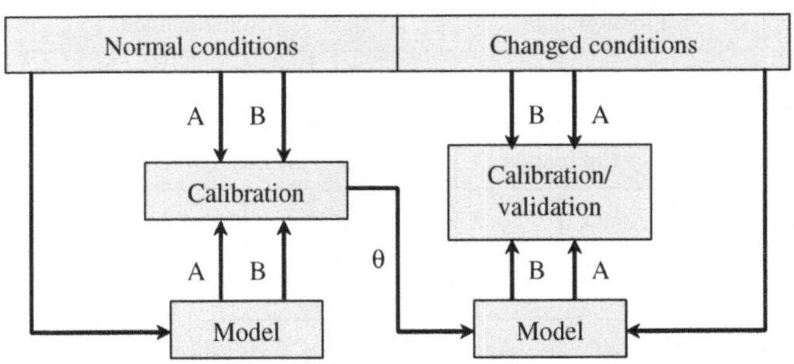

Figure 8.2. Illustration of differential split-sample validation strategy using two model responses (A and B) (adapted from Mroczkowski *et al.*, 1997)

8.1.3. Proxy-basin test

In the proxy-basin test no direct calibration of the model of the considered river basin is allowed. Experience from calibration of the model of river basins with similar characteristics and auxiliary information in terms of recommended values from literature in combination with geological profile descriptions, soil type maps, land use maps etc. are used to assess the model parameters at the river basin. The model is then validated on data collected at the river basin. The proxy-basin test is an important test in developing methodologies for model prediction at ungauged river basins.

8.1.4. Proxy-basin, differential split-sample test

The proxy-basin, differential split-sample test is a combination of the proxy-basin and differential split-sample tests. No direct calibration of the model of the considered river basin is allowed. The model parameters for the river basin are assessed to represent the "normal" and changed conditions. The model is then validated on the two periods.

Refsgaard and Knudsen (1996) presented a rigorous application of the validation test scheme by Klemes (1986). They compared the performance of a lumped, conceptual rainfall-runoff model, a semi-distributed model, and a distributed model using catchment runoff as the only data type. They found that all models performed equally well when one year or more of runoff data were used for model calibration. The distributed model performed only marginally better when no calibration was performed. Their results illustrate the problem of insufficient data to properly identify model structure and model parameters.

Mroczkowski *et al.* (1997) compared different validation strategies for a combined flow and salinity model. They showed that the power of a validation test depends on the kind of data available and the way the data are used for testing. In their application example they showed that the traditional split-sample test using catchment runoff data alone is not very powerful. The most powerful validation test is the use of multi-response data for a river basin that has experienced a shift in hydrological regime (a differential multi-response, split-sample test).

8.2. Validation tests based on model residuals

To formally validate a model it is necessary to test if the observed data used for validation are statistically consistent with the simulated model responses. For this, statistically derived model prediction and confidence limits can be used (see the section Uncertainty assessment). In this case, if a significant part of the observed validation data fall outside the model prediction limits, the model should be rejected. Mroczkowski *et al.* (1997) propose a simpler validation test based on analysing the model residuals. Similar analyses of residuals or weighted residuals are recommended by Hill (1998) to evaluate the model fit. If the model residuals are not consistent with the assumed error model, the model should be rejected.

9. UNCERTAINTY ASSESSMENT

Summary

Uncertainty is an unavoidable and inherent element in river basin modelling. The main uncertainty sources can be grouped into (i) perceptual model uncertainty, (ii) data uncertainty, (iii) parameter uncertainty, and (iv) model structural uncertainty. The disagreement between observations and model predictions is a combined effect of these different error sources, and if not appropriately accounted for in the model calibration can lead to biased model parameter estimates. One way to explicitly deal with uncertainties in model calibration is to apply weights that reflect the relative uncertainties of the data used for calibration. Parameter non-uniqueness may be a problem in model calibration. Non-uniqueness in terms of equifinality refers to a situation where different parameter sets may provide virtually similar model responses with respect to a given performance measure. This basically means that the available information used in the model calibration is insufficient to distinguish between different parameter sets. The problem of non-uniqueness can be reduced by properly constraining the parameter estimation using different types of calibration data. The effect of the different uncertainties on the model predictions should be quantified as part of the modelling. However, how to properly include all the different uncertainty sources is a complex and yet unsolved problem. Classical statistical non-linear regression techniques and Monte-Carlo based procedures can be used to quantify the effect of some of these uncertainties on the model predictions.

9.1. Uncertainty sources

Uncertainty is an unavoidable and inherent element in river basin modelling, which will affect the accuracy and precision of model predictions. The main uncertainty sources can be grouped into (e.g. Wagener and Gupta, 2005):

- *Perceptual model uncertainty*, i.e. uncertainty related to the representation of the real-world system into a numerical model. This uncertainty reflects our lack of knowledge of the different processes involved in the modelling.
- *Data uncertainty*, i.e. uncertainty in the input data (physical parameters used to set up the model and model forcings) and in the output data used in the model calibration. One important error source related to the input data is the representation of precipitation in the model. Often only data

from few rain gauges in the river basin are available, which is insufficient to represent the spatial variability of rainfall and may introduce a bias in the water balance. Radar-based rainfall measurements provide a good spatial coverage but introduce uncertainties in the transformation of reflectivity to rainfall amounts. For the output data, typical errors are related to the data processing, e.g. the translation of a water level measurement to a discharge using a rating curve. In addition, uncertainties are related to the use of point measurements to be compared with model responses over a modelling unit (see discussion about scaling uncertainties in the section Scaling issues).
- *Parameter uncertainty*, i.e. the uncertainty related to the use of non-optimal parameter values and the problem of identifying a unique optimal parameter set based on the available information.
- *Model structural uncertainty*, i.e. uncertainty introduced by conceptualisations and simplifications of the descriptions of real-world processes. This includes a variety of sources, including (i) formulations and coupling of the different processes, (ii) numerical discretisation, (iii) representations of the spatial variability and sub-grid processes, and (iv) interpretations and classifications of geology, soil type, land use etc. (Butts et al., 2004).

It must be emphasised that disagreement between model responses and observations is a combined effect of all the uncertainty sources. Thus, it is important to distinguish between the different sources during the calibration. Only the uncertainty source related to the use of non-optimal parameters can be reduced via model calibration. If other uncertainty sources are not appropriately accounted for, model calibration may compensate for these uncertainties, which may lead to biased parameter estimates. One way of dealing explicitly with uncertainties in the parameter estimation process is by using weights on the observations (see the section Weighting observations and aggregating performance measures).

9.2. Parameter non-uniqueness

The problem of parameter non-uniqueness is a common problem in model calibration. This is in principle attributed to two different causes: (i) high parameter correlations, and (ii) equifinality. In the former case, one may experience that the same model response can be obtained with different parameter sets. For instance, if two parameters are completely correlated, an infinite number of combinations of the two parameters exist which will provide exactly the same model response. In this case the two parameters cannot be uniquely determined. Gradient-based search procedures provide information about parameter correlations as an integrated part of the parameter optimisation. In practice, parameter correlations larger than 0.95 (in absolute terms) indicate uniqueness problems (Hill, 1998).

This problem of parameter non-uniqueness can often be solved by using data of a different kind in the calibration. Hill and Tiedeman (2003) demonstrated this on a steady-state groundwater calibration problem including two parameters, respectively, hydraulic conductivity and groundwater recharge. If only groundwater level observations are used for model calibration, the two parameters are completely correlated, and hence cannot be uniquely estimated. Addition of flow measurements better constrain the calibration with a well-defined optimum. High parameter correlations may also indicate model structural problems. For instance, in the calibration of the Sacramento rainfall-runoff model, two parameters related to the formulation of the percolation equation were found to be highly correlated and could not be determined uniquely (Sorooshian and Gupta, 1983). A reparameterisation of the percolation equation could be made that reduced the problem of non-uniqueness (Gupta and Sorooshian, 1983).

The other type of parameter non-uniqueness is referred to as equifinality (Beven and Binley, 1992). This occurs when different parameter sets provide practically equally fit model calibrations according to the objective function considered. This is often seen in the calibration of river basin models where the response surface has numerous very similar objective function values at very different locations in parameter space (thus, in this case not related to high parameter correlations). This basically means that the available information used in the model calibration is insufficient to distinguish between different parameter sets. This has lead to a proposed change of concept in model calibration. Instead of focusing on locating a unique optimum parameter set, one should instead appreciate the concept of equifinality and use this to represent parameter uncertainty and translate it into model prediction uncertainty (Beven and Binley, 1992). A formal strategy for this is the generalised likelihood uncertainty estimation (GLUE) procedure (Beven and Binley, 1992; Beven and Freer, 2000).

Another response to parameter equifinality is the concept of multi-objective calibration (Gupta *et al.*, 1998). According to this concept, the equifinality problem is caused by the use of an aggregated measure to evaluate model performance. Similar aggregate measures may be obtained for quite different model responses, and hence is not necessarily evidence of equifinality. Multi-objective analysis may be seen as a way to unfold the equifinality problem by looking at model performance with respect to different model responses (Madsen and Khu, 2006). In general, to better constrain the parameter estimation problem and reduce the problem of parameter non-uniqueness different types of model response data should be included in the calibration (see discussions in the sections Calibration data and choice of calibration objectives and Multi-objective optimisation).

9.3. Uncertainty propagation

It is now generally accepted that uncertainties and their effect on the model predictions should be quantified as part of the modelling. A reliable estimate of the prediction uncertainty is crucial for making efficient decisions based on the model predictions. However, how to include and combine all the different uncertainty sources is a very complex and yet unsolved problem. Preferably, the joint probability distribution of all the uncertainty sources should be quantified and propagated through the model to derive probability distributions of the model predictions. However, quantification of the probability distributions for the different uncertainty sources and propagation of these through the model is in practice not feasible, and hence simplifications are needed.

In the classical statistical approach for evaluating model prediction uncertainty using non-linear, least-squares regression (gradient-based search) uncertainties are assumed to be related to observation errors. In this regard, the observation error should not be interpreted as representing only the error related to the actual observation but should represent in a lumped way all the different error sources involved in the modelling as outlined above. To derive model prediction uncertainty bounds, assumptions about the statistical distribution of the errors have to be made. Typically independent, identically normally distributed errors are assumed. Model prediction uncertainties are then obtained using a first-order analysis where the covariance of estimated model parameters (given as output from the non-linear regression) is propagated through the model using a first order Taylor series approximation of the model operator, which are then added to the observation error uncertainty (assuming an additive error structure). While the first-order approach is an efficient solution to quasi-linear models, the method may fail when applied to more non-linear models. Regression-based uncertainty estimation methods have been developed that do not rely on a linear approximation (e.g. Christensen and Cooley, 1999).

The regression-based uncertainty estimation procedures have been widely applied in groundwater modelling. Several other uncertainty estimation procedures have been used in river basin modelling, of which Bayesian and pseudo-Bayesian techniques have found most widespread application. The GLUE method has been widely applied to a variety of models, including rainfall-runoff, groundwater and integrated modelling. The GLUE procedure is easy to implement but has been criticised for not being formally Bayesian and for not adequately sampling the high probability density region of the parameter space (unless huge computational efforts are spent). More formal Bayesian procedures have been developed based on Markov Chain Monte Carlo (MCMC) sampling (e.g. Kuczera and Parent, 1998; Vrugt *et al.*, 2003a). Recently, a new methodology that combines the GLUE concept with a MCMC sampler to more efficiently elaborate the high probability density region of the parameter space has been developed and applied in river basin modelling (Blasone *et al.*, 2008).

Most of the uncertainty propagation procedures assume a correct model structure and only parameters within that structure are allowed to vary. When significant model structural errors are present, such procedures will provide unreliable uncertainty predictions. Model structural errors may be qualitatively addressed within a multi-objective calibration framework (as discussed in the section Multi-objective optimisation). Butts *et al.* (2004) proposed a general modelling framework to address the effect of different model structures on model predictions. Explicit integration and quantification of model structural errors in uncertainty propagation, however, is much more complicated, since these errors are difficult to isolate from the errors originating from parameter uncertainty. For groundwater flow and transport modelling, Neuman (2003) proposed a Bayesian model averaging procedure to combine the predictions of several alternative conceptual hydrogeological models and to assess their joint probability. The GLUE method and Markov Chain Monte Carlo sampling procedure offer other frameworks to address jointly parameter and model structural errors but so far have considered mainly parameter uncertainty.

More information on uncertainty estimation in river basin modelling can be found in the Harmoni-CA guideline Uncertainty Analysis (see part 4 of this guideline).

10. APPLICATION EXAMPLES

10.1. Lumped, conceptual rainfall-runoff model (MIKE 11/NAM)

This application example is based on Madsen (2000) and Madsen *et al.* (2002).

10.1.1. Model description

The MIKE 11/NAM model (Nielsen and Hansen, 1973; DHI, 2005b) is a typical example of a lumped, conceptual rainfall-runoff model. The NAM model represents the rainfall-runoff processes at catchment scale by continuously accounting for the water content in four mutually interrelated storages. These storages are: (1) snow storage, (2) surface storage, (3) lower zone (root zone) storage, and (4) groundwater storage. The structure of the NAM model is shown in Figure 10.1.

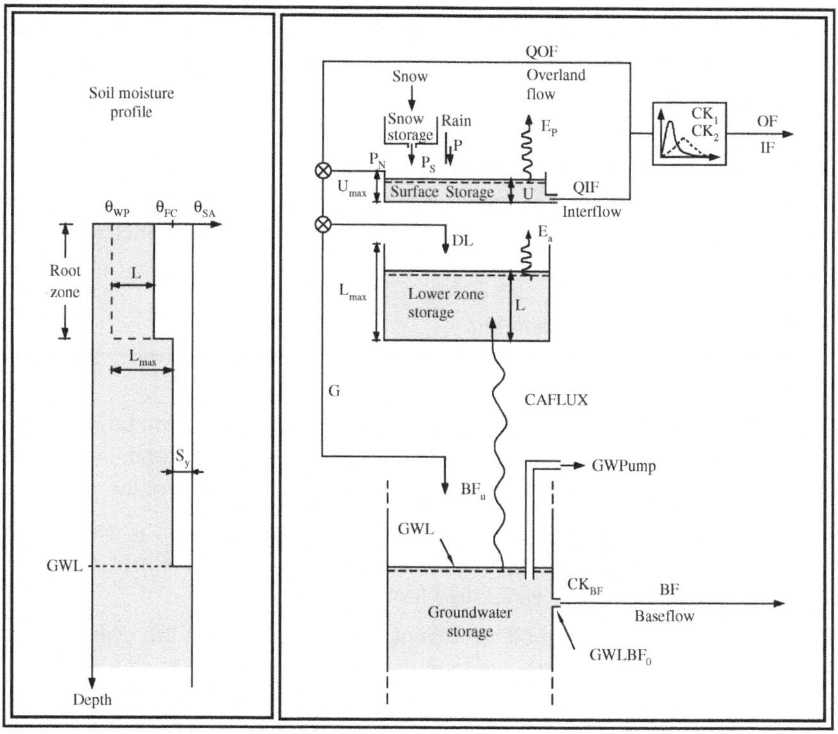

Figure 10.1. NAM model structure

The meteorological input to the model consist of precipitation, potential evapotranspiration and temperature (if snow modelling is included). On this basis, it produces as its main results catchment runoff as well as information about other elements of the land phase of the hydrological cycle, such as evapotranspiration, soil moisture content, groundwater recharge and groundwater storage. The resulting catchment runoff is split conceptually into overland flow, interflow and baseflow components, representing, respectively, the quick, intermediate and slow response modes of the hydrograph.

10.1.2. Model setup

The NAM model was applied to the Tryggevaelde catchment in Denmark. This catchment has an area of 130 km^2, an average rainfall of 710 mm/year, and an average runoff of 240 mm/year. The catchment is dominated by clayey soils, implying a relatively flashy flow regime. For the calibration, a 5-year period (1 January 1984–31 December 1988) was used where daily data of precipitation, potential evapotranspiration, mean temperature, and catchment runoff were available. For validation, data in the period 1 January 1989–31 December 1993 were used.

10.1.3. Calibration parameters

The basic version of the NAM model was applied, which includes 9 parameters to be determined by calibration. A brief description of these parameters is given in Table 10.1.

Table 10.1. NAM model parameters

Parameter	Description
U_{max} [mm]	Maximum water content in the surface storage. U_{max} can be interpreted as including the water content in the interception storage, in surface depression storages, and in the uppermost few cm's of the soil.
L_{max} [mm]	Maximum water content in the lower zone storage. L_{max} can be interpreted as the maximum soil water content in the root zone available for the vegetative transpiration.
CQOF [-]	Overland flow runoff coefficient. *CQOF* determines the distribution of excess rainfall into overland flow and infiltration ($0 \leq CQOF \leq 1$).
TOF, *TIF*, *TG* [-]	Threshold values for, respectively, overland flow, interflow and recharge. Flow is only generated if the relative moisture content in the lower zone storage is larger than the threshold value.
CK_{IF} [hours]	Time constant for interflow from the surface storage. CK_{IF} is the dominant routing parameter of the interflow because $CK_{IF} >> CK_{12}$.
CK_{12} [hours]	Time constant for overland flow and interflow routing. Overland flow and interflow are routed through two linear reservoirs in series with time constants CK_{12}.
CK_{BF} [hours]	Baseflow time constant. Baseflow from the groundwater storage is generated using a linear reservoir model with time constant CK_{BF}.

10.1.4. Calibration methods

In the study, two different calibration methods were applied and compared. These include (i) multi-objective calibration using the shuffled complex evolution (SCE) algorithm, and (ii) knowledge-based expert system.

SCE multi-objective calibration procedure

The multi-objective calibration approach presented by Madsen (2000) was applied. In this method four basic objective functions are considered:

(1) Overall volume error:

$$F_1(\theta) = \left| \frac{1}{N} \sum_{i=1}^{N} \left[Q_{obs,i} - Q_{sim,i}(\theta) \right] \right| \tag{13}$$

(2) Overall RMSE:

$$F_2(\theta) = \left[\frac{1}{N} \sum_{i=1}^{N} \left[Q_{obs,i} - Q_{sim,i}(\theta) \right]^2 \right]^{1/2} \tag{14}$$

(3) Average RMSE of peak flow events:

$$F_3(\theta) = \frac{1}{M_p} \sum_{j=1}^{M_p} \left[\frac{1}{n_j} \sum_{i=1}^{n_j} \left[Q_{obs,i} - Q_{sim,i}(\theta) \right]^2 \right]^{1/2} \tag{15}$$

(4) Average RMSE of low flow events:

$$F_4(\theta) = \frac{1}{M_l} \sum_{j=1}^{M_l} \left[\frac{1}{n_j} \sum_{i=1}^{n_j} (Q_{obs,i} - Q_{sim,i}(\theta))^2 \right]^{1/2} \tag{16}$$

In Eqs. (13)–(16), $Q_{obs,i}$ and $Q_{sim,i}$ are, respectively, observed and simulated catchment runoff, N is the total number of time steps in the calibration period, M_p is the number of peak flow events, M_l is the number of low flow events, n_j is the number of time steps in peak/low flow event no. j, and θ is the set of model parameters to be calibrated. Peak flow events are defined as periods where the observed discharge is above a given threshold level. Similarly, low flow events are defined as periods where the observed discharge is below a given threshold level.

The objective functions are aggregated into a single objective function using transformations of the original objective functions as described in Madsen (2000). These transformations are automatically calculated as part of the optimisation to provide a balanced weight between the different objectives.

The SCE algorithm is applied to optimise the aggregated objective function. As suggested by Kuczera (1997), to reduce the chance of convergence into local optima, the number of complexes in the SCE algorithm was set equal to the number of calibration parameters. For the other algorithmic parameters the recommended values given by Duan *et al.* (1994) were used.

Knowledge-based expert system

The knowledge-based calibration procedure is formulated as a step-wise procedure, focusing on different process descriptions. The procedure involves a preliminary recession analysis, followed by an iterative scheme where individual parameters are adjusted according to different response modes of the hydrograph. Numerical goodness-of-fit measures are applied at each stage of the calibration process.

In the recession analysis the three time constants in the model that determines the recession of the quick flow, intermediate flow and base flow components are estimated. These estimates are then used as initial values in the following, iterative calibration procedure. This includes the following steps:

(1) The storage parameters U_{max} and L_{max} are optimised against the overall water balance cf. Eq. (13).
(2) The overland flow runoff coefficient *CQOF* is optimised by comparing the moment around the origin of the observed and simulated hydrographs.
(3) The time constant for the quick flow component CK_{12} is fine-tuned by optimising peak flow simulations using Eq. (15).
(4) The threshold values *TOF*, *TIF* and *TG* are determined by comparing observed and simulated hydrographs in the beginning of the wet season.
(5) The time constant for interflow CK_{IF} is fine-tuned by comparing observed and simulated runoff in periods where interflow dominates.

Steps 1–5 are repeated until no further improvement can be made.

10.1.5. Results

To compare the results from the two optimisations different statistical performance measures were calculated for the validation period. These include:

(1) Water balance error (difference between average simulated and average observed runoff).

(2) Nash-Sutcliffe coefficient (*E*).
(3) Peak flow statistics for four selected peak flow events, including peak error, bias and *RMSE*.
(4) Low flow statistics for four selected low flow periods, including bias and *RMSE*.

The results are shown in Table 10.2.

Table 10.2. Performance statistics for the calibrated models

		Multi-objective SCE	Expert system
Overall measures			
Water balance error	[%]	0.2	−0.2
E	[-]	0.85	0.83
Peak flow measures			
Absolute peak error	[m^3/s]	1.10	1.00
Bias	[m^3/s]	0.45	0.48
RMSE	[m^3/s]	0.67	0.73
Low flow measures			
Bias	[m^3/s]	−0.050	−0.027
RMSE	[m^3/s]	0.091	0.078

Considering the overall performance measures (water balance error and E) the multi-objective SCE calibration has slightly better performance. For the peak flow simulations the multi-objective SCE calibration has slightly better performance for bias and RMSE, but slightly worse performance with respect to peak flow error. The low flow statistics show a slightly better performance for the calibration based on the expert system.

The results show that none of the calibration methods are superior with respect to all performance measures considered. Thus, calibration based on the automatic and generic search routine in combination with user-specified multiple calibration objectives compares favourably with the expert system that is specifically designed for the model considered and which requires substantial user intervention during the calibration process.

The results illustrate the common problem of uniqueness in model calibration due to trade-offs between the different objectives considered. Basically, the two methods put emphasis on different hydrograph behaviours, hence providing different solutions on the trade-off (or Pareto) surface. This aspect is illustrated in Figure 10.2 where the Pareto optimum solutions are shown for a calibration based on the peak flow and low flow RMSE, cf. Eqs. (15)-(16). A pronounced trade-off between the two objectives is observed; that is, no unique parameter set can be obtained that is optimal for both low flow and peak flow simulations. One should choose a compromise solution (or a number of compromise solutions) using other considerations, depending on the model application.

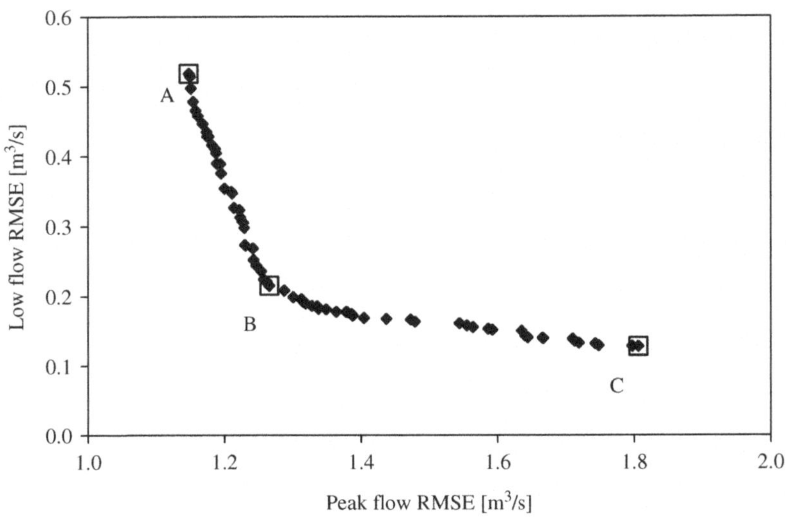

Figure 10.2. Estimated Pareto front for optimisation of peak and low flow RMSE. Marked points correspond to single objective calibrations using, respectively, peak flow RMSE (A) and low flow RMSE (C). Point B corresponds to the weighted balanced optimum found by the SCE multi-objective calibration procedure

10.2. Groundwater model (MODFLOW)

This application example is based on Harrar *et al.* (2003).

10.2.1. Model description

MODFLOW (McDonald and Harbaugh, 1989) is a three-dimensional finite-difference groundwater flow model. The model was originally developed to simulate groundwater flow only, but is developed in a modular structure, and different packages that describe different components of the water cycle can be included in the model. The model can thus be used for distributed and integrated modelling, including the major hydrological processes of the land phase. The different processes can be described in varying detail depending on the modules included in the model, e.g. physically-based descriptions or lumped/conceptual descriptions.

MODLFOW is developed by the US Geological Survey (USGS) and exists in different downloadable versions. The version used in the present example is the MODFLOW-96 code (Harbaugh and McDonald, 1996). Only steady-state groundwater flow was considered, including a relative simple expression to describe groundwater-stream interactions.

10.2.2. Model parameterisation

The model was setup for a 1000 km^2 catchment in a dissected glacial-till plain in the western part of Denmark. The geology at the study site is highly complex with interbedding clay, fine and coarse sand sediments in the Tertiary strata overlaid by Quaternary meltwater sand with till and clay lenses. The Tertiary sediments are disserted by several channels and buried valleys with depths exceeding 150 m. These channels and Valleys are filled with Quaternary deposits. Two alternative geological conceptual models were constructed (Figure 10.3). Model A resembles a relatively simple geological conceptualisation, whereas model B includes the observed heterogeneity in greater details.

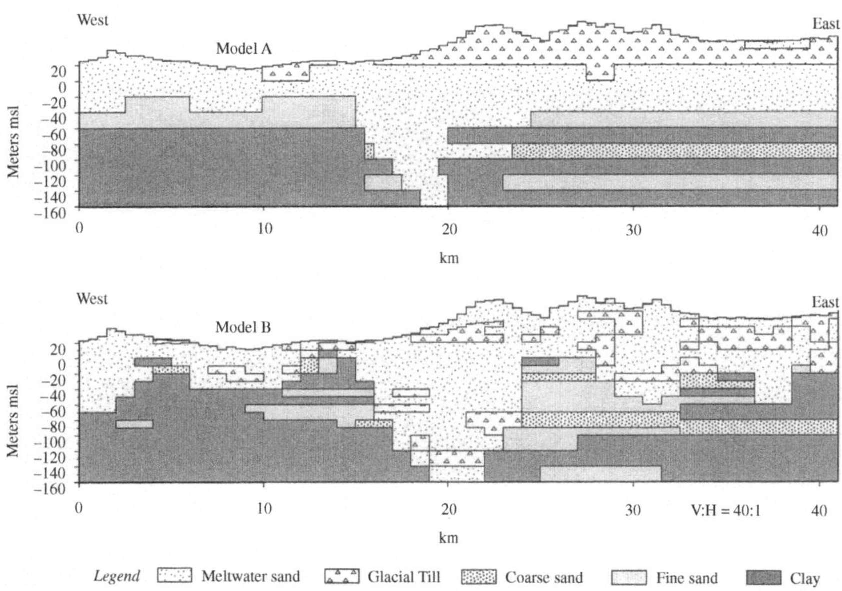

Figure 10.3. Cross sections of the two geological models used to construct the groundwater flow models

The models were set up to reflect a dry period where baseflow is expected to comprise the main contribution of the stream discharges. Only saturated flow was considered in the study, and infiltration to the groundwater table was thus estimated externally. Average precipitation was measured at precipitation stations and evaporation was estimated by the national water resource model (Henriksen *et al.*, 2003), while surface/near surface flow and drainage was estimated from discharge measurements. Infiltration to the groundwater table was thus taken to be precipitation minus the components: evaporation, shallow flow and drainage. The resulting infiltration was divided into five zones.

The river network in the catchment is shown in Figure 10.4. Only minor tributaries that could not be resolved in the numerical grid (500×500 m horizontally) were excluded in the model. Exchange between the groundwater and streams were described by a general head boundary condition, i.e. flow in the streams was not considered. The streams were divided into 36 reaches based on low flow estimates and surface geology. Unique values of leakage coefficients could potentially be assigned to the 36 reaches.

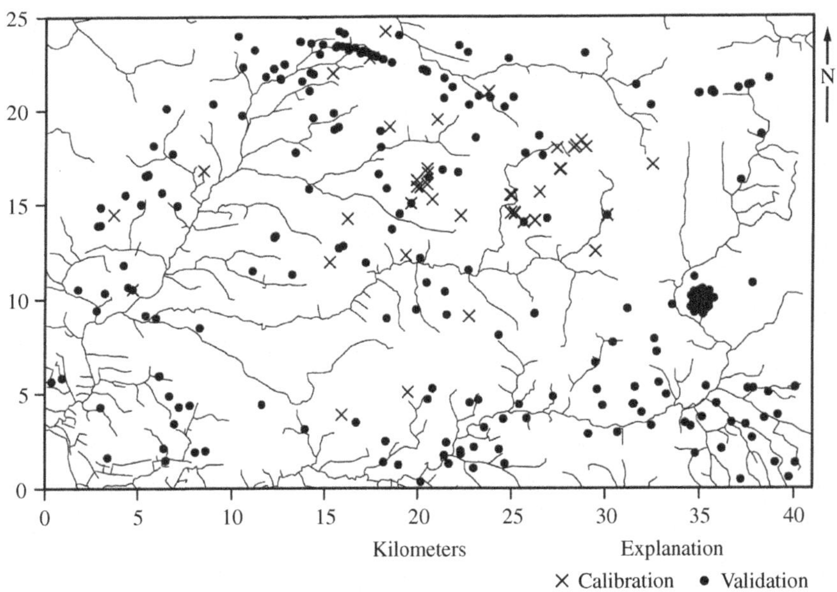

Figure 10.4. Location of the monitoring wells used for model calibration and validation

10.2.3. Calibration objectives

The models were calibrated for steady-state conditions using hydraulic heads and stream discharges as calibration targets. Since the estimated infiltration to the groundwater was given as recharge input to the model, the simulated river flow corresponds to the groundwater-stream exchange (no surface runoff or drainage), and baseflow estimates were used in the calibration, which were available at 109 locations within the catchment. The hydraulic head data used for calibration consisted of 48 observations from a synchronous round during the summer (August). Another 251 hydraulic head data were available. These observations were from different dates and not used in the model calibration but only in the model validation. The spatial distribution of the head measurements are shown in Figure 10.4.

A weighted least squares objective function was used, where the residuals were weighted by an estimated standard deviation of the observation error. All head observations included in the calibration were measured at the same time and observation uncertainty due to temporal variation was thus eliminated. The head data were considered very accurate as the location and elevation of the monitoring wells were measured precisely, and all head data were thus assigned a uniform standard deviation of 1 m. Uncertainties in the baseflow measurements were taken to be proportional to the absolute magnitude of the baseflow. Four classes were defined: 0–20, 20–100, 100–500 and above 500 l/s, where the coefficient of variations were set to 1.0, 0.5, 0.2 and 0.1, respectively. This means that the standard deviation of a measurement of 50 l/s was assumed equal to 25 l/s.

10.2.4. Calibration parameters

In both geological models the variability in the hydraulic conductivity was represented by four zones. Coarse and fine sand was defined as one zone, while a unique zone was assigned to each of the three remaining geological units, thus assuming homogenous geological units. Experiences from initial trial-and-error calibration showed that the fit to baseflow estimates for individual stream reaches could be improved by distributing streambed conductance with as few as two zones. A recharge multiplication factor was included in the calibration to account for uncertainty in the estimated infiltration to the groundwater table. A single factor was used, thus the relative spatial distribution was maintained during the optimisation.

Prior to the optimisation a sensitivity analysis was carried out to identify the most sensitive parameters to be included in the optimisation. Composite Scaled

Sensitivities (CSS) were used to evaluate the parameter sensitivities with respect to the objective functions. Only the sensitive component of the hydraulic conductivity was optimised for each zone, whereas a fixed anisotropy ratio was used to link the horizontal (K_h) and vertical (K_v) hydraulic conductivity. Anisotropy ratios (K_v/K_h) where fixed based on preliminary trial-and-error optimisation and set to 0.1 for all hydrofacies except for the Quaternary glacial till, where a value of 0.5 was specified. The sensitivity analysis showed that the regression was incapable of producing distributed streambed conductance values that were statistically significant, and a single value was thus used in the calibration.

A total of six parameters were included in the calibration:

- Horizontal hydraulic conductivity of the Quaternary meltwater sands, fixed anisotropy ratio of 0.1.
- Vertical hydraulic conductivity of the Quaternary glacial till, fixed anisotropy ratio of 0.1.
- Horizontal hydraulic conductivity of the Tertiary quartz sand, fixed anisotropy ratio of 0.1.
- Vertical hydraulic conductivity of the Tertiary clay, fixed anisotropy ratio of 0.5.
- Streambed conductance of stream system.
- Multiplication factor for recharge (one factor for the entire catchment).

Initial hydraulic conductivity values were obtained from pump tests and specific-capacity data in the area.

10.2.5. Calibration method

The two models were calibrated inversely by use of the universal optimisation code UCODE (Poeter and Hill, 1998). The weighted least-squares objective function used in the study was defined as

$$F(\theta) = \sum_{i=1}^{N_h} \left(\frac{h_{obs,i} - h_{sim,i}(\theta)}{s_{h,i}} \right)^2 + \sum_{j=1}^{N_Q} \left(\frac{Q_{obs,j} - Q_{sim,j}(\theta)}{s_{Q,j}} \right)^2 \tag{17}$$

where θ is the vector of parameters to be estimated, $h_{\text{obs,i}}$ and $h_{\text{sim,i}}(\theta)$ are the ith observed and simulated values of hydraulic head, respectively, N_{h} is the number of head observations, $s_{\text{h,i}}$ is the standard deviation of the *i*th hydraulic-head measurement, $Q_{\text{obs,j}}$ and $Q_{\text{sim,j}}(\theta)$ are the *j*th observed and simulated stream discharge, respectively, N_{Q} is the number of discharge observations, and $s_{\text{Q,j}}$ is the standard deviation of the jth discharge observation. Analysis of the objective function showed that the head levels and baseflow measurements contributed nearly equally to the objective function in the present study, and thus no weighting was applied to the head and baseflow measures in Eq. (17).

10.2.6. Results

The model calibrations were evaluated in four ways: (1) comparing optimised hydraulic conductivities to field scale estimates from pump tests and specific-capacity data, (2) evaluation of the estimated confidence intervals for the optimised parameters, (3) comparison of the calibration performance with respect to the fit to observed heads and baseflows, and (4) validation of the models by using an independent dataset of hydraulic heads.

Optimised hydraulic conductivity values compared well with the field estimates, with a slightly better agreement for model A. The estimated parameters were, in general, associated with narrow 95% confidence intervals, indicating a relatively high confidence in the estimated values. Again, model B performed worst, where the 95% confidence intervals for till and clay ranged four orders of magnitude. The fit to observed heads and estimated baseflow were expressed by the root mean squared weighted residuals (RMSW) calculated as:

$$RMSW_h = \sqrt{\frac{1}{N_h}\sum_{i=1}^{N_h}\left(\frac{h_{obs,i} - h_{sim,i}(\theta^*)}{s_{h,i}}\right)^2} \tag{18}$$

$$RMSW_Q = \sqrt{\frac{1}{N_Q}\sum_{j=1}^{N_Q}\left(\frac{Q_{obs,j} - Q_{sim,j}(\underline{\theta}^*)}{s_{Q,j}}\right)^2} \tag{19}$$

where θ^* is the vector of optimised parameters.

The models performed almost equally well in reproducing the observed quantities, but a slightly better fit was obtained for model A for both data types. The final model test was a split-sampling test. Here the 251 head observations not used in the calibration were used. These observations represent measurements made during different times of the year over a 30-year period. Although this data set was expected to reflect a stationary situation reasonable well, a larger uncertainty was assigned to the measurements, reflecting both a less precise data set, but especially accounting for temporal variations. Similarly to the previous tests, model A performed slightly better in reproducing the hydraulic heads used in the validation.

The study showed that it was possible to obtain almost similar calibration results for two different conceptual models. None of the two models could be rejected based on the calibration performance, and may thus have been accepted and used for predictive simulations. Such predictive simulations were also carried out by Harrar *et al.* (2003), where the calibrated models were applied for particle tracking and solute transport simulations. While the models gave comparable results in the delineation of a well field capture zone, the travel times and solute breakthrough varied markedly between the two models. A similar conclusion was found by Højberg and Refsgaard (2005) who calibrated three models based on different geological interpretations. While the models performed almost equally well in reproducing observed heads and stream discharges, significant variations were observed for the simulated breakthrough concentration at selected well fields. Taking the estimated 95% confidence intervals as an expression of the parameter uncertainty Højberg and Refsgaard (2005) performed an uncertainty assessment by Monte Carlo simulations and found that the parameter uncertainty contributed only little to the overall uncertainty. These findings illustrate that great care must be taken when models are used to predict variables not included in the calibration and/or validation of the model.

10.3. Distributed, integrated model (MIKE SHE)

This application example is based on Madsen (2003) and Madsen and Khu (2006).

10.3.1. Model description

MIKE SHE (Graham and Butts, 2006) is a distributed and integrated modelling system for the simulation of all the major hydrological processes occurring in the land phase of the hydrological cycle. The model combines different components, each describing the main processes in individual parts of the hydrological cycle. For these components several model descriptions are available ranging from complex, physically-based descriptions that solve the governing partial differential equations to simpler, conceptual models (see Figure 10.5).

In this application the following model descriptions were applied: (i) the Kristensen-Jensen model for evapotranspiration, (ii) 2D diffusive wave approximation of the Saint Venant equations for overland flow, (iii) Muskingum routing for flow in the river system, (iv) 3D Boussinesq equation for flow in the saturated zone, (v) 1D Richards' equation for vertical flow in the unsaturated zone, (vi) linear reservoir model for flow in drains, (vii) Darcy equation for the river-aquifer interaction, and (viii) a degree-day approach for snow melt.

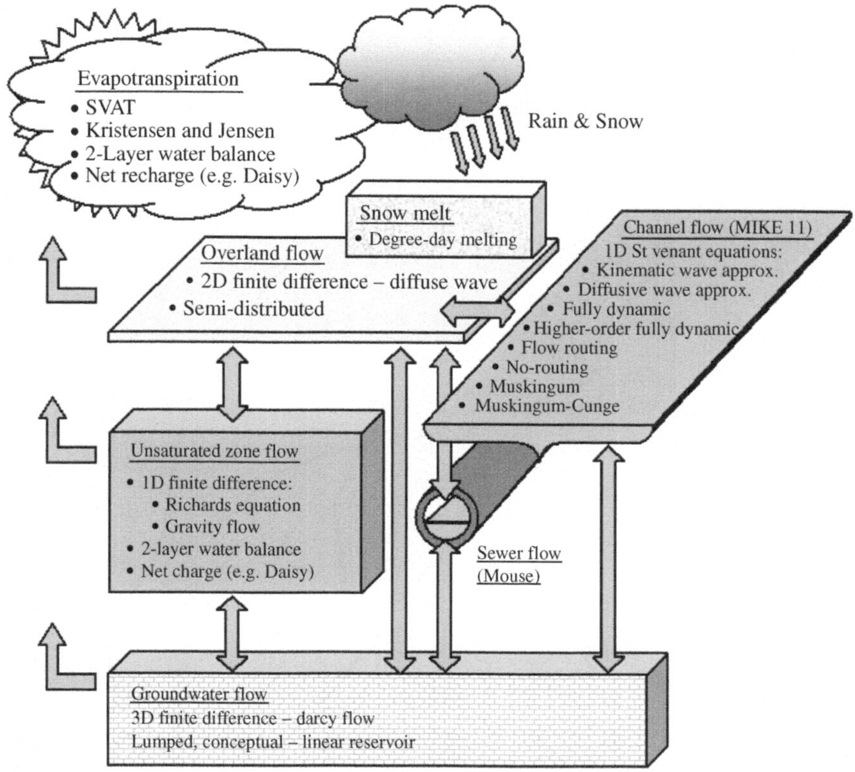

Figure 10.5. Schematic view of the processes in MIKE SHE, including the available numerical engines for each process. The arrows show the available exchange pathways for water between the different process models

10.3.2. Model parameterisation

The MIKE SHE model was applied to the Karup catchment in Denmark. The Karup catchment has an area of 440 km^2 and is drained by the Karup River and about 20 tributaries (see Figure 10.6). The topography varies from about 20 m to 100 m. The geology is relatively homogeneous with highly permeable sand and gravel deposits and small lenses of moraine clay. The aquifer is mainly unconfined and varies in thickness from about 10 m at the western and central part of the catchment to more than 90 m at the upstream eastern water divide. The depth of the unsaturated zone varies from 25 m at the eastern water divide to less than 1 m in the wetland areas along the river. The land use consists of agriculture (67%), forest (18%), heath (10%), and wetland areas (5%).

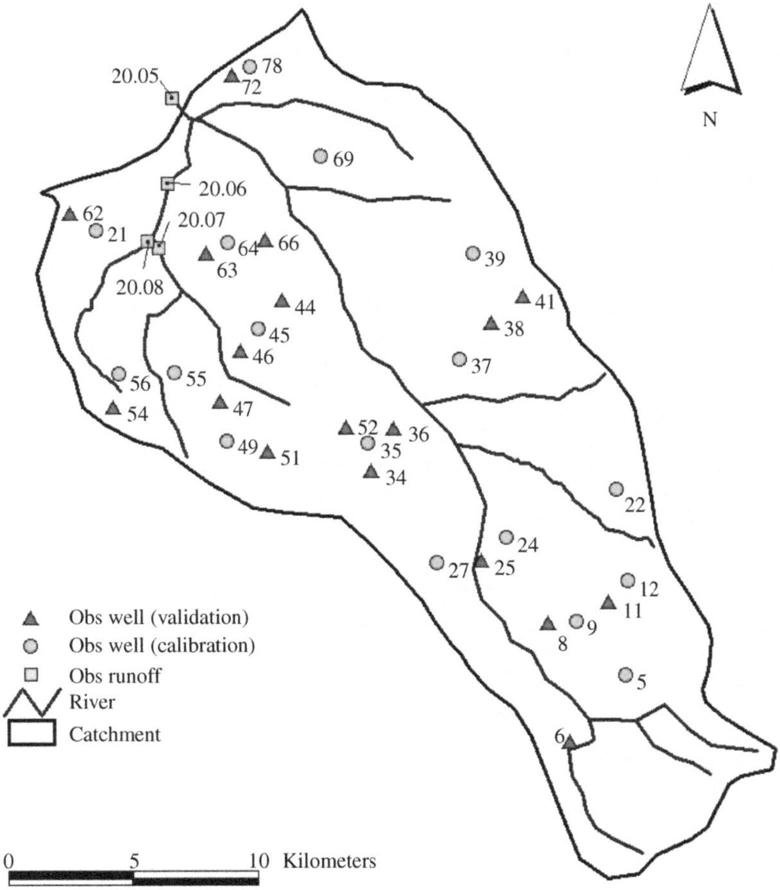

Figure 10.6. Karup catchment. Location of groundwater wells and discharge measurement stations

The geological conceptualisation of the saturated zone is taken from the Danish National Water Resources model which is interpreted in 10 meter thick layers using a grid size of 1×1 km. For each grid element a soil type is assigned with a given code and hydrogeological parameters. Six general soil units are defined (see Table 10.3).

The Karup River and the main tributaries are included in the model. For modelling the river-aquifer interaction a thin permeable layer is assumed between the river and the main aquifer. The leakage coefficient characterising this layer is assumed homogeneous in the catchment.

Table 10.3. Soil types used in the conceptualisation of the saturated zone

Soil code	Soil name	Description
1	Melt water sand	Quarternary and Post-Glacial sand and gravel
2	Clay	Glacial, Inter-Glacial and Post-Glacial clay and silt
3	Quartz sand	Miocene, medium to coarse grained sand and gravel
4	Mica sand	Miocene, fine to medium grained sand
5	Mica clay/silt	Pre-Quarternary clay and silt
6	Limestone	Limestone

The wetland areas are drained by ditches and drain pipes. Drainage is modelled conceptually using a linear reservoir model description in each cell. The drainage level (relative to ground surface) and the time constant of the linear reservoir model are assumed homogeneous in the catchment.

For the unsaturated zone two soil profiles are defined. A soil profile used for the main part of the catchment that comprises loamy sand to a depth of 100 cm and fine sand below, and a soil profile for the heath areas that consists of fine sand to a depth of 55 cm and coarse sand below. For each of the four soil types Van Genuchten retention and conductivity curves are specified.

10.3.3. Calibration objectives

The available measurements consist of groundwater level data sampled every two weeks from 35 locations in the catchment and daily discharge data from four stations in the river system, including the runoff at the catchment outlet (see Figure 10.6). In the calibration, groundwater level data from 17 wells and runoff data from the catchment outlet were used. Based on these data two objective functions were defined, respectively, the average RMSE of the groundwater levels at the 17 wells and the RMSE of the runoff at the catchment outlet.

For calibration, data in the period 1 January 1971–31 December 1974 were used. To minimise the effect from the initial conditions for calculation of the objective functions, a 2-year warm-up period was applied in the simulations. For validation of the calibrated model, data in the period 1 January 1975–31 December 1977 were used.

10.3.4. Calibration parameters

A preliminary sensitivity analysis was adopted to identify the most sensitive parameters to be included in the optimisation. Scaled sensitivity coefficients were evaluated around a nominal parameter set with respect to the two objective functions. The parameters selected for calibration are:

- Hydraulic conductivity of soil type 1 (melt water sand) and 3 (quartz sand). A constant anisotropy of 0.1 between vertical and horizontal hydraulic conductivity was assumed.
- Leakage coefficient for the river-aquifer interaction.
- Drainage level and time constant.
- Saturated hydraulic conductivity and the α and n-parameter in the van Genuchten equations of the two soil types in the soil profile covering the main part of the catchment.

Thus, in total 11 parameters were included in the calibration.

10.3.5. Calibration method

For analysing the trade-off between the two objectives Pareto optimisation was adopted. In this case the Population Simplex Evolution (PSE) method (Madsen, 2007) implemented in the AUTOCAL software (DHI, 2005a) was applied. The PSE algorithm is an evolutionary algorithm similar to genetic algorithms but uses the simplex method operators (Nelder and Mead, 1965) for evolution. For evaluating the goodness-of-fit of the different objective functions the concept of Pareto ranking according to the definition by Goldberg (1986) is applied in PSE. For the optimisation a population size of 50 was used and in total 2000 model evaluations were performed.

10.3.6. Results

The estimated Pareto front from the PSE optimisation is shown in Figure 10.7. A significant trade-off between groundwater levels and runoff simulation is observed in this case. That is, no unique parameter set can be found that provides both an optimal solution in terms of groundwater levels and catchment runoff. For the Pareto optimal solutions average RMSE of groundwater levels vary between 0.94 m and 1.04 m, and RMSE of runoff vary between 0.43 m^3/s and 0.88 m^3/s.

The modeller or decision-maker can subsequently use the results of the Pareto optimisation to choose a preferred solution (or several preferred solutions). This allows putting emphasis on certain modelling aspects (in this case runoff or groundwater level simulations) or to use other information not directly included in the calibration. In general, the final choice of preferred solutions should depend on the actual model application being considered.

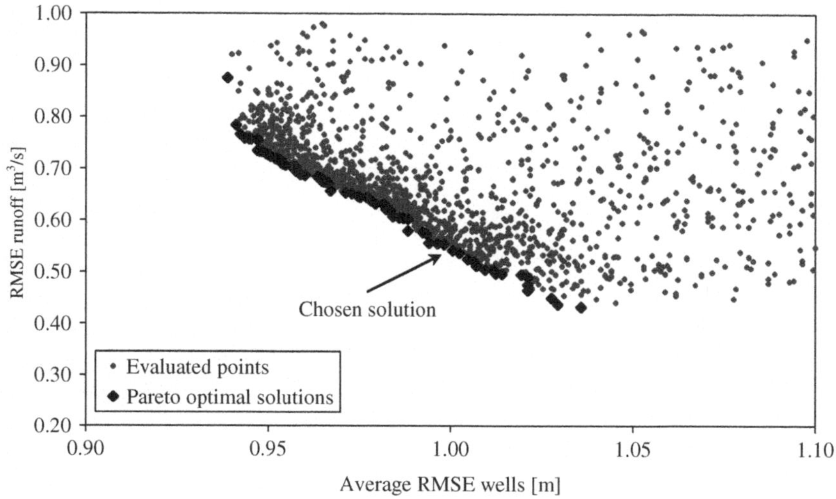

Figure 10.7. Estimated Pareto front for optimisation of the average RMSE of groundwater level and the RMSE of runoff at the catchment outlet

For model validation, three different sets of data are used, respectively, groundwater levels from the 17 calibration wells and catchment runoff in the validation period, and groundwater levels from the 18 validation wells and runoff from the 3 internal validation stations in both the calibration and validation periods. In general, model performance statistics for the different validation data sets corresponding to the Pareto optimal solutions show the same level of accuracy as the for the calibration data.

The most challenging validation test is the one that considers data not used in the calibration and applied to the validation period. To illustrate this, one of the Pareto optimal solutions as shown in Figure 10.7 is chosen for validation. In Figure 10.8, time series of simulated and observed runoff are compared for one of the internal runoff stations (station 20.06) that drains a sub-catchment in the downstream part of the catchment (see Figure 10.6).

10.4. Water quality model (ESWAT)

Water quality models have the additional complexity compared to hydrological models that they deal with multiple signals, whereby strict multi-criteria analysis, such as Pareto optimisation, is not practically applicable. An aggregation of objective functions, representing the errors of the different output signals, is hence necessary. Several aggregation approaches are possible. Here, one aggregation method is presented that is based on the work presented in van Griensven and Bauwens (2003). The method is applied to a water quality model for the River Dender, Belgium using an extension of the Soil and Water Assessment Tool (SWAT; Arnold *et al.*, 1998).

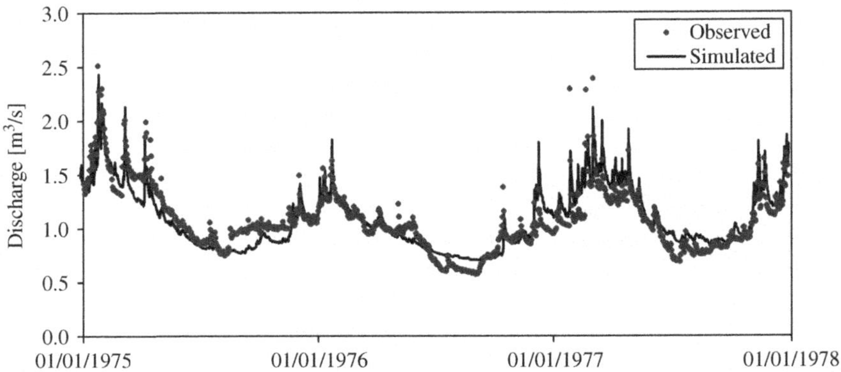

Figure 10.8. Time series of simulated and observed runoff at station 20.06 in the validation period

10.4.1. Model description

All simulations were performed using the Extended Soil and Water Assessment Tool (ESWAT; van Griensven and Bauwens, 2005), which was developed by extending the catchment delivery simulator SWAT with a detailed river water quality module. SWAT (Arnold *et al.*, 1998) is a basin-scale hydrologic/water quality model developed to predict the effects of alternative land use management decisions on water, sediment and chemical yields of large river basins. SWAT operates on a daily time step and is capable of simulating 100 years or more. Major components of the land phase of the model include hydrology, weather, erosion, soil temperature, crop growth, nutrients, pesticides, subsurface flow and agricultural management. Major components of the routing phase of the model include channel flow routing, channel sediment routing and reservoir routing.

The main developments in ESWAT consist of the incorporation of a detailed river water quality module. The model was modified to perform calculations on a sub-daily time step because sub-daily processes can be determinant for the river water quality (photosynthesis, respiration, combined sewer overflows, etc.). An hourly time step is used for the simulation of water, nutrients, pesticides and river quality processes. The simulation of the erosion processes is performed at a user-defined fraction of an hour and requires precipitation data with an hourly or sub-hourly time step.

The ESWAT model is hence able to perform integrated water quality simulations at the catchment scale. It integrates rainfall-runoff, erosion, plant growth and nutrient wash-off processes and the in-stream water quantity and quality processes in the river basin in a temporally dynamic and spatially semi-distributed way. This integration leads to a high number of parameters and output variables, aiming at the description of several observed series for water quality and quantity variables at different observation sites in the catchment. All these observations contain some information on the model parameters and should thus be used in the calibration.

10.4.2. Model setup

The river Dender is a tributary of the river Scheldt and drains an area of 1384 km^2. The river reacts quickly to precipitation events. After storms, the flow can rise to over 100 m^3/s, while during dry periods the flows are less than 1 m^3/s. These irregularities are tempered by a number of sluices built on the main channel to guarantee ship-traffic. The main channel is also partly canalised. The river is characterised by high pollution loads and long residence times in summer periods. This results in very bad water quality indices. The pollution originates from manure applications in agriculture (up to 80% of the area), industrial activities and households (the inhabitant density is more than 500 inhabitants per km^2; less than 10% was treated in 1997).

For the simulation, the Dender catchment was divided into 15 sub-basins and reaches, from which 7 form the main channel. The sub-basins are further partitioned into 80 hydrological response units (HRU), as defined by land use (impervious, agriculture, corn culture, forest, pasture) and soil type (mostly loam and sandy loam).

10.4.3. Calibration objectives

The combined calibration of the water quantity and quality variables differs from the classical approach whereby the water flow parameters are calibrated first, followed by a calibration of the water-quality parameters. This procedure is preferable since (i) the water quantity processes force the quality processes, and therefore are the key processes, and (ii) the quality of the flow data is usually much better than the quality of the pollutant data. While these issues are acknowledged, one should also consider the limitations of the calibration of the water quantity model, based on flow hydrographs only. Achieving good fits for the hydrographs does not necessarily imply a correct description of the hydrological processes, e.g., with respect to the correct subdivision of the river flow into groundwater, interflow and surface flow contributions. For water quality simulations, a correct subdivision is essential, as the flow contributions differ in the types and in the concentration levels of the pollutants they carry. Also the impression that the flow data are by far more accurate than the quality data is not always correct. For the Dender River case, e.g., the quality of the low flow observation is poor, as flows of a magnitude of about 1 m^3/s are measured in a canal of 50 m width and 5 m depth by means of water level and velocity observations. These low flow periods, with high pollutant concentrations and long residence times, are very important for the water quality processes and a correct estimation of the hydrological processes during these periods is essential for the water quality components of the model. For all these reasons, it was decided to perform a joint calibration of the water flow and the water quality parameters.

The locations where flow and water quality measurements are available are shown in Figure 10.9. In total, 13 objective functions were defined for the calibration of the water quantity and quality processes (Table 10.4). The flows were evaluated by comparing the daily average of the simulated hourly flows to the observations near the basin outlet at Denderbelle. Considering the importance of the flow data, several objective functions (OF) have been defined based on this flow series. Each OF puts its own emphasis on the calibration. The OF on the flow time series (Q_TS) minimises the daily errors expressed by the residuals, the OF on the ranked flow series (Q_RS) aims at a good statistical representation of the flow regime, and the total mass controller (Q_TM) stimulates the calibration process to respect the overall water balance.

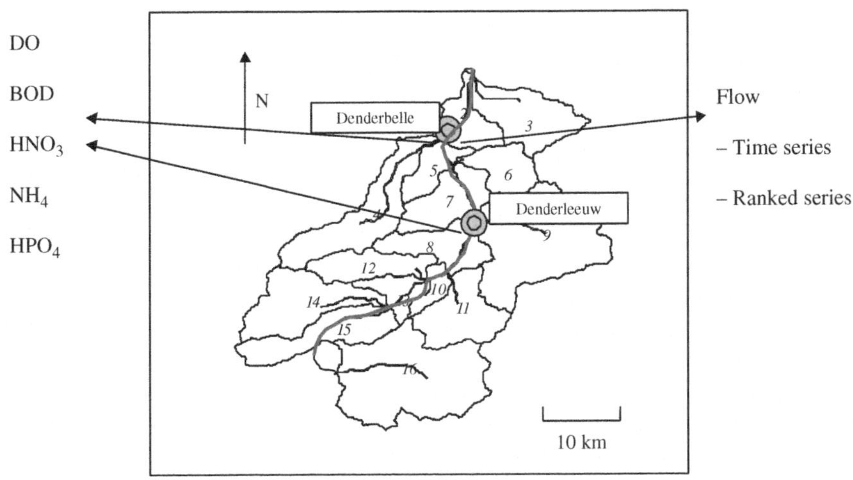

Figure 10.9. Locations of the sites for the daily flow (downstream) and water quality (middle and downstream) calibrations

Table 10.4. Defined objective functions for the optimisation

Abbreviation	Explanation
Q_TS	Sum of the squares of the residuals for the flows at Denderbelle
Q_RS	The sum of the squares of the difference of the measured and simulated values ranked series (RS)
Q_TM	Total mass controller for the flows at Denderbelle
HNO3_TS_1	Sum of the squares of the residuals for the nitrate concentrations at Denderbelle
NH4_TS_1	Sum of the squares of the residuals for the ammonia concentrations at Denderbelle
DO_TS_1	Sum of the squares of the residuals for the dissolved oxygen concentrations at Denderbelle
BOD_TS_1	Sum of the squares of the residuals for the Biological Oxygen Demand concentrations at Denderbelle
HPO4_TS_1	Sum of the squares of the residuals for the phosphate concentrations at Denderbelle
HNO3_TS_2	Sum of the squares of the residuals for the nitrate concentrations at Denderleeuw
NH4_TS_2	Sum of the squares of the residuals for the ammonia concentrations at Denderleeuw
DO_TS_2	Sum of the squares of the residuals for the dissolved oxygen concentrations at Denderleeuw
BOD_TS_2	Sum of the squares of the residuals for the Biological Oxygen Demand concentrations at Denderleeuw
HPO4_TS_2	Sum of the squares of the residuals for the phosphate concentrations at Denderleeuw

10.4.4. Calibration parameters

The number of parameters was initially reduced by sharing parameters between HRUs. For instance, all HRUs with corn have the same corn crop parameters, and all HRUs with loam soil have the same loam soil parameters. After identifying the 129 remaining parameters, a multi-variable sensitivity analysis was applied, using a One-Factor-at-a-Time design and considering the sum of squared errors for the water flow and water quality variables at Denderbelle (van Griensven *et al.*, 2002). This enabled the selection of the 32 most important parameters that are representative for the water quantity and water quality processes. The sensitivity analysis also identified that many parameters affect several variables simultaneously. From this, it can be inferred that a given output series not only provides information about the processes directly related to that variable, but also contains information that is relevant to other processes and parameters. Consequently, it is believed that complementary information on the parameters can be gained by combining the different output series during the calibration.

10.4.5. Calibration method

A multi-objective calibration procedure has been implemented in ESWAT based on the shuffled complex evolution (SCE) algorithm (Duan *et al.*, 1992). The SCE algorithm allows for the calibration of the parameters, based on a single objective function. In multi-objective calibrations, this function represents the global optimization criterion, being an aggregation of several objective functions.

In ESWAT, the individual objective functions are normalised based on the characteristics of their distributions. In addition to the normalisation, weights can be defined with the only purpose to accentuate certain output variables, e.g., observation series that are more accurate or more complete. A new method was developed, aiming at multi-objective calibrations whereby all the objectives functions are weighed in such a way that all the objectives are actively taking part in the minimisation of the aggregated objective function.

In the first step of the procedure the probability distributions of each of the objective functions in the parameter space are estimated. These distributions are approximated by random sampling of the parameter space, being the first step of the SCE method. Assuming a distribution function, its parameters can then be defined by the method of moments, given the mean and variance obtained by the random sampling. In the second step of the optimisation, each objective function is normalised to the range 0–1 by means of a transformation function $f_i(.)$ that gives the position of the objective function in its cumulative probability distribution. A value close to 0 corresponds to the best solutions and a value close to 1 corresponds to the worst solutions for the objective function (see Figure 10.10).

Considering a normal or log-normal distribution, the normalised objective functions are summed to form the aggregate objective function (global optimisation criterion, *GOC*)

$$GOC_j = \sum_{i=1}^{m} \omega_i f_i(OF_{i,j}) \tag{20}$$

or

$$GOC_j = \sum_{i=1}^{m} \omega_i f_i^*(\log(OF_{i,j})) \tag{21}$$

where $f_i(.)$ and $f_i^*(.)$ are the normalisation function using, respectively, the cumulative normal or log-normal distribution, ω_i are user-defined weights, m is the number of objective functions, and index j refers to a parameter set. The normalisation functions hereby ensure that all the objective functions are of the same order of magnitude and, therefore, will have similar influence in the search for the optimum. The method still allows for the definition of user-defined weights if this is believed to be important in view of the objectives of the model or in view of the differences in the quantity or the quality of the different data. The GOC is minimized by searching for parameter combinations that give values close to 0 for all the objective functions. Often, however, compromises may be needed, e.g., when a lower value for one $OF_{i,j}$ corresponds to an increase of one or more of the other objective functions. These trade-offs may cause relatively high normalised values for a certain objective function, indicating that the optimisation was not successful for that specific objective function.

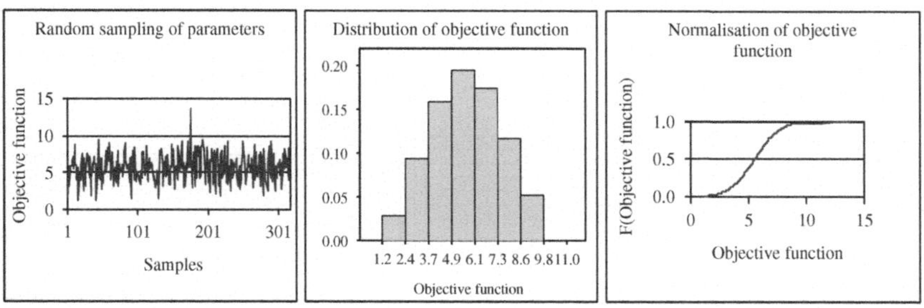

Figure 10.10. The objective function for random sampling of the parameters (left), its distribution (middle), and the normalisation function according to the cumulative distribution function (right)

10.4.6. Results

Step 1: Random parameter sampling

The initial random sampling of the parameter space consisted of 305 samples by means of the SCE method. These samples were used to calculate the major characteristics of the distributions of the objective functions (Figure 10.11). From Figure 10.11 it can be observed that most OFs have a similar slope when their percentiles are plotted. Some have, however, very high deviations for the maxima, and it was decided not to include the values above the 95% percentiles in the calculation of the means and standard deviations of the normal distributions in this study. The outliers could however indicate that the normal distribution is not well suited for these objective functions, a matter that should be further investigated.

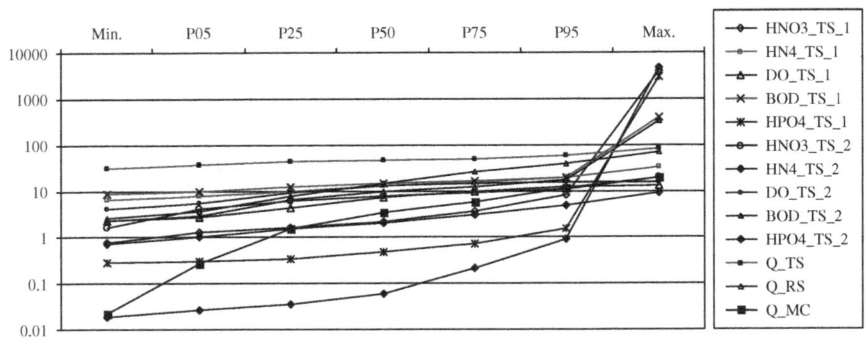

Figure 10.11. Minimum, percentiles (5% (P05), 25% (P25), 50% (P50), 75% (P75), 95% (P95)) and maximum of the different objective functions from the initial sampling

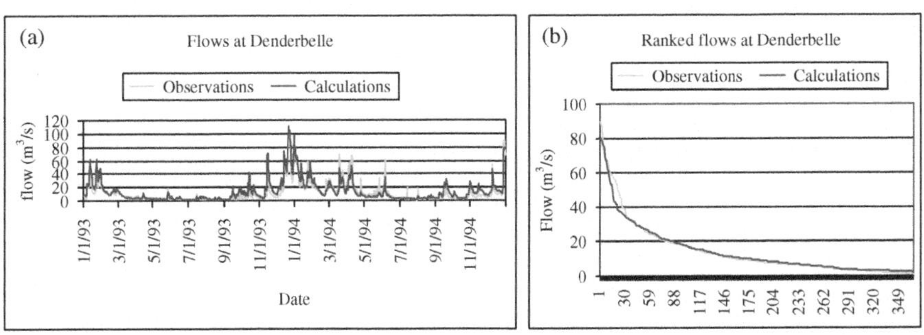

Figure 10.12. Calibration results for the daily flow at Denderbelle: (a) time series (b) ranked series

Step 2: Optimisation

The global optimisation was terminated after about 8.000 model runs. The results at Denderbelle for the optimum parameter set are plotted in Figure 10.12 and Figure 10.13.

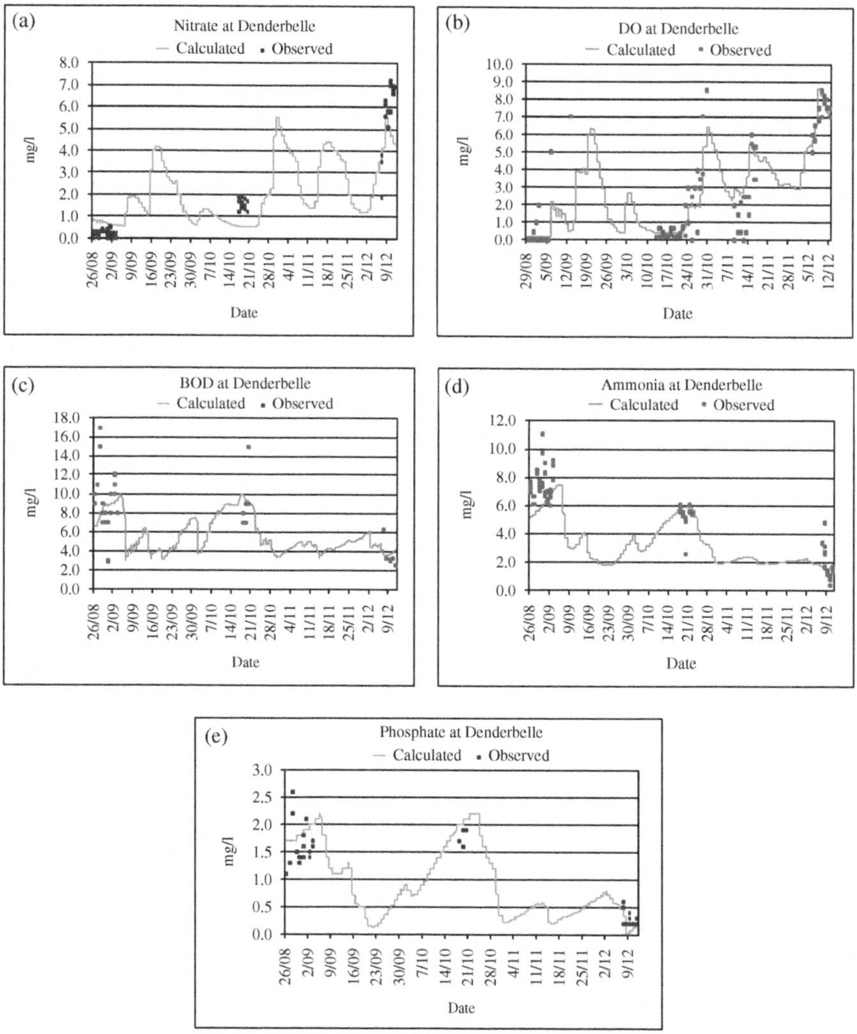

Figure 10.13. Calibration results for the water quality at Denderbelle for 1994: (a) nitrate, (b) dissolved oxygen, (c) biological oxygen demand, (d) ammonia, (e) phosphate

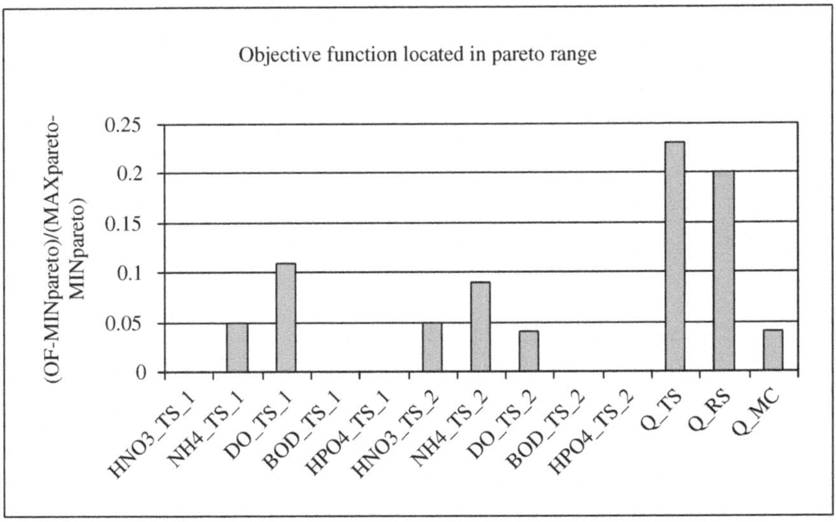

Figure 10.14. Normalised objective functions for the optimal solution

Figure 10.14 shows the different objective functions corresponding to the lowest GOC. Some OFs have a value close to zero, indicating that the results of the multi-objective optimisation are nearly as good as when a single objective optimisation would have been used. In other words, the multi-objective calibration did not require too much trade-off for that objective function. If the OFs were independent, meaning that they depend on different parameters, all would have been close to their individual minimum value. Since some parameters affect several of the OFs, a trade-off is needed in the parameter calibration, leading to an increase of some of the OFs at the global optimum. In Figure 10.14 it can be seen that this is the case for a number of OFs. For example, a large deviation can be observed for Q_TS (calibration for the time series of the flows). In terms of the Nash and Sutcliffe efficiency measure, a value of 0.93 could be reached when only Q_TS was considered as an objective. Due to the trade-off with the other objectives during the global optimization, the efficiency was reduced to 0.86. That is, however, still a quite good efficiency.

11. REFERENCES

Agyei, E. and Hatfield, K. (2006). Enhancing gradient-based parameter estimation with an evolutionary approach, *J. Hydrol.*, **316**, 266–280.

Arnold, J.G., Srinivasan, R., Muttiah, R.S. and Williams, J.R. (1998). Large area hydrologic modeling and assessment part I: model development. *J. American Water Resour. Ass.*, **34**(1), 73–89.

Barlebo, H.C., Hill, M.C., Rosbjerg, D. and Jensen, K.H. (1998). Concentration data and dimensionality in groundwater models: evaluation using inverse modeling, *Nordic Hydrol.*, **29**, 149–178.

Barlebo, H.C., Hill, M.C. and Rosbjerg, D. (2004). Investigating the Macrodispersion Experiment (MADE) site in Columbus, Mississippi, using a three-dimensional inverse flow and transport model, *Water Resour. Res.*, **40**, doi:10.1029/2002WR001935.

Beldring, S. (2002). Multi-criteria validation of a precipitation-runoff model, *J. Hydrol.*, **257**, 189–211.

Bergström, S. (1995). The HBV model, *Computer Models of Watershed Hydrology* (ed. Singh, V.P.), Water Resources Publications, Colorado, 443–476.

Bergström, S. and Forsman, A. (1973). Development of a conceptual deterministic rainfall-runoff model, *Nordic Hydrol.*, **4**, 147–170.

Beven, K. and Binley, A.M. (1992). The future of distributed models: Model calibration and uncertainty prediction, *Hydrol. Proces.*, **6**, 279–298.

Beven, K. and Freer, J. (2001). Equifinality, data assimilation, and uncertainty estimation in mechanistic modelling of complex environmental systems using the GLUE methodology, *J. Hydrol.*, **249**, 11–29.

Blasone, R.S., Madsen, H. and Rosbjerg, D. (2007). Parameter estimation in distributed hydrological modelling: comparison of global and local optimization techniques, *Nordic Hydrol.*, **38**, 451–476.

Blasone, R.S., Vrugt, J.A., Madsen, H., Rosbjerg, D., Robinson, B.A. and Zyvoloski, G.A. (2008). Generalized likelihood uncertainty estimation (GLUE) using adaptive Markov Chain Monte Carlo sampling, *Adv. Resour. Res.*, **31**, 630–648.

Boyle, D.P., Gupta, H.V. and Sorooshian, S. (2000). Toward improved calibration of hydrologic models: Combining the strengths of manual and automatic methods, *Water Resour. Res.*, **36**(12), 3663–3674.

Brath, A., Montanari, A. and Toth, E. (2004). Analysis of the effects of different scenarios of historical data availability on the calibration of a spatially-distributed hydrological model, *J. Hydrol.*, **291**, 232–253.

Brun, R., Reichert, P. and Künsch, H.R. (2001). Practical identifiability of large environmental simulation models, *Water Resour. Res.*, **37**(4), 1015–1030.

Burnash, R.J.C. (1995). The NWS river forecast system – catchment modeling, *Computer Models of Watershed Hydrology* (ed. Singh, V.P.), Water Resources Publications, Colorado, 311–366.

Burnash, R.J.C., Ferral, R.L. and McGuire, R.A. (1973). *A generalized streamflow simulation system – conceptual modeling for digital computers*, US Department of Commerce, National Weather Service and State of California, Department of Water Resources.

Butts, M.B., Payne, J. T., Kristensen, M. and Madsen, H. (2004). An evaluation of the impact of model structure and complexity on hydrological modelling uncertainty for streamflow prediction, *J. Hydrol.* **298**, 242–266.

Cheng, C.T., Ou, C.P. and Chau, K.W. (2002). Combining a fuzzy optimal model with a genetic algorithm to solve multi-objective rainfall-runoff model calibration, *J. Hydrol.*, **268**, 72–86.

Christensen, S. and Cooley, R.L. (1999). Evaluation of confidence intervals for a steady-state leaky aquifer model, *Adv. Water Resour.*, **22**(8), 807–817.

Contractor, D.N. and Jenson, J.W. (2000). Simulated effect of vadose infiltration on water levels in the Northern Guam Lens Aquifer, *J. Hydrol.*, **229**, 232–254.

Cooper, V.A., Nguyen, V.T.V. and Nicell, J.A. (1997). Evaluation of global optimization methods for conceptual rainfall-runoff model calibration, *Water Sci. Tech.*, **36**(5), 53–60.

D'agnese, F.A., Faunt, C.C., Hill, M.C. and Turner, A.K. (1999). Death valley regional ground-water flow model calibration using optimal parameter estimation methods and geoscientific information systems, *Adv. Water Res.*, **22**(8), 777–790.

Deb K., Agrawal S., Pratap A. and Meyarivan T. (2000). A fast elitist non-dominated sorting genetic algorithm for multi-objective optimization: NSGA-II. *Proc. Parallel Problem Solving form Nature VI Conf.*, 849–858.

DHI, 2005a, AUTOCAL – Autocalibration tool. User Guide, DHI Software 2005, DHI Water and Environment, Denmark.

DHI, 2005b, MIKE 11 – A modelling system for rivers and channels, Reference Manual, DHI Software 2005, DHI Water and Environment, Denmark.

Doherty, J. (2003). Ground water model calibration using pilot points and regularization, *Ground Water*, **41**, 170–177.

Doherty, J. (2005). PEST: Model Independent Parameter Estimation. Edition 5 of user manual, *Watermark Numerical Computing*.

Doherty, J. and Skahill, B.E. (2006). An advanced regularization methodology for use in watershed model calibration, *J. Hydrol.*, **327**, 564–577.

Duan, Q., Sorooshian, S. and Gupta, V. (1992). Effective and efficient global optimization for conceptual rainfall-runoff models, *Water Resour. Res.*, **28**(4), 1015–1031.

Duan Q., Sorooshian, S. and Gupta, V.K. (1994). Optimal use of the SCE-UA global optimization method for calibrating watershed models, *J. Hydrol.*, **158**, 265–284.

Eckhardt, K. and Arnold, J.G. (2001). Automatic calibration of a distributed catchment model, *J. Hydrol.*, **251**, 103–109.

Francés, F., Vélez, J.I. and Vélez, J.J. (2006). Split-parameter structure for the automatic calibration of distributed hydrological models, *J. Hydrol.*, **332**, 226–240.

Franchini, M. and Galeati, G. (1997). Comparing several genetic algorithm schemes for the calibration of conceptual rainfall-runoff models, *Hydrol. Sci. J.*, **42**(3), 357–379.

Franchini, M., Galeati, G. and Berra, S. (1998). Global optimization techniques for the calibration of conceptual rainfall-runoff models, *Hydrol. Sci. J.*, **43**(3), 443–458.

Gan, T.Y. and Biftu, G.F. (1996). Automatic calibration of conceptual rainfall-runoff models: Optimization algorithms, catchment conditions, and model structure, *Water Resour. Res.*, **32**(12), 3513–3524.

Goldberg, D.E. (1989). Genetic Algorithms in Search, Optimisation and Machine Learning, Addison-Wesley, Reading, Mass.

Graham, D.N. and Butts, M.B. (2006). Flexible, integrated watershed modelling with MIKE SHE, *Watershed Models* (Eds. V.P. Singh and D.K. Frevert), Water Resources Publication, Highlands Ranch, Colorado, 245–272.

Gupta, V.K. and Sorooshian, S. (1983). Uniqueness and observability of conceptual rainfall-runoff model parameters: the percolation process examined, *Water Resour. Res.*, **19**(1), 269–276.

Gupta, H.V., Sorooshian, S. and Yapo, P.O. (1998). Toward improved calibration of hydrological models: Multiple and noncommensurable measures of information, *Water Resour. Res.*, **34**(4), 751–763.

Gupta, H.V., Sorooshian, S. and Yapo, P.O. (1999). Status of automatic calibration for hydrological models: comparison with multilevel expert calibration, *J. Hydrol. Eng., ASCE*, **4**(2), 135–143.

Heidari, M. and Ranjithan, S.R. (1998). A hybrid optimization approach to the estimation of distributed parameters in two-dimensional confined aquifers, *J. Amer. Water Resour. Ass.*, **34**(4), 909–920.

Harbough, A.W. and McDonald, M.G. (1996). User's documentation for MODFLOW-96, an update to the U.S. Geological Survey Modular Finite-Difference Ground-Water Flow Model, U.S. Geological Survey, Open-File Report 96–485.

Harlin, J. (1991). Development of a process oriented calibration scheme for the HBV hydrological model, *Nordic Hydrol.*, **22**, 15–36.

Harrar W., Sonnenborg T.O. and Henriksen, H.J. (2003). Capture zone, travel time, and solute-transport predictions using inverse modeling and different geological models, *Hydrogeol. J.*, **11**(5), 536–548.

Havnø, K., Madsen, M.N. and Dørge, J. (1995). MIKE 11 – a generalized river modelling package, *Computer Models of Watershed Hydrology* (ed. Singh, V.P.), Water Resources Publications, Colorado, 733–782.

Henriksen, H.J. and Sonnenborg, T.O. (2005). Performance criteria, Chapter 12, *Handbook in groundwater modelling* (In Danish), Geological Survey of Denmark and Greenland, GEUS Report 2005/80.

Henriksen, H.J., Troldborg, L., Nyegaard, P., Sonnenborg, T.O., Refsgaard, J.C. and Madsen, B. (2003). Methodology for construction, calibration and validation of a national hydrological model for Denmark, *J. Hydrol.*, **280**(1–4), 52–71.

Hill, M.C. (1998). Methods and guidelines for effective model calibration, U.S. Geological Survey, Water Resources Investigations Report 98–4005.

Hill, M.C., Middlemis, H., Hulme, P., Poeter, E., Riegger, J., Neuman, S.P., Williams, H. and Anderson, M. (2004). Brief overview of selected groundwater modeling guidelines, Proceedings of the conference Finite-element models, MODFLOW and more – Solving groundwater problems (Eds. Kovar, K. and Hrkal, Z.), Carlsbad, Czech Republic, September 13–16, 2004, 105–120.

Hill, M.C. and Tiedeman, C.R. (2003). Weighting observations in the context of calibrating groundwater models, Calibration and Reliability in Groundwater Modelling: A Few Steps Closer to Reality (Proceedings of ModelCARE 2002, Prague, Czech Republic, 17–20 June), IAHS Publ. no. 277, 196–203.

Højberg, A.L. and Refsgaard, J.C. (2005). Model uncertainty – parameter uncertainty versus conceptual models, *Water Sci. Tech.*, **52**(6), 177–186.

Khu, S.T., Madsen, H. (2005). Multi-objective calibration with Pareto preference ordering. An application to rainfall-runoff model calibration, *Water Resour. Res.*, **41**, 10.1029/2004WR003041.

Khu, S.T., Madsen, H. and di Pierro, F. (2008). Incorporating multiple observations for distributed hydrologic model calibration: an approach using a multi-objective evolutionary algorithm and clustering, *Adv. Water Resour.*, **31**, 1387–1398.

Klemes, V. (1986). Operational testing of hydrological simulation models, *Hydrol. Sci. J.* **31**(1), 13–24.

Kuczera, G. (1997). Efficient subspace probabilistic parameter optimization for catchment models, *Water Resour. Res.*, **33**(1), 177–185.

Kuczera, G. and Parent, E. (1998). Monte Carlo assessment and parameter uncertainty in conceptual catchment models: the Metropolis algorithm, *J. Hydrol.* **211**, 69–85.

LaVenue, A.M. and Pickens, J.F. (1992). Application of a coupled adjoint sensitivity and kriging approach to calibrate a groundwater flow model, *Water Resour. Res.*, **28**(6), 1543–1569.

Legates, D.R. and McCabe, G.J. (1999). Evaluating the use of goodness-of-fit measures in hydrologic and hydroclimatic model validation, *Water Resour. Res.*, **35**(1), 233–241.

Leijnse, A., Pastoors, M.J.H. and Roelofsen, F.J. (2003). The effect of weighting the observations on parameter estimates in groundwater modelling, Calibration and Reliability in Groundwater Modelling: A Few Steps Closer to Reality (Proceedings of ModelCARE 2002, Prague, Czech Republic, 17–20 June), IAHS Publ. no. 277, 204–211.

Madsen, H. (2000). Automatic calibration of a conceptual rainfall-runoff model using multiple objectives, *J. Hydrol.*, **235**, 276–288.

Madsen, H. (2003). Parameter estimation in distributed hydrological catchment modelling using automatic calibration with multiple objectives, *Adv. Water Resour.*, **26**, 205–216.

Madsen, H., Evaluation of an evolutionary algorithm based on the simplex method for hydrological model calibration, In preparation.

Madsen, H. and Khu, S.T. (2006). On the use of Pareto optimization for multi-criteria calibration of hydrological models, Calibration and Reliability in Groundwater Modelling: From Uncertainty to Decision Making, Proceedings of ModelCARE 2005, The Hague, The Netherlands, June 2005 (Eds. M.F.P. Bierkens, J.C. Gehrels and K. Kovar), IAHS Publ. No. 304, 93–99.

Madsen, H., Wilson, G. and Ammentorp, H.C. (2002). Comparison of different automated strategies for calibration of rainfall-runoff models, *J. Hydrol.*, **261**, 48–59.

Mayer, A.S. and Huang, C. (1999). Development and application of a coupled-process parameter inversion model based on the maximum likelihood estimation method, *Adv. Water Resour.*, **22**(8), 841–853.

McDonald M.G. and Harbaugh, A.W. (1989). A modular three-dimensional finite-difference ground-water flow model, U.S. Geological Survey Open-File Report 83–875.

Medina, A. and Carrera, J. (1996). Coupled estimation of flow and solute transport parameters, *Water Resour. Res.*, **32**(10), 3063–3076.

Meixner, T., Bastidas, L.A., Gupta, H.V. and Bales, R.C. (2002). Multicriteria parameter estimation for models of stream chemical composition, *Water Resour. Res.*, **38**(3), 10.1029/2000WR000112.

Moore, C. and Doherty, J. (2006). The cost of uniqueness in groundwater model calibration, *Adv. Water Resour.*, **29**, 605–623.

Morris, M.D. (1991). Factorial sampling plans for preliminary computational experiments. *Technometrics*, **33**, 161–174.

Mroczkowski, M, Raper, G.P. and Kuczera, G. (1997). The guest for more powerful validation of conceptual catchment models, *Water Resour. Res.*, **33**(10), 2325–2335.

Muleta, M.K. and Nicklow, J.W. (2005). Sensitivity and uncertainty analysis coupled with automatic calibration for a distributed watershed model, *J. Hydrol.*, **306**, 127–145.

Nash, J.E. and Sutcliffe, J. (1970). River flow forecasting through conceptual models, Part I A discussions of principles, *J. Hydrol.*, **10**, 282–290.

Nelder, J.A. and Mead, R. (1965). A simplex method for function minimization, *Comput. J.*, **7**(4), 308–313.

Nielsen, S.A. and Hansen, E. (1973). Numerical simulation of the rainfall runoff process on a daily basis, *Nordic Hydrol.*, **4**, 171–190.

Neumann, S. (2003). Accounting for conceptual model uncertainty via maximum likelihood averaging. Calibration and Reliability in Groundwater Modelling: A Few Steps Closer to Reality (Proceedings of ModelCARE 2002, Prague, Czech Republic, 17–20 June), IAHS Publ. no. 277, 303–313.

Poeter, E.P. and Hill, M.C. (1997). Inverse models: A necessary next step in groundwater modeling, *Ground Water*, **35**, 250–260.

Poeter, E.P. and Hill, M.C. (1998). Documentation of UCODE: a computer code for universal inverse modelling, U.S. Geological Survey, Water Resources Investigations Report 98–4080, Denver, Colorado, USA.

Poeter, E.P., Hill, M.C., Banta, E.R., Mehl, S. and Christensen, S. (2005). UCODE_2005 and six other computer codes for universal sensitivity analysis, calibration, and uncertainty evaluation, U.S. Geological Survey Techniques and Methods 6-A11, 283p.

Reed, P., Minsker, B. and Goldberg, D.E. (2000). Designing a competent simple genetic algorithm for search and optimisation, *Water Resour. Res.*, **36**(12), 3757–3761.

Refsgaard, J.C. (1997). Parameterisation, calibration and validation of distributed hydrological models, *J. Hydrol.*, **198**, 69–97.

Refsgaard, J.C. and Henriksen, H.J. (2004). Modelling guidelines – terminology and guiding principles, *Adv. Water Resour.*, **27**, 71–82.

Refsgaard, J.C., Henriksen, H.J., Harrar, W.G, Scholten, H. and Kassahun, A. (2005). Quality assurance in model based water management – review of existing practice and outline of new approaches, *Env. Model. Software*, **20**, 1201–1215.

Refsgaard, J.C. and Knudsen, J. (1996). Operational validation and intercomparison of different types of hydrological models, *Water Resour. Res.*, **32**(7), 2189–2202.

Saltelli, A., Chan, K., Scott, E.M. (2000). Sensitivity Analysis, Wiley Publishers.

Schoups, G., Hopmans, J.W., Young, C.A., Vrugt, J.A., Wallender, W.W. (2005). Multicriteria optimization of a regional spatially-distributed subsurface water flow model, *J. Hydrol.*, **311**, 20–48.

Skahill, B.E., Doherty, J. (2006). Efficient accommodation of local minima in watershed model calibration, *J. Hydrol.*, **329**, 122–139.

Solomatine, D.P., Dibiki, Y.B., Kukuric, N. (1999). Automatic calibration of groundwater models using global optimisation techniques, *Hydrol. Sci. J.*, **44**(6), 879–894.

Sonnenborg, T.O., Christensen, B.S.B, Nyegaard, P., Henriksen, H.J. and Refsgaard, J.C. (2003). Transient modelling of regional groundwater flow using parameter estimates from steady-state automatic calibration, *J. Hydrol.*, **273**, 188–204.

Sonnenborg, T.O. and Henriksen, H.J. (eds) (2005). *Handbook in groundwater modelling* (In Danish), Geological Survey of Denmark and Greenland, GEUS Report 2005/80.

Sorooshian, S. and Gupta, V.K. (1983). Automatic calibration of conceptual rainfall-runoff models: the question of parameter observability and uniqueness, *Water Resour. Res.*, **19**(1), 251–259.

Sorooshian, S., Duan, Q. and Gupta, V.K. (1993). Calibration of rainfall-runoff models: Application of global optimization to the Sacramento soil moisture accounting model, *Water Resour. Res.*, **29**(4), 1185–1194.

Sorooshian, S. and Gupta, V.K. (1995). Model calibration, *Computer Models of Watershed Hydrology* (ed. Singh, V.P.), Water Resources Publications, Colorado, 23–68.

Tang, Y., Reed, P. and Wagener, T. (2006). How effective and efficient are multiobjective evolutionary algorithms at hydrologic model calibration?, *Hydrol. Earth Syst. Sci.*, **10**, 289–307.

Thyer, M., Kuczera, G. and Bates, B.C. (1999). Probabilistic optimisation for conceptual rainfall-runoff models: A comparison of the shuffled complex evolution and simulated annealing algorithms, *Water Resour. Res.*, **35**(3), 767–773.

van Griensven, A., Francos, A. and Bauwens, W. (2002). Sensitivity analysis and auto-calibration of an integral dynamic model for river water quality, *Water Sci. Tech.*, **45**(9), 325–332.

van Griensven, A. and Bauwens, W. (2003). Multi-objective auto-calibration for semi-distributed water quality models, *Water Resour. Res.*, **39**(10), 1348 doi: 10.1029/2003WR002284.

van Griensven, A. and Bauwens, W. (2005). Application and evaluation of ESWAT on the Dender basin and the Wister lake basin, *Hydrol. Processes*, **19**(3), 827–838.

Vrugt, J.A., Gupta, H.V., Bouten, W. and Sorooshian, S., 2003a, A shuffled complex evolution Metropolis Algorithm for optimisation and uncertainty assessment of hydrological model parameters, *Water Resour. Res.* **39**(8), 1201, doi:10.1029/2002WR001642.

Vrugt, J.A., Gupta, H.V., Bastidas, L. A., Bouten, W. and Sorooshian, S., 2003b, Effective and efficient algorithm for multi-objective optimization of hydrologic models, *Water Resour. Res.*, **39**(8), 1214, doi:10.1029/2002WR001746.

Wagener, T. and Gupta, H.V. (2005). Model identification for hydrological forecasting under uncertainty, *Stoch. Environ. Res. Risk. Assess.*, doi: 10.1007/s00477-005-0006-5.

Wang, Q.J. (1991). The genetic algorithm and its application to calibrating conceptual rainfall-runoff models, *Water Resour. Res.*, **27**(9), 2467–2471.

Willmott, C.J., Ackleson, S.G., Davis, R.E., Feddema, J.J, Klink, K.M., Legates, D.R., O'Donnell, J. and Rowe, C.M. (1985). Statistics for the evaluation and comparison of models, *J. Geophys. Res.*, **90**, 8995–9005.

Yapo, P.O., Gupta, H.V. and Sorooshian, S. (1996). Automatic calibration of conceptual rainfall-runoff models: sensitivity to calibration data, *J. Hydrol.*, **181**, 23–48.

Yapo, P.O., Gupta, H.V. and Sorooshian, S. (1998). Multi-objective global optimization for hydrological models, *J. Hydrol.*, **204**, 83–97.

Zhang, X. and Lindström, G. (1997). Development of an automatic calibration scheme for the HBV hydrological model, *Hydrol. Proces.*, **11**, 1671–1682.

12. SOFTWARE

12.1. PEST

Optimisation Algorithms

- Gauss Marquardt Levenberg (GML).
- Truncated Singular Value Decomposition.
- "SVD-Assist".
- LSQR.
- Support software includes SCE and multi-start GML inverse methods.
- Linear predictive error analysis (in both over- and under-determined calibration contexts).
- Nonlinear predictive error analysis using constrained maximisation/minimisation in both under-determined and over-determined calibration contexts.
- Support for efficient calibration-constrained Monte-Carlo analysis.

Key Features

- Single or multi-objective optimisation.
- Broyden Jacobian upgrading.
- Cyclic parameter sensitivity computation.
- Manual and "automatic" user intervention.
- Ability to re-start from any stopping point.
- Inclusion of prior information.
- Sensitivity and uncertainty analysis.
- Utilities for checking and verification of input datasets, and for automatic construction of new datasets based on existing datasets.
- Parallel processing.
- Tikhonov, SVD, SVD-assist and hybrid regularisation for stable estimation of hundreds or even thousands of parameters for spatially distributed models.
- Linear and nonlinear uncertainty and predictive error analysis in both under-determined and over-determined parameter estimation contexts.
- Null space analysis.
- Calibration constrained uncertainty analysis based on "null-space Monte Carlo", "stochastic field warping" and "variogram regularisation" methods.
- Optimisation of data acquisition based on maximum potential reduction of predictive error variance.
- Computation of comprehensive descriptors of inversion process including:
    - estimated parameter uncertainty statistics
    - estimated parameter covariance, correlation coefficient and eigvector/eigenvalue matrices

- composite parameter and observation sensitivities
- influence statistics such as Cooke's D and DFBETA
- resolution matrix
- calibration and null space specifications

- Supported by over 100 utility programs which automate PEST setup in the groundwater and surface water modelling contexts based on complex multi-component objective functions and/or regularised inversion of large parameter ensembles.
- Matrix manipulation utilities support singular value decomposition, Jacobian matrix manipulation, and definition of optimal "super parameters" for efficient calibration and predictive error analysis in highly parameterised contexts.

Model Interface

- PEST is "model-independent" in that it can be linked to any model. It communicates with a model through the model's own input/output files. The model can be a single executable, or many executables run in succession through a batch or script file.
- "Parameters" can be any model inputs; observations can be matched to any model outputs.
- Input files which specify PEST-to-model linkages can be created using a text editor, or written automatically using utility support software.
- Utility software automates linkages to MODFLOW, MT3D, FEFLOW, MICROFEM, HSPF and other models.
- Complex parameter preprocessing and model output postprocessing carried out by software supplied with PEST allows the user to creatively tailor the inversion process to meet the requirements of each specific calibration problem.
- Supported by commercial and public-domain graphical model user interfaces including Groundwater Vistas, PMWIN, GMS, Visual MODFLOW and BASINS.
- Interface with MODFLOW-2005 adjoint process for highly parameterised inversion based on one or multiple regularisation techniques.
- Interface software supports use of pilot points as a parameterisation device for spatial models.

References

Manuals

Doherty, J. (2005). PEST: Model Independent Parameter Estimation. Edition 5 of user manual. Watermark Numerical Computing.

Doherty, J. (2007). Addendum to PEST manual. Watermark Numerical Computing.

Doherty, J. (2006a). Manual for PEST Groundwater Data Utilities. Watermark Numerical Computing.

Doherty, J. (2006b). Manual for PEST Surface Water Utilities. Watermark Numerical Computing.

Scientific Journals

Doherty, J. (2003). Groundwater model calibration using Pilot Points and Regularisation. *Ground Water*, **41**(2), 170–177

Doherty, J. and Johnston, J.M. (2003). Methodologies for calibration and predictive analysis of a watershed model. *J. American Water Resources Association*, **39**(2), 251–265.

Gallagher, M.R. and Doherty, J. (2007). Parameter estimation and uncertainty analysis for a watershed model. *Environmental Modelling and Software*, doi:10.1016/j.envsoft.2006.06.007.

Gallagher, M.R. and Doherty, J. (2007). Predictive error analysis for a water resource management model. *Journal of Hydrology*, **34**(3–4), 513–533.

Gallagher, M.R. and Doherty, J. (2007). Parameter interdependence and uncertainty induced by lumping in a hydrological model. *Water Resources Research*, **43**, W05421, doi:10.1029/2006WR005347.

Moore, C. and Doherty, J. (2005). The role of the calibration process in reducing model predictive error. *Water Resources Research*, **41**(5), W05050.

Moore, C. and Doherty, J. (2006). The cost of uniqueness in groundwater model calibration. *Advances in Water Resources*, **29**(4), 605–623.

Tonkin, M. and Doherty, J. (2005). A hybrid regularised inversion methodology for highly parameterised models. *Water Resources Research*, **41**, W10412, doi:10.1029/2005WR003995, 2005.

Tonkin, M, Doherty, J. and Moore, C. (2007). Efficient nonlinear predictive error variance for highly parameterized models. *Water Resources Research*, **43**, W07429, doi:10.1029/2006WR005348.

Doherty, J. and Skahill, B. (2006). An advanced regularization methodology for use in watershed model calibration. *Journal of Hydrology*, **327**(3–4), 564–577.

Skahill, B. and Doherty, J. (2006). Efficient accommodation of local minima in watershed model calibration. *Journal of Hydrology*, **329**(1–2), 122–139.

Download

PC and Unix versions available; source code provided.
http://www.sspa.com/pest

12.2. UCODE_2005

Includes the programs RESIDUAL_ANALYSIS, RESIDUAL_ANALYSIS_ADV, MODEL_LINEARITY, MODEL_LINEARITY_ADV, LINEAR_UNCERTAINTY, CORFAC_PLUS, OPR-PPR, and MMA.

Graphical packages: GW_Chart and ModelViewer

Optimisation algorithms

- Modified Gauss-Newton (also called Levenberg-Marquardt)
- Includes options for (a) quasi-Newton updates and (b) two trust region approaches called (i) double dogleg and (ii) hookstep.

Key features

- Flexible definition of parameters and simulated equivalents to observations using a powerful equation module.
- Tools to accommodate observation types with non-detects and observations for which the forward model cannot obtain a solution for some situations.
- Prior information on parameters.
- Observation weighting based on observed or simulated values.
- Full weight matrix for observations and prior information.
- Single objective optimisation.
- Local sensitivity analysis methods, including scaled sensitivities, parameter correlation coefficients, and measures of leverage and influence. The sensitivity analysis methods address the importance of observations to parameters and predictions, and the importance of parameters to predictions. The OPR-PPR program calculates the Observation-PRediction and Parameter-PRediction statistics.
- Uncertainty analysis methods include intervals that can be confidence or predictions intervals, linear or nonlinear, and individual or simultaneous. Nonlinear intervals require two executions of UCODE_2005 and one execution of CORFAC_PLUS.
- Uncertainty resulting from aspects of the system other than parameter values and embodied in alternate models of a system can be evaluated using MMA.
- Dynamic omission of insensitive parameters.
- Constrained parameter values.
- Residual analysis based on summary statistics and graphical analyses using output files from UCODE_2005, RESIDUAL_ANALYSIS and RESIDUAL_ANALYSIS_ADV (the graphical analyses can be achieved using GW_Chart).

- Analysis of model linearity using MODEL_LINEARITY, MODEL_LINEARITY_ADV, and GW_Chart.
- Easy to use, largely self-documented input structure based on JUPITER API input blocks.
- Robust distributed computing capabilities.
- Data-exchange files for transferring data between programs.

Model interface

- The model used for calibration needs to be capable of running in batch mode based on text input and output files. Even programs commonly run through graphical user interfaces often are capable of meeting these criteria.
- UCODE_2005 is linked to the process model using template files to define parameters and instruction files to extract simulated equivalents to observations. Template and instruction files are the same as those constructed for PEST except in the first line "ptf" is replaced by "jtf". This results from John Doherty's generous involvement in the JUPITER API, from which UCODE_2005 is constructed.
- In addition, simulated equivalents to observations can be extracted using a standard files option if the simulated values are listed uninterrupted in a column of a file produced by the model.
- For observations, weights can be defined using any of six user-specified statistics: the variance, standard deviation or coefficient of variation of the observation error; the weight or the square root of the weight for one observation; or a variance-covariance matrix needed when observation errors are correlated. The same options are available for prior information.

References

Key references:

- Poeter, E.P., Hill, M.C., Banta, E.R., Mehl, Steffen and Christensen, Steen (2005). UCODE_2005 and six other computer codes for universal sensitivity analysis, calibration, and uncertainty evaluation: U.S. Geological Survey Techniques and Methods 6-A11, 28 3 p.
- Tonkin, M, Tiedeman, C.R., Ely, D.M. and Hill, M.C. (in press). OPR-PPR, a computer program for assessing data importance to model predictions using linear statistics: U.S. Geological Survey Techniques and Methods 6-E2.
- Poeter, E.P. and Hill, M.C. (2007). MMA, a computer code for multi-model analysis: U.S. Geological Survey Techniques and Methods 6-E3.

Applications in different model domains

- Foglia, L., S.W.Mehl, M.C. Hill, Perona, P., Burlando, P. (in press). Testing alternative ground water models using cross validation and other methods: Ground Water.

Download

Source code and executable available freely at
http://www.mines.edu/igwmc/freeware/ucode/
http://water.usgs.gov/nrp/gwsoftware/GW_Chart/GW_Chart.html
http://water.usgs.gov/nrp/gwsoftware/modelviewer/ModelViewer.html
When they are published, sites for OPR-PPR and MMA will be listed at http://water.usgs.gov/software/ground_water.html

12.3. AUTOCAL

Optimisation algorithms

- Shuffled Complex Evolution (SCE) algorithm.
- Population Simplex Evolution (PSE) algorithm, which is a revised version of SCE optimised for distributed computing.

Key features

- A general multi-objective optimisation framework in which several objectives can be defined and optimised simultaneously. AUTOCAL supports both single objective optimisation using a weighted aggregate of the objective functions and multi-objective Pareto optimisation.
- Sensitivity analysis tool for calculating local scaled sensitivity coefficients around a nominal parameter set.
- Definition of feasible parameter space using box constraints as well as inequality constraints supporting any linear or non-linear relationship between the parameters. In addition equality constraints can be defined to tie parameters in the optimisation using a general linear or non-linear expression.
- Different comparison statistics for definition of objective functions, including average error, root mean square error, standard deviation of residuals, and event-based statistics.
- Distributed computing facilities based on grid technologies for execution on multi-processor PCs and office grids.

Model interface

AUTOCAL includes a graphical user interface to define the simulation model, model parameters, and objective functions. The only requirements for the model to be linked to AUTOCAL are:

- The model can be assessed via an executable file (.exe) or batch file (.bat).
- Model parameters are defined in ASCII files. To define model parameters AUTOCAL uses templates of the model input files in which parameter ID values are specified for the parameters to be manipulated by AUTOCAL.
- Model output can be transformed to a specified output format using available software tools, or the objective functions can be directly calculated by the model engine and saved in an ASCII file to be read by AUTOCAL.

References

Description of the SCE and PSE algorithms:

- Duan, Q., Sorooshian, S. and Gupta, V. (1992). Effective and efficient global optimization for conceptual rainfall-runoff models. *Water Resour. Res.*, **28**(4), 1015–1031.
- Madsen, H., Evaluation of an evolutionary algorithm based on the simplex method for hydrological model calibration, In preparation.

Applications to lumped, conceptual rainfall-runoff modelling:

- Madsen, H. (2000). Automatic calibration of a conceptual rainfall-runoff model using multiple objectives. *J. Hydrol.*, **235**, 276–288.
- Madsen, H., Wilson, G. and Ammentorp, H.C. (2002). Comparison of different automated strategies for calibration of rainfall-runoff models. *J. Hydrol.*, **261**, 48–59.

Applications to distributed and integrated hydrological modelling:

- Madsen, H. (2003). Parameter estimation in distributed hydrological catchment modelling using automatic calibration with multiple objectives. *Adv. Water Resour.*, **26**, 205–216.
- Mertens, J., Madsen, H., Feyen, L., Jaques, D. and Feyen, J. (2004). Including prior information in the estimation of effective soil parameters in unsaturated zone modelling. *J. Hydrol.*, **294**, 251–269.

- Madsen, H. and Khu, S.T. (2006). On the use of Pareto optimization for multi-criteria calibration of hydrological models, Calibration and Reliability in Groundwater Modelling: From Uncertainty to Decision Making, Proceedings of ModelCARE 2005, The Hague, The Netherlands, June 2005 (Eds. M.F.P. Bierkens, J.C. Gehrels and K. Kovar), IAHS Publ. No. 304, 93–99.
- Madsen, H. and Vinter, B. (2006). Parameter optimisation in complex hydrodynamic and hydrological modelling systems using distributed computing, Proceedings of the 7th International Conference on Hydroinformatics (Eds. P. Gourbesville, J. Cunge, V. Guinot and S.Y. Liong), 4–8 September 2006, Nice, France, Vol. 4, 2489–2496.

Download

AUTOCAL is developed by DHI and is included in their basic software package MIKE Zero.
A demo version can be downloaded from http://www.dhigroup.com/Software.aspx

12.4. GLOBE

Optimisation algorithms

- Adaptive cluster covering (ACCO)
- Combination of ACCO with multiple local searches (ACCOL)
- Adaptive cluster covering combined with downhill simplex descents (ACD)
- Evolutionary and genetic algorithms
- Controlled random search
- Multi-start version of Powell-Brent algorithm (Multis)
- Multi-start version of Simplex descent algorithm (M-Simplex)

Key features

GLOBE is an optimization tool that can find the minimum of a function of multiple variables which value is given by an external program or a dynamic-link library (DLL). It is also possible to obtain N next best estimators of the minimum which are at a certain (user-controlled) distance from each other. It is possible to impose box constraints (bounds) on the variables' values. No special properties of the function are assumed. GLOBE performs the "global" minimization; that is, the function may have multiple local minima. There are seven (with variations – nine) algorithms implemented that the user can tune to his/her problem and that can be run in a batch for the same function. GLOBE includes advanced visualization features. No sensitivity or uncertainty analysis is performed. It is run on a single PC.

Model interface

User can specify the external program (EXE) or a dynamic-link library (DLL) name which GLOBE will run to obtain the objective function value via a standard file (in case of EXE) or directly via a procedure call (in case of DLL). This external program in its turn can run any software (model to be calibrated) and calculate the calibration error.

References

Description of software and implemented algorithms:

Solomatine, D.P. (1998). Genetic and other global optimization algorithms – comparison and use in model calibration. *Proc. 3rd Intern. Conf. Hydro-informatics*, Copenhagen, pp. 1021–1028.

Solomatine, D.P. (1999). Two strategies of adaptive cluster covering with descent and their comparison to other algorithms. *Journal of Global Optimization*, **14**(1), 55–78.

Applications in different model domains:

Maskey, S., Jonoski, A. and Solomatine, D.P. (2002). Groundwater remediation strategy using global optimization algorithms. *ASCE Journal of Water Resources Planning and Management*, **128**(6), 431–440.

Nishida, W., Noguchi, M. and Solomatine, D.P. (2005). Study on Numerical Prediction of Tidal Current using Forecasted Weather Data. *JSCE Ann. J. Hydraulic Engineering*, 49(2), 1279–1284 (in Japanese).

Solomatine, D.P., Dibike, Y. and Kukuric, N. (1999). Automatic calibration of groundwater models using global optimization techniques. *Hydrological Sciences Journal*, 44(6), 879–894.

Vdovin, G., Middelhoek, S., Bartek, M., Sarro, P.M. and Solomatine, D.P. (1995). Technology, characterization and applications of adaptive mirrors fabricated with IC-compatible micromachining. *Proc. SPIE's Int. Symposium on Optical Science, Engineering and Instrumentation, Conference 'Adaptive Optical Systems and Applications'*, vol. 2534/13, San-Diego, USA.

Download

http://www.data-machine.com

Guidance Report I.3

Review of sensitivity analysis methods

Florian Pappenberger, Marco Ratto and Veronique Vandenberghe

Abstract

This document is a review of sensitivity analysis methods and should help practitioners, who may be unfamiliar with the benefits and want to evaluate the added value of sensitivity analysis. It also acts as a quick reference guide for any reader who is confronted with sensitivity analysis and needs to understand the advantages and disadvantages of a methodology.

The document centers around a decision tree to suggest a sensitivity method according to a set of questions as it is impossible to suggest a universal sensitivity analysis method, which is optimal for each situation. Questions include: how long does the model run; are the model equations simple and access to the model code available; or how many parameters and factors need to be investigated; or what is the aim of the sensitivity analysis (model simplification, prioritizing research, mapping, calibration).

This is followed by an introduction into different aspects and methodologies of sensitivity analysis (more mathematical details are given in an appendix). The chapter begins with a section on experimental design which describes ways parameters and factors can be varied within a sensitivity analysis framework. Fundamentally, sensitivity analysis can be grouped in four major subdivisions: *Graphical methods* include different ways to plot results of model outcome and parameters (e.g. scatter plots) or of other sensitivity analysis's themselves (e.g. in the form of histograms); *Screening methods* are often preliminary numerical experiments whose purpose is to isolate the most important factors that may affect a particular model response; *Local methods* investigate how small changes of parameters affect model output; and *Global methods* evaluate the effect on the model output by varying all parameters at the same time, in a multivariate fashion, usually by a Monte Carlo approach. An important approach to sensitivity falls within the emulation context. The basic idea is to represent the relationship between model output and model parameters or factors without the need to run the model multiple times.

Another way to make a decision on the type of methodology to be used is based on case studies. For example, a global sensitivity analysis of a sub-surface model highlights the complexities inherent in the sensitivities of the model factors, but demonstrates that the upslope pressure boundary condition and the Brooks Corey soil algorithm 'air entry' value are particularly important. Understanding the controls on the numerical simulation of hydrological processes allow informed decision making, for example, on where next to deploy field resources or on pollution transport and groundwater mixing problems. Other examples in this document include flood inundation models, a weather prediction system and comparisons between different river water quality modeling concepts.

Keywords

Monte Carlo Sampling, Morris Screening, Sensitivity Analysis, Sobol' method, Uncertainty Analysis

By Peter A. Vanrolleghem. ISBN: 9781843392231. Published by IWA Publishing, London, UK.

1. INTRODUCTION

This document reviews sensitivity analysis which analyses how the output of a model can be apportioned to variations in the model input and how such an analysis is beneficial to the user of any model.

This introduction outlines the aims and objectives of this document, specifies the target audience, explains the benefits of sensitivity analysis and briefly introduces sensitivity analysis. It further puts sensitivity analysis in context of the Water Framework Directive and gives an overview of the content of this guidance report.

1.1. Aims and objectives

The report aims to promote sensitivity analysis tools and highlights the cost effectiveness of such methodologies. It has the objective to guide the reader to perform a sensitivity analysis. This is achieved in four main chapters:

(1) A decision tree to choose a methodology
(2) Examples to learn through proxy
(3) A glossary to understand the terminology
(4) A detailed introduction into each method

1.2. Target audience

The document is aimed at *practitioners*, who may be unfamiliar with the benefits and want to evaluate the added value of sensitivity analysis. The document also allows to choose a methodology if necessary. It especially targets applications within the Water Framework Directive (see section 1.4).

Moreover, it can serve as a quick reference for *any reader* who is *confronted with sensitivity analysis* and needs to understand the advantages and disadvantages of a methodology. It further allows them to evaluate the justification given to use a particular methodology.

Finally, it gives an introduction to *scientists* who want to understand their models in more detail.

1.3. Benefits of sensitivity analysis

Sensitivity analysis can be used as an aid in **identifying the important uncertainties** for the purpose of prioritizing additional data collection or research (Frey *et al.*, 2004). In addition, sensitivity analysis can play an important role in model verification and validation throughout the course of **model development and refinement** (e.g., Kleijnen and Sargent, 2000; Fraedrich and Goldberg, 2000; Wagner *et al.*, 2002). Sensitivity analysis also can be used to provide **insight into the robustness of model results when making decisions** (Saltelli *et al.*, 2000a).

Modelers conduct sensitivity analysis for a number of reasons including the desire to determine:

(1) Input parameters which contribute the most to output variability, thereby requiring additional research to increase knowledge of parameter behavior in order to reduce output uncertainty;
(2) If parameter interactions are present, which (group of) parameters interact with each other;
(3) Which parameters are insignificant and can be held constant or eliminated from the final model; and
(4) The optimal regions within the parameter space for use in subsequent calibration studies.

Sensitivity analysis methods have been applied in various research fields, including complex engineering systems, economics, physics, social sciences, medical decision making, and others (e.g., Helton, 1993).

1.4. What is sensitivity analysis?

Sensitivity analysis (SA) is "The study of how the uncertainty in the output of a model (numerical or otherwise) can be apportioned to different sources of uncertainty in the model input" (Saltelli *et al.*, 2000b). This definition is general enough to cover a variety of strategies for sensitivity analysis, while committing the strategy to some sort of quantitative partition of the output uncertainty (however this uncertainty is defined) into factors-related components.

Originally, SA was created to deal simply with uncertainties in the input variables and model parameters. Over the course of time the ideas have been extended to incorporate model conceptual uncertainty, i.e. uncertainty in model structures, assumptions and specifications. As a whole, SA is used to increase the confidence in the model and its predictions, by providing an understanding of how the model response variables respond to changes in the inputs, be they data used to calibrate it, model structures, or factors, i.e. the model independent variables. SA is thus closely linked to uncertainty analysis (UA, see part 4 in this volume, Beven 2007), which aims to quantify the overall uncertainty associated with the response as a result of uncertainties in the model input.

Often the terms sensitivity and uncertainty analysis are mixed as both are sometimes using the same techniques. One can say that SA focuses on the inputs and studies the influence of variations on parameters, initial conditions and inputs on model outputs, UA is more related to uncertain aspects in the modelling process and its influence on the model outputs.

1.5. Sensitivity analysis and the Water Framework Directive

The WFD poses new challenges to river basin managers. The traditional physical domain specific and sectoral approaches need to be combined and extended to fulfill the WFD requirements. The models become more and more extended and complex. This way these models contain a lot of parameters that need calibration or these are models realised by the coupling of different submodels. Typical sub-models might treat: ground water flow; surface and near surface flow; rivers; rainfall runoff; abstraction wells; and transport of agricultural and industrial contaminants. Coupling these sub-models is non-trivial and needs a lot of standardisation, input output coupling, conversions, etc. In addition to the problems posed by coupling various models together, there is the additional difficulty that many of the parameters that appear in the sub-models are subject to uncertainty. Here SA can play a key role to identify the most influential parameters and to detect processes of less relevance that can be left out of those coupled models. The important parameters and processes need to be focussed on and need to be well identified either by experiments (physically based modelling where the parameters represent a physical value) or calibration of the model.

Models used in river basin management provide answers on management questions and these can serve for European environmental policy at the river basin scale. The river basin management and planning process prescribed in the WFD focuses on integrated management, involving all physical domains in water management, all sectors of water use, socio-economics and stakeholder participation. The planning process should move from a more rational-instrumental type of planning to an interactive planning with an open eye for the power of fundamental debate. The uncertainties present in such planning processes are judged of an increasing importance (EC, 2004a). An iterative planning process can deal with these uncertainties, for instance by revising the programmes of measures according to the circumstances (EC, 2004b). As one of the goals of sensitivity analysis is to point to origins of uncertainties, the SA is here again an important step in the modelling process (presented in Figure 1.1). Further SA is preferably performed at the following steps:

- assessment of uncertainties
- model construction
- model calibration
- evaluation of scenarios

1.6. Content and overview

This document is structured according to the needs of the target audience.

It therefore starts with a **decision tree** on choosing a methodology by a set of questions or limitations of the modelling process as the overwhelming amount of methods for sensitivity analysis makes it difficult for a user to decide which method will be most fit for purpose (especially as there is a steep learning curve associated with most methods).

In the next chapter **details of each sensitivity analysis** are introduced. It has a section on experimental design, graphical methods, local sensitivity analysis and global sensitivity analysis (further details are given in an appendix). All methodologies are set in context of two simple examples. Unfortunately, in many applications the user prefers to use complex models which have a long run time and many parameters. A solution to gain information about the sensitivity of such models can be the usage of model emulators, which approximate the response surface. The latter are also introduced in this chapter.

Understanding and applying sensitivity analysis is very often easier if it is set in the context of **examples of applications**, as highlighted in the next chapter.

A **glossary** has been added to the document in order to maximize the use of this document and support the understanding of the specific terminology. This should help to prevent misunderstanding and improve communication.

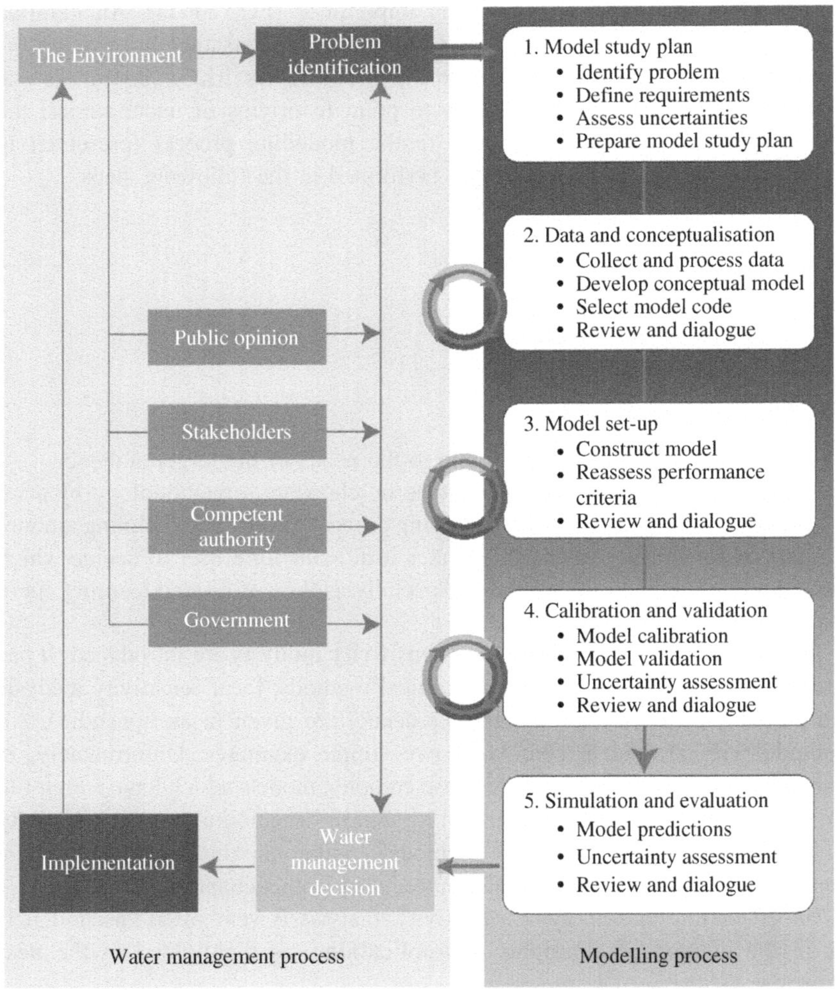

Figure 1.1. The modelling process

2. DECISION TREE FOR CHOOSING A SENSITIVITY ANALYSIS

It is impossible to suggest a universal sensitivity analysis methodology. Discussion about what might be the 'right' methodology continues. Here we give guidelines that consider two aspects: (i) the sensitivity aim/setting which the analyst is targeting, and (ii) the computational cost of the analysis.

2.1.1. Sensitivity aim/setting

Saltelli *et al.* (2004) have argued that the effectiveness of a sensitivity analysis is greater if the purpose of the analysis is specified unambiguously beforehand. Over time, practitioners have identified cogent questions for sensitivity analysis. These questions define the setting, which in turn allows for the selection of the strategy. The settings associated to different global methods will be succinctly mentioned in the next sub-sections. The most typical settings are:

Factors prioritization (FP)
Assume that, in principle, the uncertain input factors can be 'discovered', i.e. determined or measured, so as to find their true value. One legitimate question is then "which factor should one try to determine first in order to have the largest expected reduction in the variance of the model output"? This defines the 'factors prioritization' setting. Saltelli and Tarantola (2002) have shown that the variance-based *main effect* provides the answer to the Factor Prioritization setting.

Factors fixing (FF)
Another aim of sensitivity analysis is to simplify models. If a model is used systematically in a Monte Carlo framework, so that input uncertainties are systematically propagated into the output, it might be useful to ascertain which input factors can be fixed, anywhere in their range of variation, without sensibly affecting a specific output of interest. This may be useful for simplifying a model in a larger sense, because we may be able then to condense entire sections of our models if all factors entering in a section are non-influential. Saltelli and Tarantola (2002) also showed that the variance-based *total effect* provides the answer to the Factor Fixing setting. A null total effect is a sufficient condition for an input factor to be irrelevant, and therefore to be fixed.

Factors mapping (FM)
In this case, the analyst is interested to as many information as possible, either global and local, i.e. which values of an input factor (or of group of factors) are

responsible for driving the model output in a given region? Which conditions are able to drive a specified model behaviour? In this case, a full array of methods, from local ones, to Monte Carlo Filtering, to model emulators, to variance-based and entropy-based methods can provide useful insights about model properties.

Calibration
In this case, the analyst aims at identifying the portions of the input factors space that allow the model to behave according to available information (data, constraints) about the modelled system. FP and FF settings are linked to calibration and can be used to prepare a calibration step: with FP and FF one can say what can be effectively calibrated (those parameters having large main effects) and what cannot (those having null total effect). Parameters in the grey area (i.e. small main effect and large total effect) fall under the so-called equifinality conditions. An effective calibration can be performed by first concentrating on the input factors highlighted by the FP setting.

Arguments and research efforts on sensitivity analysis methodologies continue. Therefore, any attempt to produce guidance in choosing a methodology for a particular application will necessarily reflect the view of the authors. The following decision tree is such an attempt to provide guidance on the choice of methods, based on the most widely agreed approaches. This decision tree is presented before a detailed introduction into the methodologies. The reader is referred to the next chapter for more details.

The decision tree is a pragmatic approach to sensitivity analysis. For a scientific approach we suggest to use as many methods as possible to highlight differences and have a maximum understanding of the processes, in particular when the aim of the analysis is a 'mapping'. This can be easily achieved as most methods can be based on the same sampling design. However, they all have individual learning curves, which makes this approach prohibitive for practical applications.

Therefore, specifying the aim for the sensitivity analysis becomes crucial for concentrating on specific sensitivity methods:

(1) FP? Variance based (VB) main effects provide the answer. This setting is also connected to Type I errors, i.e. when one is 'scared' of labelling as important something that is not: if one gets a high main effect, he can be sure that this parameter is really relevant for the model. As far as the computational cost is concerned, the classical Monte Carlo algorithms

(Sobol') to estimate main effects require a few thousands of model runs: this might be unaffordable in some cases. In such cases, we point the readers to recently developed smoothing methods (RS-HDMR, SDP, spline) that are not particularly difficult to code and that allow to reliably estimate main effects with few hundreds of model runs.

(2) FF? Variance based total effects provide the answer. This is also linked to Type II errors, i.e. one is 'scared' of fixing something that can have an effect on the model. In this case, therefore, one is mostly interested at the zero sensitivity indices. From the computational point of view the only reliable route to compute total effects is the classical Monte Carlo algorithm (Sobol'). However, it has been shown that the screening method based on elementary effects (Morris, 1991, as extended by Campolongo *et al.*, 2007), provides an excellent proxy to total effects at a much smaller computational cost: whenever a zero index is found using the Morris approach, it is extremely difficult that any other method detects any effect for such an input factor.

(3) Both the FF and the FP settings can be treated at a quite small computational cost by taking a few hundred runs to perform the screening test by Campolongo *et al.* (2007) plus another few hundred runs to apply a smoothing technique for estimating the main effects.

(4) FM? In this case, the suggestion of using as many methods as possible is justified by the fact that one is interested to as many information, both global and local, i.e. which parameter values are responsible for driving the output in a given region? It has to be noted that many methods can be applied using the same sample, for example Monte Carlo Filtering (MCF), regression analysis, smoothing and emulation approaches can be applied on the same Monte Carlo sample.

(5) In the case of numerical instabilities, MCF can be very useful to map the occurrence of such numerical problems. Moreover, most of the sensitivity methodologies (MCF, Morris, Sobol', smoothing, etc.) can be applied also in such cases when model simulation fails for a portion of the sample.

2.1.2. Computational cost

Figure 2.1 shows a decision tree based on the computational cost and learning curve of performing a sensitivity analysis. Explanations for each field are given below.

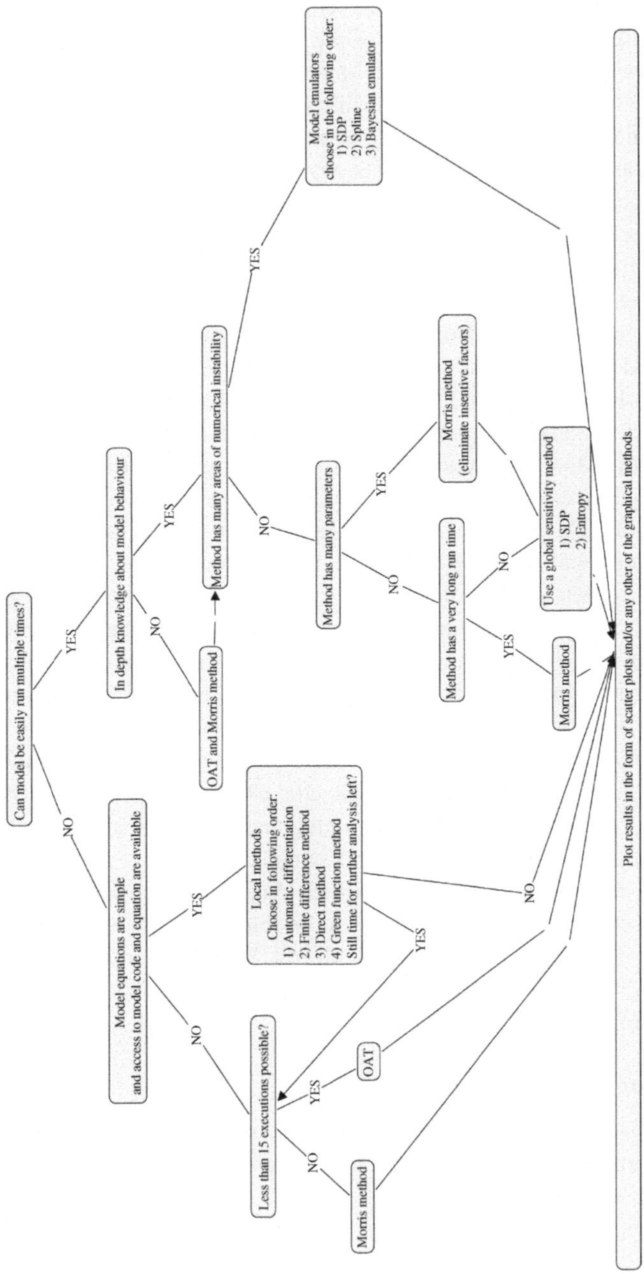

Figure 2.1. Decision tree to choose a sensitivity analysis method. We recommend plotting the model results always. Either in the form of scatter plots and/or some of the graphical methods

Explanation of fields

Can model be run easily multiple times?
The model can be executed externally e.g. from another programming language without the use of a Graphical User Interface. The model output is parsed back or the output files are sufficiently documented so they can be read from an external controller (e.g. another programming language).

Model structure is 'simple' e.g. Manning equation
No numerical analysis or complex direct solution schemes are required to compute these equations.

In depth model knowledge
The user is greatly familiar with the model and has used it many times before.

Model has many areas of numerical instability
For many parameter set, the model does not converge to a solution. 'Many' has to be defined in respect to the model run time and the number of model parameters.

Model has a long runtime and/or many parameters
The more parameters a model has the more runs are necessary to describe the response surface adequately.

For example, following a full factorial design, a two parameter model would need four simulations to sample a two by two grid (two-level full factorial design). 100 simulations would be already necessary to sample a 10 by 10 grid (ten-level full factorial design, see Figure 2.2). The amount of subdivision needed for each parameter would depend on the nonlinearity of the response surface. The sample dimension of the full factorial design increases exponentially with the number of parameters:

$$N_S = D^{N_p} \tag{2.1}$$

N_S: Number of samples
D: Number of subdivisions in the parameter space (the levels)
N_p: Number of parameters.

Using two levels, 10 parameters would require $2^{10} = 1024$ simulations, and 20 parameters would need more than a million. It has to be noted that this is just a crude example as most sensitivity techniques have more efficient techniques to quantify the response surface: for example it is possible to select a fraction of a full factorial design to generate a smaller, feasible design that can still produce useful results (see *sampling methods*).

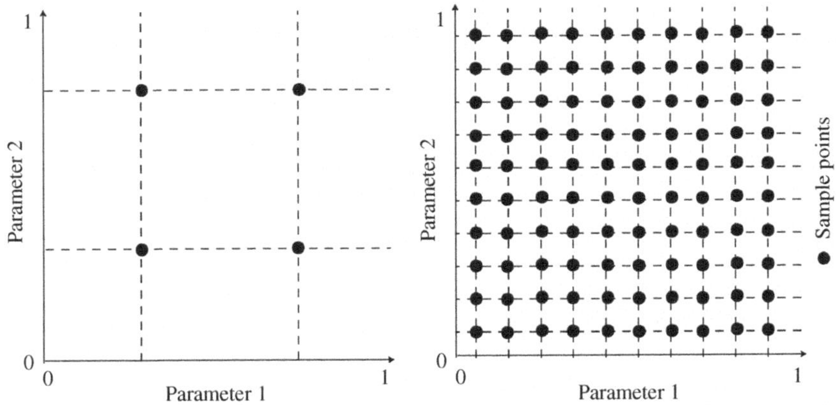

Figure 2.2. Example of sampling the parameter space following a full factorial design of level two (left) and ten (right)

In summary, the question about the sample dimension depends highly on the type of model used and should be based on previous experience. In general, a few hundreds/one thousand of model runs is in many cases sufficient for deriving useful information about the most important model characteristics. This is due to the fact that there is always an upper bound to the number of important input factors. This implies that there is always a small subset of input factors (usually not larger than 10) that produce most of the output variation, which makes the 'effective' dimension of the problem much smaller than the *k*-dimensional input space considered by the modeller. As a very bold statement, we argue that a model with more than 20 parameters and an execution time of more than 10 minutes should be considered as computer intensive (if executed on a single CPU).

3. DETAILED METHODOLOGY

In order to apply sensitivity analysis, it is paramount to understand in which ways parameters and factors can be varied. For example, it maybe necessary to run the model(s) with multiple set of parameters (e.g. a floodplain model with different surface roughness parameters) to observe the change in the prediction variable (e.g. flood inundation extent). In sensitivity analysis this is understood as *Experimental Design*, which will be discussed in the first section of this chapter.

Fundamentally, Sensitivity analysis can be grouped in four major subdivisions, which are graphical methods, screening methods, local methods and global methods:

Graphical methods include different ways to plot results of model outcome and parameters (e.g. scatter plots) or of other sensitivity analysis's themselves (e.g. in the form of histograms). They should be the companion of any sensitivity analysis in order to visualize the results and complement communication. Moreover, graphical methods can also be used for screening prior to the analysis and detect possible floors (e.g. non-monotonic relationships, which would prohibit a straightforward linear regression analysis).

Screening methods are often preliminary numerical experiments whose purpose is to isolate the most important factors amongst a large number that may affect a particular model response. By using screening methods, factors can also be ranked in order of their importance. However, the percentage of the output variation that each factor is accounting for cannot be quantified.

Local methods investigate how small changes of parameters affect the model output. Therefore, these analysis techniques evaluate this partial derivative at one specific set of parameter values, also called the nominal parameter set. Most local methods require complex manipulations of the model equations. However, this is not practically feasible because the models are too complicated or the model equations are not directly accessible (e.g. because they are compiled in executable commercial code).

Global methods evaluate the effect on the model output by varying all parameters at the same time, in a multivariate fashion. Properties of global methods should be the inclusion of the influence of range and shape of the probability distributions of the input factors; the sensitivity estimates of individual factors are evaluated varying all other factors as well (multi-dimensional averaging); and global methods should be model-independent, i.e. they should work irrespective to the nonlinearity, interaction structure and possible discontinuity of the model at hand.

An important approach to sensitivity, which has developed in the last few years, falls within the *emulation* context. The basic idea is to represent in a direct way the relationship between model output and model parameters or factors without the need to run the model multiple times. If the emulation exercise is successful, one can obtain a simple relationship between the model parameters or factors and the model output that fits well the original model and is less computationally demanding. Given the emulator, this can be used to compute any measure of interest, including sensitivities.

3.1. Examples

The methodologies, which will be introduced in what follows, have been demonstrated on two examples. They are purely mathematical functions, which have been used earlier to demonstrate sensitivity analysis techniques.

3.1.1. Example 1

The first example is a simple additive combination of two factors. The example is taken from Saltelli *et al.* (2004):

$$Y = X_1^2 + X_2^2 \tag{3.1}$$

with

$$X_j \sim U(-0.5, 0.5)|_{0.2<Y<0.25} \tag{3.2}$$

Y: Predicted variable
X_j: Factors j
U: Uniform distribution

3.1.2. Example 2

This example uses the Sobol g-function which is a nonlinear and non-additive model. It has been used in previous studies to test global sensitivity analysis models (Ratto *et al.*, 2004; Saltelli *et al.*, 2004).

$$Y = \prod_{j=1}^{k} g_j(X_j) \tag{3.3}$$

where

$$g_j(X_j) = \frac{|4X_j - 2| + a_j}{1 + a_j} \tag{3.4}$$

with $a_j \geq 0$ and $X_j \sim U(0, 1)$
Y: Predicted variable
X_j: Factors j
U: Uniform distribution

This allows tuning the importance of factors by changing a. The smaller a is the more important it should be. For this structure eight input factors have been considered ($k = 8$) with $a_j = [0,1,4.5,9,99,99, 99,99]$.

3.2. Experimental design of SA (sampling)

Different sampling procedures exist, the most important ones are: screening design (one-at-a-time, Morris, fractional factorial), random sampling, Latin Hypercube sampling,

3.2.1. Screening design

Probably the simplest way to sample parameters for doing a sensitivity analysis is to use a 'one-at-a-time' (OAT) design, where only one parameter changes values between consecutive simulations. If there is any change in value of the output between two consecutive simulations, it can only be attributed to a change in the parameter x_i that has been changed. OAT sampling is inefficient when the number of parameters k is large and only a few of them are influential. Each simulation changes the value of one parameter. If only a few parameters are influential, most of the simulations would be devoted to determining the very small effects of non-influential parameters. These simulations would be duplicates as far as the values of influential parameters are concerned. Very little new information would be generated.

It is possible that there are no influential parameters for sensitivity analysis to reveal. This happens when sensitivity analysis indicates that many parameters have similar small effects, but cannot give further insight into this model's behaviour. All of the parameters are equally non-influential. Usually, however, there are a few parameters that stand out from the rest. So with large numbers of parameters, it is desirable to find a more efficient approach than changing one parameter at a time.

In fractional factorial design (FF), a fraction of a full factorial design (mentioned earlier in this handbook) is 'cleverly' selected to generate a smaller, feasible design that can still produce useful results. In general, if one starts with a full factorial design on two levels for n parameters, requiring $k = 2^n$ simulations, it can be converted into a fractional factorial (FF) design for $k - 1$ parameters.

Morris design is perhaps one of the most efficient ways to sample parameters for screening purposes. It allows estimating sensitivity measures at the cost of $r(k+1)$ model runs, where k is the number of parameters and $r \approx 8$ (see dedicated Section on Morris screening method).

3.2.2. Random sampling

Here the input factor is sampled randomly according to the joint probability distribution. This is a technique easy to explain, provides unbiased estimates for means, variances and distribution functions and is easy to implement.

The problem is that the sample has to be sufficiently large and that the term 'sufficiently' is not well defined. When the underlying models need a long time to evaluate then the required sample size to achieve a certain purpose may be too large to be computational practicable.

3.2.3. Latin Hypercube sampling (LHS)

The Latin Hypercube sampling method uses a stratified sampling approach (McKay *et al.*, 1979) to ensure the full coverage of the range of each variable. It subdivides the distribution of each parameter into N ranges, each with a probability of occurrence equal to 1/N. Random values of the parameters are generated such that each range is sampled only once (Weijers and Vanrolleghem, 1997; Manache, 2001). It is possible to combine two-level FF and LH sampling to secure the advantages of both (see e.g. Saltelli *et al.*, 2007: Chapter 2).

3.2.4. Correlation control

Control of correlation within a sample can be very important. If two or more variables are correlated, then it is necessary that the appropriate correlation structure be incorporated into the sample if meaningful results are to be obtained in subsequent uncertainty/sensitivity studies. For most uncertainty/sensitivity analysis problems, rank correlation is probably the most natural measure of congruent variable behaviour. What is present in most situations is some idea of the extent to which variables tend to move up or down together and more information about variable linkage is not available.

3.2.5. Quasi-random sampling with low-discrepancy sequences

Using a pseudo-random generator for simple Monte Carlo sampling is much simpler than using a complex design: as many values as necessary can be generated, and if more parameters or more simulations are desired, it is a simple matter to generate more. Unfortunately samples generated randomly tend to have clusters and gaps. Where a cluster occurs, function values in that vicinity are overemphasized in statistical analysis. Where a gap arises, function values within that gap are not sampled for statistical analysis. The net effect is that mean values estimated with random samples have an uncertainty that diminishes slowly as $1/\sqrt{N}$. To reduce an estimated uncertainty by a factor of 10, the analyst must increase N by a factor of $10^2 = 100$.

A mathematical measure called discrepancy characterizes the lumpiness of a sequence of points in a multidimensional space. Smaller discrepancy values are better for sensitivity analysis (the distribution is less lumpy).

Random sequences (and also LH samples) of k-dimensional points have a relatively high discrepancy (see e.g. Saltelli *et al.*, 2007, Chapter 2). But there are infinite sequences of k-dimensional points that behave much better with respect to this measure. They are called low-discrepancy sequences.

They have the property that as the sequence length N gets very large, the discrepancy shrinks at the theoretically optimal rate. As a result, an estimated mean for a function $y = f(x_{i1},\dots,x_{ik})$ evaluated on points $(x_{i1},\dots,x_{ik})$, $i = 1,\dots,N$ from such a sequence will converge much more quickly than would an estimated mean based on the same number of random points. While the rate of stochastic convergence for pure pseudo-random Monte Carlo methods is $N^{-1/2}$, the actual rate of convergence of quasi-Monte Carlo methods can be N^{-c} with $c \leq 1$ (Sobol' 1998). This means that the rate of convergence can be doubled, in such cases where c approaches one.

Samples made from a finite subset of such sequences are called quasi-random samples. These samples are not random, in the sense of being completely unpredictable. In fact, to maintain an even spread of points, an algorithm that generates low-discrepancy sequences must somehow bias the selection of new points to keep them away from the points already present. But they are like random points in the sense that they are uniformly distributed across the entire sample space.

3.3. Methods of sensitivity analysis

This section introduces different methodologies to perform sensitivity analysis. Categories of sensitivity analysis are introduced in the first section which is followed by a more detailed explanation of each methodology.

3.3.1. Graphical methods and visualisation

Graphical methods have to be the companion of any sensitivity analysis and have just recently attracted more attention by authors such as Cooke and Noortwijk (2000), Bedford and Cooke (2002) or McCamley and Rudel (1995). They are an intrinsic part of most analysis in order to visualize the results and complement communication. Moreover, they can be also used for screening prior to the analysis and detect possible floors (e.g. non-monotonic relationships, which would prohibit a straight forward linear regression analysis). Similar to any of the other techniques, graphical methods have to be carefully explained and accompanied by the relevant documentation. Both are neglected in many reports and scientific articles, despite being obvious and clear to the authors. Every graph should have a clear legend and label, which makes it understandable on its own.

Moreover, it should be accompanied by metadata. Metadata is structured, encoded data that describe characteristics of information-bearing entities to aid in the identification, discovery, assessment, and management of the described entities (Durrell, 1985). In what follows, we will concentrate on four types of typical graphical methods, namely scatter plots, histogram type plots, generalized reachable sets and finally cobweb plots.

Scatter plots

In scatter plots (sometimes also termed dotty plots) each model simulation is plotted usually as a dot of a model factor *vs* a model outcome, the result of a sensitivity analysis or any other transformation of the former (e.g. ranks, logarithm etc.) (Frey and Patil, 2002). These plots are representations of the response surface and can be two (Figures 3.2 and 3.3) or three dimensional (Figures 3.1 and 3.4). In the three dimensional case usually a marginal distribution has to be plotted. For example, a model with more than two factors can have multiple outcomes in which the two factors depicted are the same (e.g. only a third one changed). In such cases, the maximum, minimum or median has to be displayed. The two dimensional plot (Figure 3.1) indicates that several combinations of both factors fit this equation (concept of equifinality see Beven, 1993). This equifinality of model factors is also visible in Figure 3.1 however, this figure allows a better illustration of the response surface. The factors form a circle pattern with higher response values of Y towards the edges of each circle band. The Figures 3.3 and 3.4 illustrate very clearly a common problem in the analysis of multidimensional models: visual display is limited to a small number of dimensions and many models have often more than 7/8 factors. It is very difficult to understand and analyse multidimensional relationships in these cases. However, two-dimensional relationships are shown (e.g. top left corner in Figure 3.4) in these figures.

The method of generalized reachable sets by Bushenkov *et al.* (1994) is a variation on the three dimensional scatter plots in which the response variable of multiple dots is aggregated with a box. The density of point patterns has to be enough to observe a pattern, but not too many to obstruct the analysis. Multiple sensitivity techniques have been developed based on scatter plots. Helton *et al.* (2006) analyse scatterplots of non-random patterns based on common means, common distributions and common medians. It is assumed that any non-random pattern indicates an influence of this particular factor on the model output. Another method which is used by Helton *et al.* (2006) is entropy, which indicates that such sensitivity analysis are in many cases the application of global methods on local scale. In fact, this could be seen as factor mapping. Pappenberger *et al.* (in submission) has shown how factor mapping can be used to identify areas of desirable and undesirable model behaviour, which aids the modelling process. The latter links the methods proposed by Helton *et al.* (2006) with the tree structured density estimation (TSDE) technique (Spear *et al.*, 1994), which clusters areas of high point density based on the same argument as Helton *et al.* (2006).

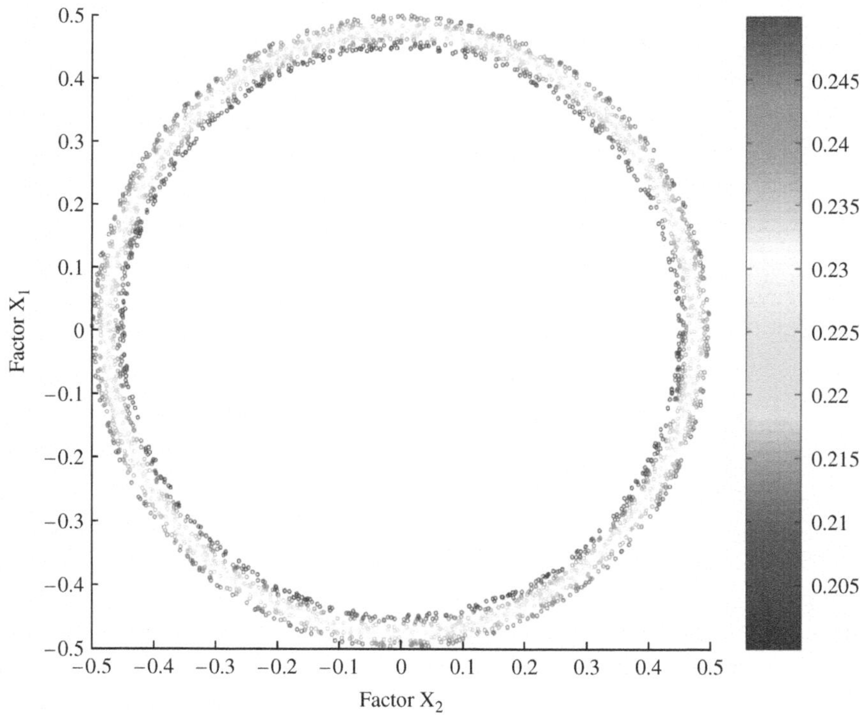

Figure 3.1. Two dimensional scatter plot of Example 1, which plots each factor against the response variable Y (see also Saltelli *et al.*, 2004)

Bar/Histogram/Tornado/Box-Whisker plots/Roses and radar graphs
Bar type graphs come in many forms and shapes and are part of many presentations. Our list is far from being exhaustive and multiple other variations exist. This section is only indented to give an overview.

Bar/Histogram/Tornado: A bar graph displays the values in a vector or matrix as horizontal or vertical bars. A histogram is the most commonly used bar type diagram and shows the distribution of data values and are in many cases used to accompany scatter plots (Figure 3.5). The Figure 3.5 shows the skewed response of Example 2.

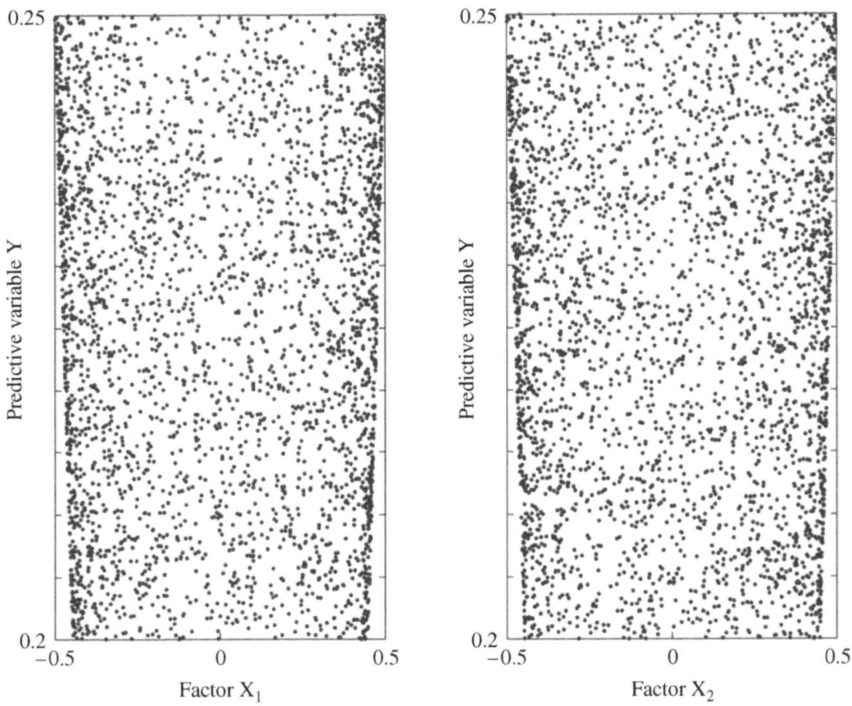

Figure 3.2. Scatter plot of Example 1, which plots each factor against the response variable Y (see also Saltelli *et al.*, 2004)

The tornado graph is a type of bar graph, which is arranged vertically (Figure 3.6).

Roses/feather/radar graphs: Rose are angle histograms, which is a polar plot showing the distribution of values grouped according to their numeric range. Each group is shown as one bin. A feather plot is a variation of the tornado plot and displays vectors emanating from equally spaced points along a horizontal axis. The vector components relative to the origin of the respective vector.

Box-Whisker plot: Box-Whisker are based on box representations (Figure 3.7). The box has lines at the lower quartile, median, and upper quartile values.

The whiskers are lines extending from each end of the box to show the extent of the rest of the data. Outliers are data with values beyond the ends of the whiskers. They are especially useful to display distributions of data without the need for a histogram (reduce the clustering).

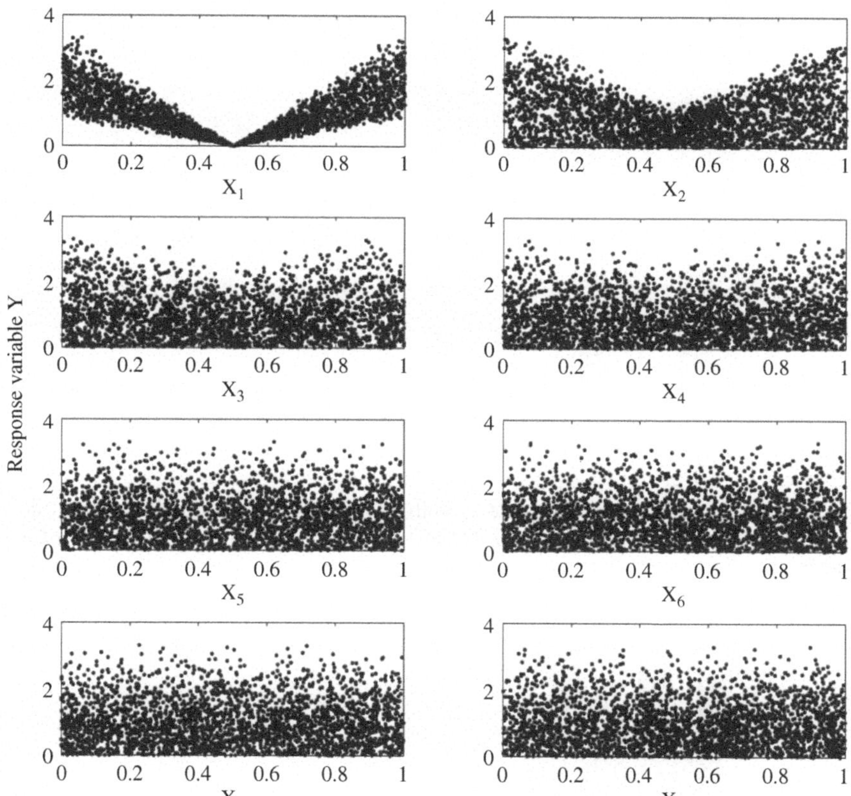

Figure 3.3. Scatter plot of Example 2, which plots each factor against the response variable Y (reproduced from Pappenberger *et al.*, 2006; Saltelli *et al.*, 2004)

Many of the plots have numerical equivalents or relatives. For example the histogram is strongly related to the entropy measure. Thus in a way all this type of techniques have been utilized numerically. However, they still can reveal additional patterns. For example a histogram could have multiple modes, which would only be represented in a limited way in the entropy value.

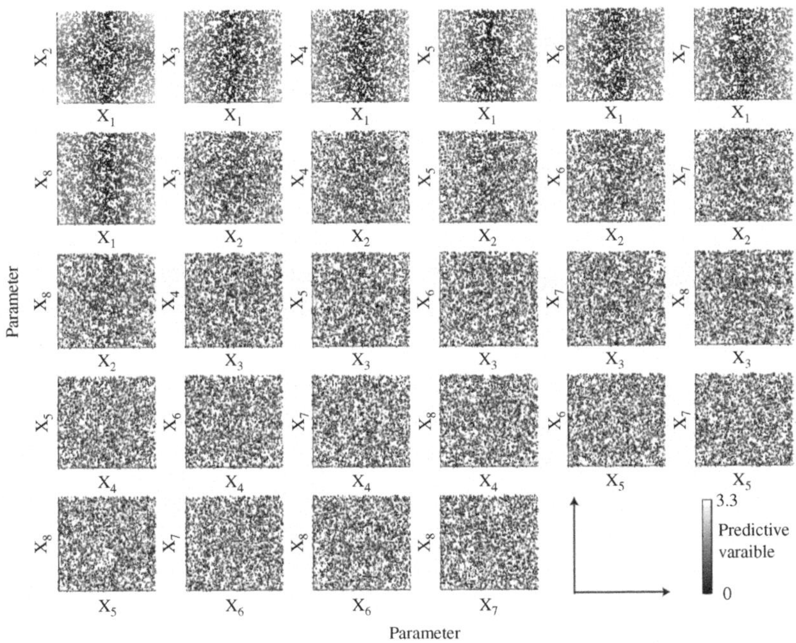

Figure 3.4. Scatter plot of Example 2, with the factors plotted against each other and the Predicted variable as grey scale. (reproduced from Pappenberger *et al.*, 2006)

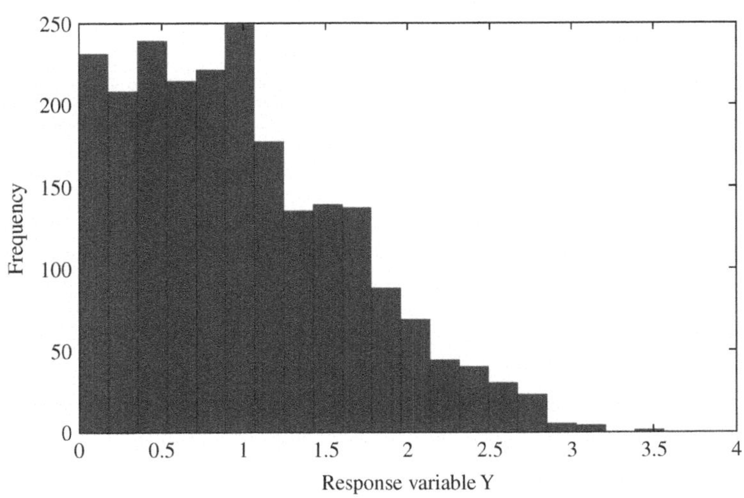

Figure 3.5. Histogram of the response variable Y of example structure 2

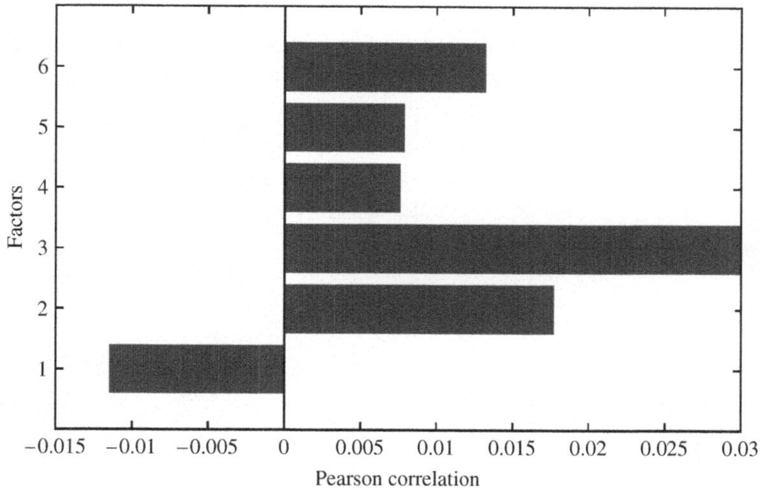

Figure 3.6. Bar diagram of the Pearson correlation between the factors of example 2 and the response variable

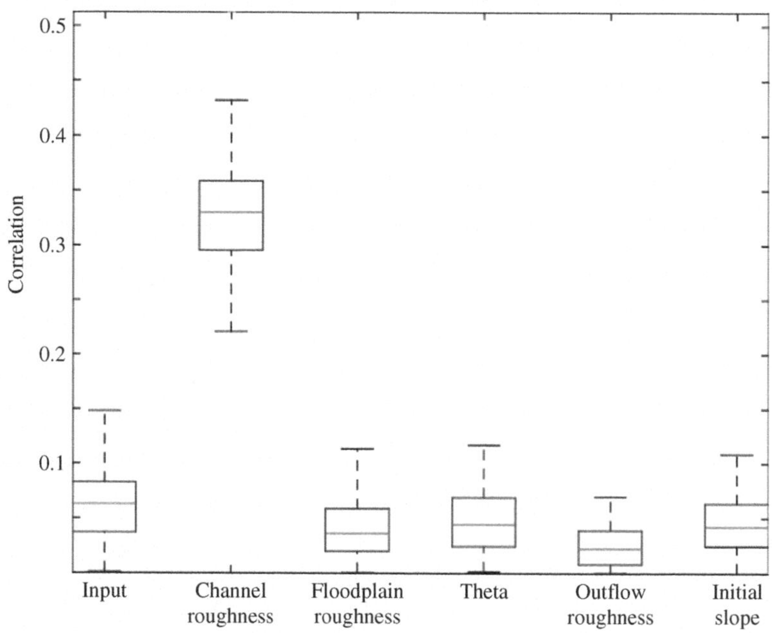

Figure 3.7. Results of correlation analysis based on bootstrap sampling of the factor space for a flood inundation model (each box-plot represents the range of correlations based on a sub-sample of the total sample) (reproduced from Pappenberger *et al.*, 2009)

Cobweb plots

Cobweb plots have been pioneered by Cooke and van Noortwijk (2000) and Wegman (1990). They represent each factor set as a line on (Figure 3.8). The lines are colour coded according quantiles of one variable. Cooke and van Noortwijk (2000) argue that this aids greatly the visual inspection. They are usually accompanied by cross-densities, which show a histogram of lines crossing midway between the vertical axis. The plot is very depended on the way in which the factors are sorted.

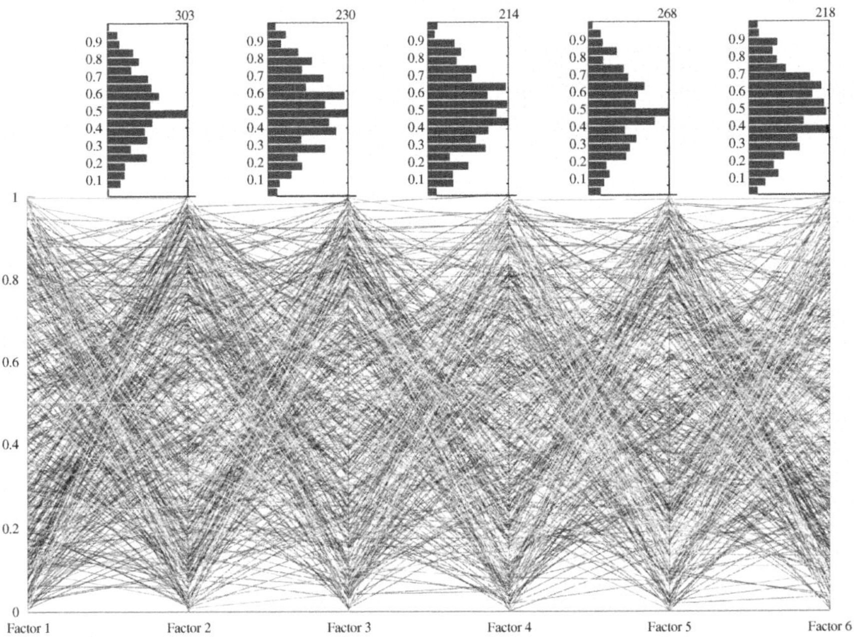

Figure 3.8. Cobweb figure of Example 2. The factors are normalized between 0 and 1. Each factor set is plotted as a line. Different quantiles of the response variable are colour coded (e.g. the top 20% of the simulations are plotted in yellow). The bar diagrams display the 'cross-density', which is the density at the line crossing midways between two factors

Time series plots

All plots above are shown for a static example and can be adapted for time series analysis by relating one of the axes to time. For example, time could be shown instead of different factors on the x axis of plot 3.8. An example of such an analysis through time is given by Wagener *et al.* (2004).

Advantages

Graphical methods allow an identification of potentially complex relationships by utilizing the human brain for pattern recognition. The latter is advantageous and disadvantageous as some patterns can be detected better and others worse (in many cases this is related to factors such as information quantity, graph familiarity etc.). Plotting in different ways very often allows getting a 'feel' for the data, which is vital on pre and post analysis.

Disadvantages

Graphical methods usually lead to a large numbers of plots and figures and in many cases one cannot see the forest for the trees. Moreover, interpretation is qualitative (which is also a strength), but especially in multidimensional space any judgement can be clouded.

Key references

For scatter plots the reader is referred to:

Helton, J.C., Johnson, J.D., Sallaberry, C.J. and Storlie, C.B. (2006). Survey of sampling-based methods for uncertainty and sensitivity analysis. *Reliability Engineering and System Safety*, 91(10–11): 1175–1209.

Spear, R.C., Grieb, T.M. and Shang, N. (1994). Parameter Uncertainty and Interaction in Complex Environmental- Models. *Water Resources Research*, 30(11): 3159–3169.

Bar type plots are explained and introduced in many standard statistical text books or part of software package descriptions such as R, Matlab, SPSS, Unicorn etc.

For Cobweb plots we recommend:

Cooke, R.M. and van Noortwijk, J.M. (2000). Graphical methods. In: A. Saltelli, K. Chan and E.M. Scott (Editors), *Sensitivity Analysis*. John Wiley and Sons, Chichester, pp. 245–264.

3.3.2. Screening methods

Screening methods are preliminary numerical experiments whose purpose is to isolate the most important factors from amongst a large number that may affect a particular model response. By using screening methods, factors can be ranked in order of their importance. However, the percentage of the output variation that each factor is accounting for cannot be quantified.

One-factor-At-a-Time (OAT)

This is the simplest class of screening designs. Here a set of simulations is done in which only one parameter is modified between two consecutive simulations.

Therefore, changes in the model outcome can for sure be assigned to that modified parameter.

OAT can be classified into five categories (Saltelli *et al.,* 2000a)

- standard, which vary one factor from a standard condition (from nominal value to an extreme value);
- strict, which vary one factor from the condition of the last preceding experimental run;
- paired, which produce two observations and hence one simple comparison at a time;
- free, which make each new run under new conditions;
- curved, which produce a subset of results by varying only one easy-to-vary factor.

Many OAT experiments are local, that is, the factors are changed over small intervals around their nominal values. Results of such a local experiment are thus dependent on the choice of this point and the model behaviour is identified only locally in the input space. If the model results show strong nonlinearity then a change in the selected nominal values provide totally different results.

Morris

A sampling design that is not dependent on the choice of the specific point in the input space is that proposed by Morris (1991). The sampling design takes a number r of incremental changes for each input factor randomly selecting the initial point on a grid of p-levels in the k-dimensional space of input factors. This provides a set of r 'elementary effects' (EE) for each input factor. In the original Morris work, taking the average (μ) and the standard deviation (σ) of the set of EE's allows to determine which factors have negligible effects, linear and additive effects or nonlinear or interaction effects. Morris' computational cost is of $r(k+1)$ model runs for k input parameters. Campolongo *et al.* (2007) have developed and refined the original Morris work. The sampling strategy of Morris is improved to avoid oversampling of some levels and undersampling of others that often occurs in the standard Morris procedure. Moreover, only the mean of the absolute values of the elementary effects is computed (μ^*). For screening purposes, this is actually the only measure needed, since this alone is able to provide negligible input factors ($\mu^* \approx 0$). Indeed, μ^* is a good proxy for the Sobol' total effect, with a much cheaper computational cost. Moreover, it is rather resilient against type II errors, i.e. if a factor is seen as non-influential by μ^*, it is unlikely to be seen as influential by any other measure (see Saltelli *et al.*, 2004, 2007; Campolongo *et al.*, 2007).

Advantages and disadvantages

Screening procedures do not give any quantitative information about the sensitivity, so they are very useful as a first screening when the number of parameters is too high to perform a quantitative analysis. The low computational cost is one of the main advantages of the screening methods. The Campolongo *et al.* extension of the Morris method provides a proxi to the variance-based total effect, allowing to treat the FF setting at a much smaller computational cost with respect to Sobol' algorithm.

Key References

Saltelli, A., Ratto, M., Andres, T., Campolongo, F., Cariboni, J., Gatelli, D., Saisana, M., and Tarantola, S. (2007). Global sensitivity analysis. Gauging the worth of scientific models. Wiley.

Morris, M.D. (1991) Factorial sampling plans for preliminary computational experiments, *Technometrics*, 33: 161–174.

Campolongo, F., Cariboni, J. and Saltelli, A. (2007). An effective screening design for sensitivity analysis of large models. Environmental Modelling and Software 22: 1509–1518.

3.3.3. Local methods

Local sensitivity analysis methods refer to small changes of parameters of example variations around an optimum. A factor is called sensitive if small changes produce significant changes in the output. On the other hand, a factor is called insensitive if changes of the factor produce insignificant changes in the model output. This variation can be analytically solved if the analytical solution of the model equation is known. Unfortunately, this is rarely the case and numerical methods have to be used in order to approximate the sensitivity function. Since sensitivity functions are used in several applications like optimal experimental design, model reduction, etc., it is of great importance that a correct approximation of the sensitivity function is calculated to produce accurate results. Various techniques for local sensitivity analysis will be described here, which are (a) the finite difference method, (b) the direct differential method, (c) the Green's function method, (d) the polynomial approximation method and (e) automatic differentiation.

The finite difference method

The simplest way of calculating local sensitivities is to use the finite difference approximation. This technique is also called the brute force method or indirect method. It is very easy to implement because it requires no extra code beyond the original model solver. In this method an infinitesimal number of variations (perturbation) of the parameters, inputs or initial conditions is applied. It should be noted that only one parameter is perturbed at a time while all others are kept at their nominal value. The finite difference technique was found to be too computationally intensive, especially in cases with a large number of input factors.

Direct method for sensitivity analysis

Atherton *et al.* (1975) developed the direct method for sensitivity analysis. Differentiation of the equations with respect to the model factors yields to a set of sensitivity differential equations. The solution of these equations results in the sensitivity functions. The direct differential method sensitivity analysis method relies on the fact that the sensitivity differential equations (obtained after differentiating the ordinary differential equation (ODE) system equations with respect to the parameters) are coupled with the equations of the ODE system. Both systems of differential equations are coupled through the jacobian matrix of the right hand side of the ODE system with respect to the states and the parameters.

Green function's method

The difficulties of solving the large sets of differential equations led to the development of the Green's function method, also called the variational method (Hwang *et al.*, 1978). This method makes use of the fact that the sensitivities can be expressed in integral, rather than differential form. There are a number of variations of the Green's function method a. Probably the most used method is the GFM/AIM method (Analytically Integrated Magnus). It is beyond the scope of this document to go into more detail on these solving techniques. In all Green's function methods, the numerical effort is proportional to the number of variables and not the number of parameters. So this method should be preferred when the number of parameters is large compared to the number of variables. When the number of variables is much larger than the number of parameters, direct differential methods should be used.

Automatic differentiation

Recently, a technique called automatic differentiation has gained a lot of attention. Automatic differentiation techniques are based on the fact that every function, no matter how complicated, is executed on a computer as a sequence of elementary functions. By applying the chain rule of differentiation repeatedly to the composition of these elementary operations, one can compute the derivative information exactly and in a completely automated fashion. This method produces compliable code that evaluates derivatives up to machine precision with a minimum of human effort.

Advantages and disadvantages
All the above techniques except for the finite difference technique have one thing in common. They all require complex manipulations of the model equations. In many studies this is not practically feasible because the models are too complicated or the model equations are not directly accessible (e.g. because they are compiled in executable commercial code). This is the basic reason why the finite difference method, although inefficient, is still used very often.

Key references
Atherton, R.W., Schainker, R.B. and Ducot, E.R. (1975). Statistical sensitivity analysis of models for chemical-kinetics. *AIChE Journal*, **21**, 441–448.
Valko, P. and Vajda, S. (1984). An extended ODE solver for sensitivity calculations. *Computers and Chemistry*, **8**, 255–271.
Grievank, A. (2000). Evaluating derivatives, Principles and techniques of algorithmic differentiation. SIAM.

3.3.4. Global methods

Two key global properties are essential for global sensitivity methods (Saltelli *et al.*, 2000a):

(1) the inclusion of influence of range and shape of the probability distributions of the input factors;
(2) the sensitivity estimates of individual factors are evaluated varying all other factors as well (multidimensional averaging).

Finally, global methods should be model-independent, i.e. they should work irrespective to the nonlinearity, interaction structure and possible discontinuity of the model at hand.

Variance based methods
Variance-based methods quantify sensitivity by decomposing the variance of model outputs into factors related components. In particular, the variance is decomposed into main effects and interaction effects. The main effect of a parameter quantifies the portion of the variance of the model output which is explained by that parameter, by allowing all other parameters to be varied at the same time. It can also be interpreted as the expected reduction in the variance (i.e. in the uncertainty) of the model output that could be obtained if we were able to fix that parameter to its 'true', albeit unknown, value. For this reason, the main effect provides the answer to the *Factors prioritization* setting. The total effect of a parameter measures the residual variance of the

model output that remains by removing the portion explained by all other parameters, i.e. quantifies the variance (i.e. the uncertainty) in the model output that would be left by fixing any other factor to its 'true', albeit unknown, value. The total effect of a parameter can also be seen as the sum of its main effect and all its interaction terms with any other model parameter. When a parameter has a null total effect, this provides a sufficient condition for assessing its irrelevance in the model. For this reason, the total effect provides the answer to the *Factors Fixing* setting. The total effect, main effect and interaction effects are normalized by the unconditional variance to obtain sensitivity indices. The variance based analysis allows for an analytical solution of example 2:

Table 3.1 shows that the main effects (i.e. the additive part of the model) account for about 93% of the entire output variation, implying that only 7% is due to interaction terms. Analysing Total effects, one can see that interaction terms involve parameters X_1, X_2 and X_3.

Table 3.1. Sensitivity indices of Example 2

Factor	Main effect	Total effect
X_1	0.7165	0.7871
X_2	0.1791	0.2422
X_3	0.0237	0.0343
X_4	0.0072	0.0105
$X_{5,6,7,8}$	0.0001	0.0001
sum	0.927	–

There are several methods available for the estimation of variance-based indices: the 'classic' ones are the Sobol' method, FAST and extended FAST. Moreover, reliable estimates of main effects and low order interaction effects can be obtained by applying parametric and non-parametric smoothing techniques (RS-HDMR, SDP, spline, see the section on *emulation*).

FAST

FAST is a methodology which allows to estimate the entire set of main effect sensitivities by Fourier transformation (Koda *et al.*, 1979; McRae *et al.*, 1982), using a single sample of size *N*. Extensions of the FAST method are described in Saltelli *et al.* (1999) and Tarantola *et al.* (2006). In classic FAST only the main effect terms are computed. Extended FAST (Saltelli *et al.*, 1999) allows the computation of higher order terms, in particular it allows to compute the entire set of main and total effects, at the cost of kN model runs.

The Sobol' method
The Sobol' method is a Monte Carlo procedure that allows to compute any term of the variance decomposition, each at the cost of N model runs (Sobol', 1993). Following Saltelli (2002), the cost of estimating the entire set of main and total effects is of $(2+k)N$ model evaluations, which roughly halves the computational cost with respect to the original Sobol' algorithm.

Advantages
Extremely robust, they work with any type of discontinuous (even randomised) mapping between input factors and the output. The Sobol' estimator is unbiased. They do not rely on any hypothesis about the smoothness of the mapping. The only key assumption is that variance (i.e. the second moment) is an adequate measure for quantifying the uncertainty of the model output.

Computing main effects and total effects for each factor, while still being far from a full factors mapping, gives a fairly instructive description of the system. Moreover, they provide unambiguous and clear answers to well specified sensitivity settings (prioritisation and fixing).

Disadvantages
The computational cost is relatively high, which implies that these methods cannot be applied to computationally expensive models.

They do not provide any mapping, i.e. they decompose the output uncertainty but they do not provide information about, e.g., the input factors responsible for producing Y values in specified regions, such as extreme high/low or any behavioural classification.

Key references
Cacuci, Dan G. (2003). In Sensitivity and Uncertainty Analysis, Volume I: Theory; Chapman and Hall.
Cacuci, Dan G., Ionescu-Bujor, M. and Navon, M. (2005). Sensitivity and Uncertainty Analysis: Applications to Large-Scale Systems (Volume II), Chapman and Hall.
Saltelli, A. Tarantola, S., Campolongo, F. and Ratto, M. (2004). Sensitivity Analysis in Practice. A Guide to Assessing Scientific Models, John Wiley and Sons publishers.
Saltelli, A., Ratto, M., Tarantola, S. and Campolongo, F. (2005). Sensitivity analysis for chemical models. *Chemical Reviews*, 105(7): 2811–2827.

Regression analysis
In a regression analysis a relationship between the conditional expected variable y given some input variables x is established (see e.g. Iman *et al.*, 1985). The simplest form of regression analysis is correlation analysis. In Table 3.2 a correlation analysis for example 1 is presented together with the p-value which is testing the hypothesis of no correlation against the alternative that there is a non-zero correlation. It can be seen that this simple analysis fails to identify sensitivities and predominantly suggests insensitive model factors.

Nonlinear regression approaches include approaches such as Neural Networks (Zeng and Yeung, 2003) or random forests (Pappenberger *et al.*, 2006). Many other models can be postulated as a relationship, however, more complex approaches should be treated with care as they usually exhibit larger uncertainties in the identification (for a discussion on that topic see Young *et al.*, 1996). Smoothing procedures to estimate the decomposition of the model output into additive factors related components (RS-HDMR, SDP and spline, see later) can be seen as nonlinear extensions of the regression analysis into the sensitivity analysis context.

The regression coefficient can be used as a measure of sensitivity. In general, the sensitivity of a factor can be quantified by computing a measure of fit (such as the correlation coefficient, Kendall's coefficient of concordance or top down coefficient of concordance see Helton *et al.*, 2005), the distribution of this measure of fitness and the significance level. The measures of fit are estimated from a random sample and are themselves random variables, therefore, if a measure of fit is not significantly different from zero, then there is no statistical significant relationship between the variables (typical tests include the Student's t-test or the Kolmogorov-Smirnov test). If the measure of fit is statistically significant then there is evidence that a relationship exists. The lower the significance level (usual levels are 0.05 or 0.1) for which the hypothesis that the measure of fit is zero can be rejected, the stronger is the evidence of a relationship (for a more detailed discussion see Frey and Patil, 2002; Patil and Frey, 2004). If linear regression is used nonlinear monotonic data can be transformed into linear relationships by for example taking the logarithm or rank transforming the data. The usage of ranked transformed correlation coefficients such as Spearman can significantly improve the resolution of the sensitivity analysis (Helton *et al.*, 2005; Helton *et al.*, 2006). However, nonlinear methods should be preferred if such transformation does not improve the analysis.

The coefficients of a regression analysis are often identified by least squares optimization and thus have to fulfil assumptions such as the independence of input variables and a normal distribution of errors (both are discussed further in the caveats).

Table 3.2. Correlation analysis of Example 2 using Spearman, Pearson and Kendall correlation. The PVal is the p-value testing the hypothesis of no correlation against the alternative that there is a non-zero correlation

	Factor 1		Factor 2		Factor 3		Factor 4		Factor 5		Factor 6		Factor 7		Factor 8	
	Corr	Pval	Corr	Pval	Corr	Pval	Corr	Pval	Corr	Pval	Corr	Pval	Corr	Pval	Corr	Pval
Spearman	0.02	0.30	0.01	0.50	0.01	0.68	0.00	0.73	0.00	0.99	0.00	0.98	0.00	0.98	0.00	0.23
Pearson	0.02	0.27	0.02	0.64	0.02	0.46	0.00	0.90	0.00	0.78	0.00	0.79	0.00	0.76	0.00	0.76
Kendall	0.01	0.27	−0.01	0.43	0.00	0.71	0.00	0.73	0.00	0.99	0.00	0.99	0.00	0.99	0.00	9.99

Helton *et al.* (2006) advocate that any regression analysis should be performed in a step-wise manner, with a regression model first constructed with the most influential variable followed by the next most important variable. The importance is identified by computing the coefficient of determination for a regression model containing only a single variable. Variable importance is then indicated by the order by which the factors are chosen, the changes of the coefficient of determination which are caused by added additional factors and a standardized regression coefficient (e.g. the Pearson correlation coefficient).

Advantages
Regression analysis is conceptually simple and is part of many software packages (Helton *et al.*, 2006). The technique allows the evaluation of the sensitivity of individual model inputs without neglecting the influence of other factors (Cullen and Frey, 1999). The methodology has a large degree of flexibility (multiple ways of transforming data and assumption of different model relationships).

Disadvantages
If the key assumptions of the regression analysis are not met, then, the results will not be robust. Moreover, a functional relationship between the input and the selected outputs has to be assumed (not model-free).

In the case of correlated inputs, the problem of multicollinearity can affect the robustness. This is the case when two variables are highly correlated and thus essentially convey the same information. Any couple of highly correlated variables would exhibit very similar sensitivity, due to the correlation, irrespective the possibly extremely different effect onto the model output itself.

A second assumption of regression analysis is that the errors are distributed normally, which is not the case for many real world examples. Therefore, if the functional relationship used for the regression is not correct, this can give biased and misleading results. Neter *et al.* (1996) have shown that the result of regression analysis largely depends on the functional form chosen a priori and that such an analysis can lead to results which are counterintuitive or statistically insignificant.

Key references
An extensive review of regression base techniques is given by:

Helton, J.C., Johnson, J.D., Sallaberry, C.J. and Storlie, C.B. (2006). Survey of sampling-based methods for uncertainty and sensitivity analysis. *Reliability Engineering and System Safety*, **91**, 1175–1209

Storlie, C.B. and Helton, J.C. (2008a). Multiple predictor smoothing methods for sensitivity analysis: Description of techniques. *Reliability Engineering and System Safety*, **93**, 28–54.

Storlie, C.B. and Helton, J.C. (2008b). Multiple predictor smoothing methods for sensitivity analysis: Example results. *Reliability Engineering and System Safety*, **93**, 55–77.

Regionalised sensitivity analysis

The Regionalised Sensitivity analysis has been originally developed in the context of environmental models by Spear and Hornberger (1980) and further developed by Beven and Binley (1992). The paper by Spear and Hornberger (1980) is one of the first landmark papers in Monte Carlo analysis and thus this method requires to execute the model multiple times using factor set sampled from their joint distributions. It has been applied to many environmental and water resource system applications (Chang and Delleur, 1992; Hornberger and Cosby, 1985; Lence and Takyi, 1992; Osidele and Beck, 2004) such as Seasonal Discharge Programs (Lence and Takyi, 1992). The Regional Sensitivity Analysis procedure involves three fundamental tasks: (1) Running of the model in a Monte Carlo framework (2) a qualitative definition of system behaviour and (3) a binary classification of model output (Osidele and Beck, 2004). The definition of system behaviour involves the setting of thresholds and boundaries in which the model output has to lie. These expected behaviour can be derived from hard data such as measurements or soft data such as empirical information (for discussion on hard and soft data see e.g. Seibert and McDonnell, 2002). For example, a predicted flow hydrograph is expected to lie within certain range of limits (Pappenberger and Beven, 2004). Such a constrain should take account of all model errors (e.g. input, parameters, model structure), the errors in the measurements and the commensurability errors (Beven, 2006; van Straten and Keesman, 1991). All model simulations which fall outside the qualitative definition of the system behaviour are called nonbehavioural (NB) and all models which fulfil the constraints are termed behavioural (B) (Spear and Hornberger, 1980). The subjectivity within this separation is based on an understanding of the system and is inherited in most other methodologies (Freer *et al.*, 2003). The distributions of the factors which are classified as NB and the ones which are classified as B can be compared by a Kolmogrov-Smirnov test (Saltelli *et al.*, 2004, p. 154).

The importance of the uncertainty of each factor is inversely related to the significance level for rejecting H_0. A low significance level means that there is stronger evidence for a significant discrepancy between the factor distributions for N and NB. On the contrary, a high significance level indicates that the H_0 hypothesis cannot be rejected and the possibility that this factor or process is redundant in the model structure. Input factors can be grouped into three sensitivity classes, based on the significance level for rejecting H_0:

[1] critical (<1%); [2] important (>1% and <10%); and [3] possibly insignificant (>10%). (Osidele and Beck, 2004; Saltelli *et al.*, 2004).

In Figure 3.9 the behavioural (0.25> y >0.2), non-behavioural (0.2>y) and prior distribution of example 1 is plotted. The hypothesis that the two distributions are the same cannot be rejected (significance level, 0.05).

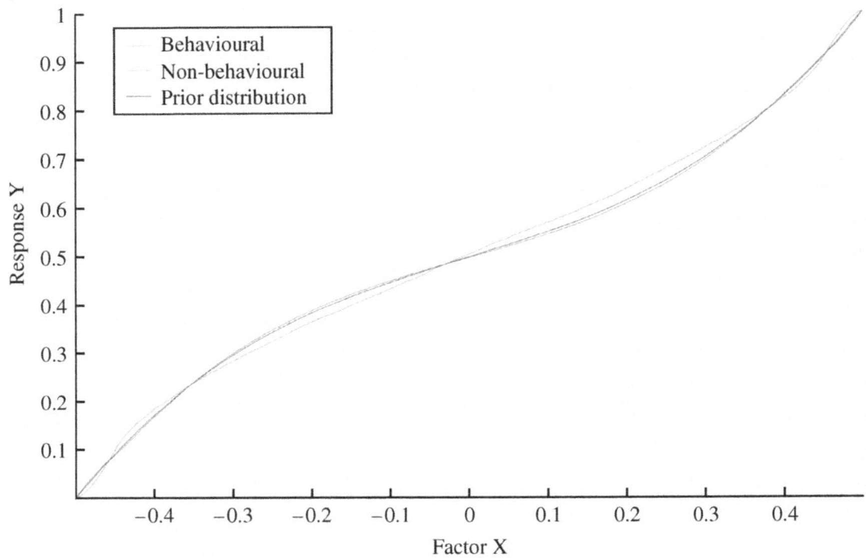

Figure 3.9. Regional Sensitivity Analysis of Example 1. Behavioural is defined as 0.25> y >0.2, non-behavioural as 0.2>y and the prior distribution is defined as the entire sampled space. For more details see Saltelli *et al.* (2004)

The method has been further developed by Beven and Binley (1992) within the Generalized Likelihood Uncertainty Estimation (GLUE) framework. In the RSA application all behavioural and nonbehavioral factor sets are regarded as equally likely representations of the modelled system whereas within the GLUE framework the performance of each simulation is taken into account. Similar to RSA the Monte Carlo runs of the model are classified into behavioural and nonbehavioural. However, a model performance (e.g. a Nash-Sutcliffe of a flow hydrograph) is computed for all behavioural runs. These model performances are binned according to their ranked performance (for example in 10% steps, meaning that the best 10% of all models are treated together as well as the 10%–20% 'best' models etc.). The performances of each bin are normalized (by for example dividing through the sum) and a cumulative distribution function against the factor values plotted. The larger the area such an ensemble of cdfs encloses the more sensitive a factor can be considered. The sensitivity measure can be quantified either by the area or the significance level on which the hypothesis that the area is equal to zero can be rejected (see Pappenberger *et al.*, 2008). Figure 3.10 shows the result of the analysis for Example 2 (assuming that all factors are behavioural and that the response variable is a performance measure). It can be seen that factor 1 and 2 show sensitivity by having a spread in the cumulative distribution functions.

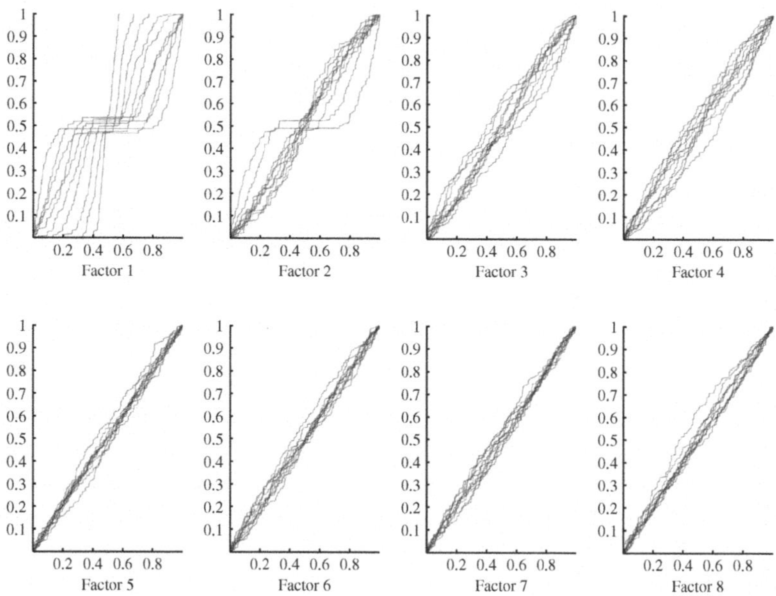

Figure 3.10. Regional Sensitivity Analysis of example structure 2. It can be seen that factor 1 and 2 are sensitive as the cumulative distribution functions show a spread. The different cdfs are based on subdividing the model response in bins of 10% (e.g. the lowest 10% of model response, all response with values between 10% and 20% etc.)

Advantages

The method has the advantage that it is conceptually simple and easy to implement. The result has an easy to understand qualitative meaning and the methodology is model free (no assumptions of a model have to be made for the sensitivity analysis).

Disadvantages
The methodology in its current form cannot quantify higher order effects (problem of multicollinearity, see previous chapter) or search for interacting structures, except for stylised structures detectable through correlation analysis withn the B or NB samples. It has also to be noted that only a hypothesis rejection is sufficient to make one argument. The acceptance of a hypothesis in this kind of test does not imply that there is no importance of that particular factor.

Saltelli *et al.* (2004) argue that the fraction of behavioural models is too small for large models and that this implies a lack of statistical power. However, one could argue that this is due to over parameterization /complexity combined with the difficult task to search the entire model response space adequately or simply that the model is inadequate for the task in question.

Key references
The reader is referred to:

Spear, R.C. and Hornberger, G.M. (1980). Eutrophication in Peel Inlet, II, Identification of critical uncertainties via Generalised Sensitivity Analysis. *Water Resources Research*, **14**, 43–49.

Beven, K.J. and Binley, A. (1992). The future of distributed models – model calibration and uncertainty prediction. *Hydrological Processes*, **6**(3), 279–298.

Entropy
Krykacz-Hausmann (2001) argued that sensitivity should be measured as deviation of the factor distribution from being uniform (if the factor has been sampled from a uniform prior distribution). The author developed a sensitivity measure which is based on the Kullback-Leibler Entropy (Kullback and Leibler, 1951) and was extended by Liu *et al.* (2004).

Liu *et al.* (2004) point out that this entropy shares many properties of a metric measure such as non-negativity, additive property and convexity despite not being a metric measure. The K-L entropy can be understood as a measure which expresses the lack of overlapping between probability density functions (within given limits). Liu *et al.* (2004) define total and main effect indices in analogy to the Sobol method. The concept of entropy and sensitivity is also exploited (in a slightly different formulation) in methodologies such as the dynamic identifiability analysis (Wagener *et al.*, 2003) for rainfall-runoff models and its numerical expression by Horritt (2006) for flood inundation models.

Advantages
It has been argued that variance-based methods rely too heavily on the assumption that the second moment is sufficient to describe the uncertainties and

sensitivities encountered. These assumptions may be invalid if the distribution is highly skewed due to nonlinear functions or inputs (Liu *et al.*, 2004). This methodology proposes a way to overcome this limitation by taking account of the full probability density function.

Disadvantage

The mathematical framework on which first and higher order effects as well as total effects are calculated is less rigid than for the variance based methods (despite being designed to be an improvement to those methods). It can be argued that the main order effect as formulated by Liu *et al.* (2004) actually contains properties of the total effect (if the definitions of the Sobol method are used). The method relies on a structured sampling design which may be difficult to achieve in some practical applications, e.g. due to numerical instabilities.

Key references

The key reference for this methodology is:

Liu, H., Chen, W. and Sudjianto, A. (2004). Relative entropy based method for global and regional sensitivity analysis in probabilistic design, Proceedings of DETC'04 ASME 2004 International Design Engineering Technical Conferences and Computers and Information in Engineering Conference Salt Lake City, Utah

3.3.5. Sensitivity and model emulators

An important approach to sensitivity analysis, which has developed in the last few years, falls within the emulation context. The basic idea is to represent in a direct way the relationship between the model factors and the model results, whose form is usually unknown to the analyst. If the emulation exercise is successful, one can obtain a simple relationship between the model factors and the model output that fits well the original model and is less computationally demanding. Given the emulator, this can be used to compute any measure of interest, including sensitivities.

There is a vast literature on this framework. Local approximation methods provide a first class of emulator that matches the properties of the model at a base point and in the nearby region (Taylor series). Interpolation methods look at 'nice' functions that go through a set of data points spanning the entire domain of the mapping. The approximation is then identified by fixing a set number of parameters (e.g. the coefficients of the polynomials) using a set number of data points (Lagrange, Chebyshev interpolation). The cut-HDMR expansion (see later) is an example of this. Regression methods differ from interpolation whereby a set of more data points than number of parameters is used to identify the approximating function. For univariate models, the interpolation approach can be extended by applying piecewise polynomials, constructing functions that are only piecewise smooth. Splines (cubic spline is the most popular) are powerful and widely used piecewise polynomial interpolations, which are smooth where the polynomial pieces connect. In the multivariate case, which is the most interesting for us, radial basis function networks can be seen as the equivalent of univariate piecewise interpolation.

All the above approaches can be referred to as model approximation methods: see Storlie and Helton (2008ab) for a review of smoothing methods for sensitivity analysis purposes and Storlie *et al.* (2009b) for algorithmic implementations in R.

cut-HDMR

The papers by H. Rabitz and co-workers (e.g. Rabitz *et al.,* 1999; Rabitz and Aliş, 2000) on High Dimensional Model Representation (HDMR) can be seen as precursory applications of the emulation approach in SA. There, the emulator is given by the so-called cut-HDMR expansion, obtained by evaluating the model on quadrature points falling on lines, planes and hyper-planes passing through a 'base' point in the input factor space, and truncating the order of the hyper-planes. Then, all the required integrals are computed on the truncated cut-HDMR function. This helps to reduce the computational cost of the analysis with respect to variance-based methods, albeit still not sufficiently to treat expensive models. Moreover, the cut-HDMR approach still depends significantly on the dimensionality of the problem.

RS-HDMR

The same research group who worked on cut-HDMR presented the so-called Random Sampling-HDMR (RS-HDMR, Liu *et al.,* 2002; Liu *et al.,* 2006), where the HDMR terms are estimated by smoothing a Monte Carlo sample of the mapping with orthogonal polynomials. Due to the curse of dimensionality problem, a truncated HDMR is estimated, usually up to order three, using polynomials of order up to 2–3 (see Liu *et al.*, 2006 for more details). RS-HDMR can be seen as the first implementation of the emulation concept within the sensitivity context. RS-HDMR proved to be good in the applications presented by the authors and allows to significantly reducing the computational cost of sensitivity analysis.

Bayesian and kriging emulators

The theoretical foundations of model emulation are grounded in statistical theory of stochastic processes. The general framework is originally due to Sacks *et al.* (1989). Oakley and O'Hagan (2004) have demonstrated its application in SA, where a Bayesian approach is applied to produce a Gaussian process emulator of the computational model and, subsequently, to compute sensitivity indices in an extremely efficient way, with a computational cost of only a few hundreds of runs for a reasonable number of input variables. Kriging emulators (Kleijnen, 2007a,b) are similar to Gaussian, except that they do not rely on Bayesian interpretation.

Their computational efficiency is linked to certain limitations and hypotheses: namely, that the model response needs to be smooth and the number of input variables should not be very high (<30). Moreover, Gaussian/kriging emulators can be prone to the curse of dimensionality and to the smoothness assumptions of the function under analysis. This is because, instead of trying to identify the best predictors of Y based on a subset of input factors or on low-order ANOVA-HDMR terms, Gaussian emulators try to interpolate and predict the $f(\cdot)$ mapping by applying a Gaussian kernel of the same k-dimensionality as the input parameter space. Therefore, as k increases, the number of hyper-parameters to be estimated (linked to the covariance structure of the k-dimensional Gaussian kernel), increases strongly, possibly implying problems with identification and over-parameterization. Under such hypotheses and caveats, the GP emulator works in an extremely efficient way and a few hundred of runs are sufficient to describe with good accuracy most of the model behaviour. More details on Gaussian emulator can be found in Kennedy and O'Hagan (2001) and Oakley and O'Hagan (2004).

SDP

This is a non-parametric methodology that is based on an approach to State-Dependent Parameter (SDP) modelling first suggested in Young (1993) and applied for sensitivity analysis and emulation by Ratto *et al.* (2004, 2007). The estimation is performed with the help of the 'classical' recursive (non-numerical) Kalman filter (Kalman, 1960) and associated fixed interval smoothing algorithms.

As for the RS-HDMR, the emulator is based on a truncated ANOVA-HDMR expansion up to the order 3 and, as in the previously published approaches, it normally estimates all the main effects with only a few hundred Monte Carlo realisations ($<1{,}000$), almost independently of the dimensionality k of the problem. A MATLAB Toolbox for the estimation of Smoothing Spline ANOVA models with Recursive algorithms can be found at http://eemc.jrc.ec.europa.eu/Software-SS_ANOVA_R.htm.

Splines

Splines can be effectively used to identify ANOVA models. The spline approach is similar to the SDP method introduced above. A recent methodology estimates the relationship of the state dependent parameters over hermite cubic splines (Pappenberger and Stauch, 2006). The shape preserving algorithm by Fritsch and Carslon (1980) is applied and the values of the knots of the cubic spline optimised using a Levenberg-Marquardt nonlinear least square minimisation (Stauch and Jarvis, in press). The optimisation procedure duplicates the backfitting algorithm for the SDP procedure. The number of nodes and location has been fixed after several initial tests, but could be optimised (Bitterlich *et al.*, 2004). However, initial tests suggest that this is not always necessary.

Smoothing splines are also successfully applied for sensitivity analysis (see recent contributions by Storlie *et al.*, 2009ab).

Advantages and disadvantages

All emulation approaches allow performing quantitative global sensitivity analysis, like variance-based analysis, with a much lower computational effort compared to classical Monte Carlo estimation approaches (FAST, Sobol'). This is because they try to make best use of the smoothness properties of the function under analysis. These methods are less simple to code with respect to classical methods, although they are conceptually simple. The RS-HDMR (polynomial regression), SDP (non-parametric, recursive regression) and spline approaches can also be seen as extensions of the regression approach for the estimation of low order HDMR expansions. As such (and irrespective to the emulation context), they can be used for estimating main effect variance based sensitivity indices, for such models where the number of model evaluations required by classical Monte Carlo methods are unaffordable.

4. CASE STUDIES

4.1. Modelling floodplain hydrological processes

Type of sensitivity analysis: Multi-method Global Sensitivity Analysis.

Application area: Understanding the controlling factors of subsurface flow of a flood inundation event on a floodplain.

Type of model: Finite element subsurface flow model (ESTEL-2D).

Location: River Severn (UK).

Reference: Cloke, H.L., Pappenberger, F. and Renaud, J.-P.' (2007). Multi-Method Global Sensitivity Analysis (MMGSA) for Modelling Floodplain Hydrological Processes, Hydrological Processes, DOI: 10.1002/hyp.6734.

Short description

In this application different methods of calculating global sensitivity (a multi-method global sensitivity analysis – MMGSA) are applied to a finite element subsurface flow model (ESTEL-2D finite element Richards equation model code). The model is applied on a well-studied floodplain and flood event on a lowland floodplain of the River Severn, UK. This work is one of the few successful attempts to apply Global Sensitivity Analysis to a complex and numerically challenging model with limited computational resources (i.e. restriction on feasible number of simulations), whilst still adequately covering the parameter/factor (hyper)space. The objectives of this work have been twofold: (i) to demonstrate multi-method GSA for complex distributed sub-surface models; (ii) to demonstrate how information on the sensitivity or importance of model factors can inform the investigation of hydrological processes and the prediction of state variables, such as Darcian velocity vectors.

The suite of results generated highlight the complexities inherent in the sensitivities of the model factors, but demonstrated that for this particular flooding case, the upslope pressure boundary condition and the Brooks Corey soil algorithm 'air entry' value were particularly important. This work demonstrates the utility of the MMGSA method for model understanding, in particular for the prediction of state variables, such as Darcian velocity vectors directed towards the hillslope. This will allow an improved estimate of the likely range of velocities possible. Understanding the controls on the numerical simulation of hydrological processes allows informed decision making, for example, where next to deploy field resources. As ESTEL and other similar models are used routinely for understanding pollution transport and groundwater mixing problems, such a decision making capability is an essential component.

4.2. Flood inundation models

Type of sensitivity analysis: Sensitivity analysis based on regional splits and regression trees and Scatter Plots.

Application area: Improving the predictions of flood inundation models.

Type of model: 1D flood inundation Model (HEC-RAS).

Location: River Alzette (Luxembourg).

Reference: Pappenberger, F., Iorgulescu, I., Beven, K.J. (2006). Sensitivity analysis based on regional splits and regression trees (SARS-RT). *Environmental Modelling and Software*, 21(7), 976–990.

Similar applications:

Hall, J.W., Tarantola, S., Bates, P.D. and Horritt, M.S. (2005). Distributed sensitivity analysis of flood inundation model calibration. *Journal of Hydraulic Engineering-Asce*, 131(2), 117–126.

Pappenberger, F., Beven, K.J., Horritt, M. and Blazkova, S., (2005). Uncertainty in the calibration of effective roughness parameters in HEC-RAS using inundation and downstream level observations. *Journal of Hydrology*, 302(1–4), 46–69

Pappenberger, F., Matgen, P., Beven, K. J., Henry, J.-B., Pfister, L. and Fraipont de, P., (2006). Influence of uncertain boundary conditions and model structure on flood inundation predictions. *Advances in Water Resources*, 29(10), 1430–1449

Short description

In this application the one dimensional flood routing model HEC-RAS which solves the one-dimensional Saint Venant equations to route flood flow is used. HEC-RAS has been applied on the Alzette catchment (Luxembourg) with 7 different parameters which consist of: 3 different surface roughnesses, 3 parameters which control the magnitude of the model input and one parameter controlling the accuracy of the numerical solution (see Table 4.1). The Monte Carlo based Generalized Likelihood Uncertainty Estimation (GLUE) has been performed with these model parameters. The model has been evaluated on a satellite picture of flood extent with a performance measure. The performance measure has been normalized and is scaled from 0 to 1 (with 1 being the best performing models). All model results which perform worse than a linear interpolation from the upstream to the downstream boundary have been classified as non-behavioural and excluded from further analysis.

Table 4.1. Parameters of the HEC-RAS model

No.	Parameter	Sample range
1	Manning surface roughness of channel	0.01–0.2
2	Manning surface roughness of right flood plain	0.01–0.2
3	Manning surface roughness of left flood plain	0.01–0.2
4	Theta (controlling accuracy of numerical implicit scheme)	0.5–1.0
5	Rating Curve Parameter (controlling magnitude of inflow)	0.05–0.2
6	Rating Curve Parameter (controlling magnitude of inflow)	0.2–0.7
7	Rating Curve Parameter (controlling magnitude of inflow)	0.001–0.01

The sampling range of all parameters is uniform over the whole range shown (only the behavioural parameter range is shown). Parameter 4 could only have the values of 0.6, 0.7, 0.8, 0.9 and 1.0 due to constraints in the model set-up. Around ~4000 model realisations have been computed.

The study reveals that higher order effects and thus combinations of parameters rather than individual parameters dominate the response of this model. The Manning roughness of the channel is the most important factor and clearly interacts with the left floodplain friction. In contrast, the right flood plain friction seems to be of minor importance when it is evaluated against a global inundation measure (it still can be of importance on a local scale). The analysis shows further an interplay between channel roughness, left floodplain roughness and input parameters 5 and 6. This reflects the physical understanding which would suggest that input and roughness have to be related in order to achieve behavioural model results. This study seems to indicate that parameter 4 (Theta) plays only a minor role in controlling the performance measure (it still plays an important role in controlling the accuracy of the numerical solution). This case study allows to quantify the sensitivity in flood inundation controlling factors and resources could be allocated according to the sensitivity of the individual parameters.

4.3. Water quality modelling

Type of sensitivity analysis: Regression analysis, parameters ranked on the basis of the normalised regression coefficient. The sampling method is Latin Hypercube.

Application area: Predictions of water quality for management decisions.

Type of model: Modified version of the SWAT model.

Location: Dender River, Belgium.

Reference: Vandenberghe, V., van Griensven, A., Bauwens, W. and Vanrolleghem, P.A. (2006). Effect of different river water quality model concepts used for river basin management decisions. *Water Science and Technology*, 53(10), 277–284.

Short description

For integrated water quality management, a holistic approach is necessary at river basin scale. As diffuse pollution sources are increasingly responsible for water quality problems, water quality modelling entered the field of catchment modelling. Upscaling of agricultural field-scale modelling tools or the inclusion of erosion and nutrient equations in catchment hydrological models has led to a number of tools that enable the calculation of the contributions of water, nutrients and sediments from drained areas. In integrated river water quality modelling, the in-stream processes play a key role, as it is here that the pollution of different origins are added and are transformed to what finally determines the water quality.

Within the Soil and Water Assessment Tool (SWAT) (Arnold *et al.*, 1996), the original water quality module – based on QUAL2E (Brown and Barnwell 1987) – appeared to be erroneously implemented. Time steps of only 1 day were possible that cannot be used in evaluating river water quality processes that change on a sub daily time base. Therefore, two alternative formulations using hourly time steps based on the QUAL2E and the more elaborated River Water Quality Model nr. 1 (RWQM1) (Reichert *et al.*, 2001) were incorporated in the SWAT model codes and applied on the highly polluted Dender river basin (Belgium) (van Griensven and Bauwens, 2001). Since these concepts represent different processes or different formulations of the processes (see Table 4.2), they may give rise to different results. This is revealed when the two models are applied for pollution abatement scenarios. When using water quality models for management purposes it is important to have knowledge of the key processes in the river system. To this end a sensitivity analysis (SA) on the parameters of both concepts was performed. With the results of this SA, one is able to define the restrictions of use of a certain river water quality model, e.g. when the model results are not sensitive towards sediment processes, this model cannot be used for the evaluation of anti-erosion measures.

Table 4.2. Comparison between QUAL2E and RWQM1 based water quality models

QUAL2E	RWQM1
Effect focused	Based on activated sludge model concept
No microbial masses modelled	Cause focused
Simple	Microbial masses modelled
Few parameters and variables	Complex
Not closed mass balance	Many parameters and variables
	Closed mass balance

To show the effects on management decisions, a specific scenario is evaluated in which the pollution load to the river Dender originating from agricultural fertilizer use is diminished. Both river water quality concepts were used to evaluate the decrease in diffuse pollution input and a comparison is made in this study.

The capabilities of the two different concepts of river water quality modelling to predict and assess the effects of future scenarios for pollution abatement are explored by studying a reduction in diffuse pollution load towards the river Dender. The ESWAT model was calibrated for the two water quality model concepts with two weekly measurements taken in 1994. The calibration was done with the multi-objective calibration method described in (van Griensven and Bauwens, 2003). The calibration of the flow was also done by multi-objective calibration. The parameters calibrated were hydraulic conductivity of the soils, canopy index infiltration runoff time lagging, groundwater parameters and routing parameters. The calibration led to an efficiency of 0.9 for the flow. For both concepts, the results are given for the time series of NO_3 and DO. Sensitivity analysis (SA) on the models is performed on the model results for 1994 as well as the model results obtained after decreasing the diffuse pollution input. This SA is done following a global sensitivity analysis to see the most influential parameters of the water quality model. The method used is a regression and correlation technique with Latin Hypercube Monte Carlo sampling. Because one of the problems in the river Dender is oxygen shortage during some periods of the year, due to eutrophication, and the diffuse pollution is influencing the nitrate content of the river, nutrient for the algae, the critical output considered here is the amount of hours that the oxygen concentration drops below 5 mg/l and the nitrate concentration is higher than 3 mg/l.

Ranking of the parameters is done according to the Standard regression coefficients (SRC). Only the parameters contributing significantly in this linear regression (90% level) are presented (Tables 4.3 to 4.6).

The sensitivity analysis on the base scenario reveals which processes will be taken into account when computing predictions for the abatement scenarios. The sensitivity analysis on the scenario shows the importance of the changed input on the modelled processes. It helps to decide what processes have to be measured and evaluated for attaining more reliable results of the model when evaluating a future scenario.

Table 4.3. Ranking of sensitivity of parameters of QUAL2E model based on standardized regression coefficient for the output NO_3 >3 mg/l and DO < 5 mg/l. (base case)

NO_3 > 3 mg/l		DO < 5 mg/l	
Parameter	SRC	Parameter	SRC
O_2 uptake/NH_3 oxidation	−0.704	O_2 uptake/NH_3 oxidation	0.521
Denitrification rate	−0.342	Rate biological oxidation of NH_4 to NO_2	0.354
Reaeration rate	0.321	O_2 uptake/algae respiration	0.279
O_2 uptake/HNO_2 oxidation	−0.211	CbOD deoxygenation rate	0.268
Rate biological oxidation of NO_2 to NO_3	−0.268	O_2 production/algae growth	−0.240
CBOD loss due to settling	0.173	Algae resp rate	0.159
O_2 uptake/algae respiration	0.121	O_2 uptake/HNO_2 oxidation	0.159
		Max algae growth rate	−0.159
		CBOD loss due to settling	−0.149

Table 4.4. Ranking of sensitivity of parameters of RWQM1 model based on standardized regression coefficient for the output NO_3 > 3 mg/l and DO < 5 mg/l. (base case)

NO_3 > 3 mg/l		DO < mg/l	
Parameter	SRC	Parameter	SRC
Growth rate first stage Nitrifiers	−0.662	Growth rate heterotrophs, aerobic	0.482
Respiration rate first stage Nitrifiers	0.341	Respiration rate algae	0.394
Respiration rate heterotrophs, aerobic	0.150	Respiration rate first stage Nitrifiers	0.319
Respiration rate heterotrophs, anoxic	−0.145	Growth rate algae	−0.313
Growth rate consumers in sediment	−0.123	Growth rate second stage Nitrifiers	−0.260
Hydration rate in sediment	0.111	Growth rate heterotrophs, aerobic in sediments	−0.073
Growth rate heterotrophs, aerobic in sediments	−0.097	Reaeration rate	−0.072
Respiration rate algae	0.088		
Reaeration rate	0.086		

Table 4.5. Ranking of sensitivity of parameters of QUAL2E model based on standardized regression coefficient for the output NO_3 > 3 mg/l and DO < 5 mg/l. (reduced diffuse pollution)

NO_3 > 3 mg/l		DO < 5 mg/l	
Parameter	SRC	Parameter	SRC
Reaeration rate	−0.728	Reaeration rate	−0.728
Benthic oxygen demand	0.360	Rate biological oxidation of NH_4 to NO_2	0.360
CBOD loss due to settling	0.229	Bethic source rate NH_4	0.229
Rate biological oxidation of NH_4 to NO_2	0.185	Benthic oxygen demand	0.185
Algae preference factor for ammonia	0.122	O_2 uptake/NH_3 oxidation	0.122
Rate org N settling	−1.00	Rate org N settling	−1.00
Half saturation constant for nitrogen	−0.083	Algae respiration rate	−0.087
Cbod deoxygenation rate	0.078	Halfsaturation constant for phosphor	−0.083
O_2 uptake/NH_3 oxidation	0.12		

Table 4.6. Ranking of sensitivity of parameters of RWQM1 model based on standardized regression coefficient for the output NO_3 > 3 mg/l and DO < 5 mg/l. (reduced diffuse pollution)

NO_3 > 3 mg/l		DO < mg/l	
Parameter	SRC	Parameter	SRC
Growth rate first stage Nitrifiers	−0.853	Growth rate heterotrophs, aerobic	0.416
Respiration rate first stage Nitrifiers	0.331	Growth rate first stage Nitrifiers	−0.376
Respiration rate heterotrophs, aerobic	0.100	Respiration rate algae	0.373
Respiration rate heterotrophs, anoxic	−0.099	Respiration rate first stage Nitrifiers	0.328
Diffusion ammonium	0.065	Growth rate algae	−0.265
Respiration rate heterotrophs, anoxic in sediments	−0.060	Growth rate second stage Nitrifiers	−0.237
Growth rate consumers in sediment	0.059	Growth rate heterotrophs, anoxic	0.101
Growth rate algae	0.048	Respiration rate heterotrophs, aerobic	−0.093
Hydration rate in sediment	0.043	Sediment boundary layer	−0.085
Diffusion nitrite	0.08		

Conclusions

The two main concepts in river water quality modelling in use today, QUAL2E and RWQM1 were here compared towards their role in management decisions. It is shown with a sensitivity analysis that the focus for the two concepts lay different. QUAL2E based water quality models are mainly relating the algae processes towards the output where as the RWQM1 is also taking into account sedimentation and stresses processes performed separately by different microbial communities. As the microbial masses change between scenarios, it can give better results to include them in the analysis. However, this only holds if sufficient measurement data are available for calibration and validation. If one needs to work with a restricted amount of data, the QUAL2E- based modelling will perform better.

When a RWQM1 model can be used that is well calibrated, it is best to use choose a RWQM1 model over a QUAL2E-based model for evaluation of a scenario of reduced diffuse pollution as it was shown that the sediment processes become more important for such a scenario.

Sensitivity of the model results towards the processes is not the same for the 2 different river water quality modelling concepts and the different models are not always able to properly answer the same management problem. This clearly shows that managers should be aware of the possibilities and limitation of the model they use and choose a model that fits their problem and expectations. Also knowing which processes will become important after execution of a scenario can ensure that extra attention is paid towards those to get more reliable results. Here expert knowledge plays also an important role. In the Dender river case study the SA on the reduction scenario showed that the sediment and diffusion processes become more important.

5. CONCLUSIONS

Current practice in modelling does not routinely involve sensitivity (and uncertainty) analysis and thus models are used deterministically without acknowledging or understanding the impact of the uncertainty in model factors on model results.

Public discussions, policy decisions and scientific discourses, results and decisions are frequently based on such deterministic model results. We proceed as if our models are true and we fully understand their behaviour, although most modellers would, after a cursory reflection, acknowledge the opposite. Sensitivity analysis is vital in understanding how uncertainty translates in model results. Certainly, sensitivity analysis is an additional complication that can only be incorporated at the cost of additional expense, understanding and training but it is a worthwhile investment, which can support active decision making. Most methods are ready to be used and have been tested in many different environments. We encourage the reader of this document to apply some of these methods.

6. GLOSSARY

(Modified from Gouldby and Samuels, 2004, Language of risk – project definitions. FLOODsite Report T34/04/01, at www.floodsite.net and Pappenberger F., Hamish H., Beven, K.J., Hall J., Romanowicz R., Smoith P. (2006). Risk and Uncertainty, Tools and Implementation, Flood Risk Research Consortium and www.floodrisk.net).

Accuracy – Closeness to reality.
Additivity – By definition, a model is additive when it is possible to separate the effects of its input variables in a variance decomposition framework. For example, $Y = \sum_i X_i^2$ is a non-linear, additive model in the input factors X_i's; $Y = \prod_i X_i$ is nonlinear and non-additive.
Behavioural – Model or system behaviour exceeds a certain threshold.
Bias – The disposition to distort the significance of the various pieces of information that have to be used.
Binary – A numbering system with only two values.
Calibration parameters/factors – which must be established to represent particular circumstances. They must be determined by calibration of model results for historical data on both input and outcome. The parameters are generally chosen to minimise the difference between model outcomes and measured data on the same outcomes. It is unlikely that the set of parameters required to achieve a "satisfactory" calibration is unique.
Conditional probability – The likelihood of some event given the prior occurrence of some other event.
Confidence interval – A measure of the degree of (un)certainty of an estimate. Usually presented as a percentage. For example, a confidence level of 95% applied to an upper and lower bound of an estimate indicates there is a 95% chance the estimate lies between the specified bounds. Confidence limits can be calculated for some forms of uncertainty (see knowledge uncertainty), or estimated by an expert (see judgement).
Correlation – Between two random variables, the correlation is a measure of the extent to which a change in one tends to correspond to a change in the other.

One measure of linear dependence is the correlation coefficient p. If variables are independent random variables then p = 0. Values of +1 and −1 correspond to full positive and negative dependence respectively. Note: the existence of some correlation need not imply that the link is one of cause and effect.
Cobweb – A type of sensitivity plot.
Dependence – The extent to which one variable depends on another variable. Dependence affects the likelihood of two or more thresholds being exceeded simultaneously. When it is not known whether dependence exists between two variables or parameters, guidance on the importance of any assumption can be provided by assessing the fully dependent and independent cases (see also correlation).
Deterministic process/method – A method or process that adopts precise, single-values for all variables and input values, giving a single value output.
Distribution – An arrangement of values of a variable showing their frequency of occurrence.
Emulator – Ii is a *statistical* approximation of a model. An emulator provides an entire probability distribution for the function $Y = f(X_1, \ldots, X_k)$, and its approximation $g(X_1, \ldots, X_k)$ is given by the mean of that distribution. However, the emulator, also provides a distribution around the mean which gives information on how close is likely to be the 'true' $Y = f(X_1, \ldots, X_k)$ to the approximation $g(X_1, \ldots, X_k)$.
Entropy – A measure of the disorder of a system.
Equifinallity – Occurs if multiple parameter/factor set can all be seen as equally acceptable representations of the environmental system in question.
Error – Mistaken calculations or measurements with quantifiable and predictable differences.
Exact parameters – Which are universal constants, such as the mathematical constant: Pi (3.14259…).
Expectation – Expectation, or "expected value" of a variable, refers to the mean value the variable takes.
Extrapolation – The inference of unknown data from known data, for instance future data from past data, by analysing trends and making assumptions.
Factors – All model parameters (see parameters) and other variables which are used in a model (such as the input).
Fixed parameters – Which are well determined by experiment and may be considered exact, such as the acceleration of gravity, g (approximately 9.81 m/s).
Hierarchy – A process where information cascades from a greater spatial or temporal scale to lesser scale and vice versa.
Hypothesis – A proposal intended to explain certain facts or observations.

Identifiability – If a unique optimal factor set can be identified (also see equifinallity).
Ignorance – Lack of knowledge.
Independence (statistically) – See dependence.
Insignificant – See significant.
Institutional uncertainty – Inadequate collaboration and/or trust among institutions, potentially due to poor communication, lack of understanding, overall bureaucratic culture, conflicting sub-cultures, traditions and missions.
Integrated Water Resource Management – IWRM is a process which promotes the co-ordinated management and development of water, land and related resources, in order to maximise the resultant economic and social welfare in an equitable manner without compromising the sustainability of vital ecosystems.
Interaction – Two or more input factors have interactions when their effect on the model output cannot be represented simply by main effects, i.e. they act jointly, in a non-separable (non-additive) way.
Joint probability – The probability of specific values of one or more variables occurring simultaneously. For example, extreme water levels in estuaries may occur at times of high river flow, times of high sea level or times when both river flow and sea level are above average levels. When assessing the likelihood of occurrence of high estuarine water levels it is therefore necessary to consider the joint probability of high river flows and high sea levels.
Knowledge – Spectrum of known relevant information.
Knowledge uncertainty – Uncertainty due to lack of knowledge of all the causes and effects in a physical or social system. For example, a numerical model of wave transformation may not include an accurate mathematical description of all the relevant physical processes.
Least-square optimization – Mathematical optimization technique which, when given a series of measured data, attempts to find a function which closely approximates the data (a "best fit"). It attempts to minimize the sum of the squares of the ordinate differences between points generated by the function and corresponding points in the data (from wikipedia).
Likelihood – A general concept relating to the chance of an event occurring. Likelihood is generally expressed as a probability or a frequency.
Linear Combination – A linear combination of two quantities means simply their sum or their difference.
Measure of sensitivity – Way sensitivity is defined.
Meta-data – Means data about data. While this definition is commonly offered, it is also commonly not helpful. An example is a library catalog card, which contains data about the nature and location of a book: It is data about the data in the book referred to by the card (from wikipedia).

Metric Measure – See performance indicator

Model – An abstract construct to represent a system for the purposes of reproducing, simplifying, analyzing, or understanding it. The definition of a model can be broadly divided into perceptual, conceptual and procedural models.

Model factor – See factor.

Model, conceptual – The mathematical description of a perceptual model is a conceptual model. It is a construct of mathematical and logical statements that describe a complex system in quantitative terms; a carefully constructed, but sharply limited simulation of nature. It includes hypotheses and assumptions to simplify the processes.

Model, perceptual – Summary of our (personal) perceptions on how a system responds. Perceptual models are frameworks representing how a given theorist views the phenomena of concern to a discipline. People receive information, process this information, and respond accordingly many times each day. This sort of processing of information is essentially a perceptual model of how things in our surrounding environment work. The perceptual understanding of systems is far greater than most material model implementations.

Model, procedural – Converts a conceptual model essentially to a computer code for example the replacement of differentials of the original equation by finite-difference or finite-volume equivalents.

Monotonic – changing in one direction only; thus either strictly rising or strictly falling, but not reversing direction.

Monte carlo (analysis) – Is performed by multiple model executions.

Multicollinearity – The presence of high correlations between predictor variables.

Nash-Sutcliffe – Type of performance measure.

Natural variability – Uncertainties that stem from the assumed inherent randomness and basic unpredictability in the natural world and are characterised by the variability in known or observable populations.

Non-additive – See additive.

Non-Behavioural – See behavioural.

Nonlinear – See linear.

Non-monotonic – See monotonic.

Outlier – For a set of numerical data, any value that is significantly smaller or larger than other values.

Over-parameterisation – A model is over-parameterised when it is built with a larger number of parameters that it would be actually identifiable from data/available observations. Equifinality is also linked to over-parameterisation, in the sense that an identifiable, non over-parameterised does not present equifinality.

Parameters – The parameters in a model are the "values", chosen to represent the chosen context and scenario.
Performance – The degree to which a process or activity succeeds when evaluated against some stated aim or objective.
Performance measure – Informal likelihood. Measures the performance of model outcome against e.g. observations.
Precision – Degree of exactness regardless of accuracy.
Probabilistic method – Method in which the variability of input values and the sensitivity of the results are taken into account to give results in the form of a range of probabilities for different outcomes.
Probability – A measure of our strength of belief that an event will occur. For events that occur repeatedly the probability of an event is estimated from the relative frequency of occurrence of that event, out of all possible events. In all cases the event in question has to be precisely defined, so, for example, for events that occur through time reference has to be made to the time period, for example, annual exceedance probability. Probability can be expressed as a fraction, % or decimal. For example the probability of obtaining a six with a shake of four dice is 1/6, 16.7% or 0.167.
Probability density function (distribution) – Function which describes the probability of different values across the whole range of a variable (for example flood damage, extreme loads, particular storm conditions etc.).
Process model uncertainty – All models are an abstraction of reality and can never be considered true. They are thus subject to process model uncertainty. Measured data versus modelled data comparisons give an insight into the extent of model uncertainty but do not produce a complete picture.
Process model uncertainty – See Knowledge uncertainty.
Quasi Monte Carlo – Quasi-Monte Carlo methods are sampling methods that do not rely on probabilistic ideas and pseudo-random sequences for constructing the sample and analysing the output. Quasi-random samples are made of a finite subset out of infinite, low-discrepancy sequences of k-dimensional points. These samples are not random, in the sense of being completely unpredictable.
Residual – The difference between the measured and predicted values of some quantity.
River Basin – The area from which water runs off to a given river
Screening – Sensitivity method based on a 'cheap' design, that allows to identify the most sensitivity ones, out of a large number of input factors.
Sensitivity Analysis – Is the study of how the variation in the output of a model can be apportioned, qualitatively or quantitatively, to different sources of variation.
Significant – See insignificant.

Stakeholders – Parties/persons with a direct interest (stake) in an issue — also Stakeowners.
Statistical model uncertainty – See Knowledge uncertainty.
Statistical model uncertainty – Uncertainty associated with the fitting of a statistical model. The statistical model is usually assumed to be correct. However, if two different models fit a set of data equally well but have different extrapolations/interpolations then this assumption is not valid and there is statistical model uncertainty.
Stratified sampling – A sampling pattern in which the site is divided into (usually) non-overlapping sub-areas. Different sampling densities and sampling patterns are used in the different sub-areas.
System – An assembly of elements, and the interconnections between them, constituting a whole and generally characterised by its behaviour. Applied also for social and human systems.
System state – The condition of a system at a point in time.
Uncertainty – A general concept that reflects our lack of sureness about someone or something, ranging from just short of complete sureness to an almost complete lack of conviction about an outcome.
Uncertainty Analysis (Model) – Assesses the uncertainty in model outputs that derives from uncertainty in structure, parameters, boundary conditions and evaluation data.
Uniform Distribution – The probability of a number falling in a particular interval is proportional only to the size of the interval.
Validation – is the process of comparing model output with observations of the 'real world'.
Variability – The change over time of the value or state of some parameter or system or element where this change may be systemic, cyclical or exhibit no apparent pattern.
Variable – A quantity which can be measured, predicted or forecast which is relevant to describing the state of the flooding system e.g. water level, discharge, velocity, wave height, distance, or time. A prediction or forecast of a variable will often rely on a simulation model which incorporates a set of parameters.
Verification – See validation.
Water Framework Directive – A European Directive that aims to establish a framework for the protection of inland surface waters (rivers and lakes), transitional waters (estuaries), coastal waters and groundwater.

7. REFERENCES

Atherton, R.W., Schainker, R.B. and Ducot, E.R. (1975). Statistical sensitivity analysis of models for chemical-kinetics. *AIChE Journal*, **21**, 441–448.

Bedford, T. and Cooke, R.M. (2002). Vines – A new graphical model for dependent random variables. *Annals of Statistics*, **30**(4), 1031–1068.

Beven, K.J. (2007). Environmental Modelling: An Uncertain Future? Wiley, London.

Beven, K.J. and Binley, A. (1992). The future of distributed models – model calibration and uncertainty prediction. *Hydrological Processes*, **6**(3), 279–298.

Beven, K.J. (1993). Prophecy, reality and uncertainty in distributed hydrological modeling. *Advances in Water Resources*, **16**(1), 41–51.

Box, G.E.P., Hunter, W.G. and Hunter, J.S. (1978). Statistics for experimenters. John Wiley and Sons, New York.

Bryson, A. and Ho, Y. (1969). Applied Optimal Control. Blaisdell, Waltham, Mass.

Bushenkov, V.A., Gusev, D.V., Kamenev, G.K., Lotov, A.V. and Chernykh, O.L. (1994). Visualization of Pareto Set in Multiattribute Choice Problem. *Doklady Akademii Nauk*, **335**(5), 567–569.

Cacuci, Dan G. (2003). In Sensitivity and Uncertainty Analysis, Volume I: Theory; Chapman and Hall.

Cacuci, Dan G., Ionescu-Bujor, M. and Navon, M. (2005). Sensitivity and Uncertainty Analysis: Applications to Large-Scale Systems (Volume II), Chapman and Hall.

Campolongo, F., Cariboni, J. and Saltelli, A. (2007). An effective screening design for sensitivity analysis of large models. *Environmental Modelling and Software* **22**, 1509–1518.

Chang, F.J. and Delleur, J.W. (1992). Systematic parameter-estimation of watershed acidification model. *Hydrological Processes*, **6**(1), 29–44.

Cooke, R.M. and van Noortwijk, J.M. (2000). Graphical methods. In: A. Saltelli, K. Chan and E.M. Scott (Editors), Sensitivity Analysis. John Wiley and Sons, Chichester, pp. 245–264.

Cukier, R.I., Fortuin, C.M., Shuler, K.E., Petschek, A.G. and Schaibly, J.H. (19730). Study of sensitivity of coupled reaction systems to uncertainties in rate coefficients 1. Theory. *Journal of Chemical Physics*, **59**(8), 3873–3878.

Cukier, R.I., Levine, H.B. and Shuler, K.E. (1977). Nonlinear sensitivity analysis of multiparameter model systems. *Journal of Physical Chemistry*, **81**(25), 2365–2366.

Cukier, R.I., Schaibly, J.H. and Shuler, K.E. (1975). Study of sensitivity of coupled reaction systems to uncertainties in rate coefficients 3. Analysis of Approximations. *Journal of Chemical Physics*, **63**(3), 1140–1149.

Cullen, A.C. and Frey, H.C. (1999). Probabilistic techniques in exposure assessment : a handbook for dealing with variability and uncertainty in models and inputs. Plenum Press, New York: London, xvi, 335p.

Durrell, W.R. (1985). Data Administration: A Practical Guide to Data Administration. McGraw-Hill, New York.

EC, (2004a). Common Implementation Strategy for the Water Framework Directive (2000/60/EC) Guidance Document No 11, Planning Processes. European Communities. Available on http://forum.europa.eu.int/Public/irc/env/wfd/library?l = /framework_directive/guidance_documents/0publishedsguidancesdocuandvm = detailedandsb = Title

EC, (2004b). Common Implementation Strategy for the Water Framework Directive (2000/60/EC) Guidance Document No 3, pressures and impactes, IMPRESS. Available on http://forum.europa.eu.int/Public/irc/env/wfd/library?l=/framework_directive/guidance_documents/0publishedsguidancesdocuandvm=detailedandsb=Title

Fraedrich, D. and Goldberg, A. (2000). A methodological framework for the validation of predictive simulations. *European Journal of Operational Research*, **124**, 55–62.

Freer, J.E., Beven, K.J. and Peters, N.E. (2003). Multivariate seasonal period model rejection within the generalised likelihood uncertainty estimation procedure. In: Q. Duan, H. Gupta, S. Sorooshian, A.N. Rousseau and R. Turcotte (Editors), Calibration of watershed models. AGU, Water Science and Application Series, Washington, pp. 69–88.

Frey, H.C. and Patil, S.R. (2002). Identification and review of sensitivity analysis methods. *Risk Analysis*, **22**(3), 553–578.

Frey, H.C., Mokhtari, A. and Zheng, J. (2004). Recommended Practice Regarding Selection, Application, and Interpretation of Sensitivity Analysis Methods Applied to Food Safety Process Risk Models, Prepared by North Carolina State University for U.S. Department of Agriculture, Washington, D.C. 148 pp.

Grievank, A. (2000). Evaluating derivatives, Principles and techniques of algorithmic differentiation. SIAM.

Hastie, T. and Tibshirani, R. (1990). Generalized Additive Models. Chapman and Hall.

Helton, J.C. (1993). Uncertainty and sensitivity analysis techniques for use in performance assessment for radioactive-waste disposal. *Reliability Engineering and System Safety*, **42**(2–3), 327–367.

Helton, J.C., Davis, F.J. and Johnson, J.D., (2005). A comparison of uncertainty and sensitivity analysis results obtained with random and Latin hypercube sampling. *Reliability Engineering and System Safety*, **89**(3), 305–330.

Helton, J.C., Johnson, J.D., Sallaberry, C.J. and Storlie, C.B. (2006). Survey of sampling-based methods for uncertainty and sensitivity analysis. *Reliability Engineering and System Safety*, **91**(10–11), 1175–1209.

Helton, J.C., Johnson, J.D., Sallaberry, C.J. and Storlie, C.B., (2006). Survey of sampling-based methods for uncertainty and sensitivity analysis. *Reliability Engineering and System Safety*, **91**, 1175–1209

Homma, T. and Saltelli, A. (1996). Importance measures in global sensitivity analysis of nonlinear models. *Reliability Engineering and System Safety*, **52**(1), 1–17.

Hora, S.C. and Iman, R.L. (1986). A comparison of maximum/bounding and Bayesian/Monte Carlo for fault tree uncertainty analysis, Sandia Laboratories Report, SAND85-2839.

Hornberger, G.M. and Cosby, B.J. (1985). Selection of Parameter Values in Environmental-Models Using Sparse Data – a Case-Study. *Applied Mathematics and Computation*, **17**(4), 335–355.

Hornberger, G.M. and Spear, R.C. (1981). An approach to the preliminary analysis of environmental systems. *Journal of Environmental Management*, **12**, 7–18.

Horritt, M.S. (2006). A methodology for the validation of uncertain flood inundation models. *Journal of Hydrology*, **326**, 153–165.

Hwang, J.T., Dougherty, E.P., Rabitz, S. and Rabitz, H. (1978). The Green's function method of sensitivity analysis in chemical kinetics. *Journal of Chemical Physics*, **69**, 5180–5191.

Iman, R.L. and Conover, W.J. (1980). Small sample sensitivity analysis techniques for computer models, with an application to risk assessment. *Communications in Statistics*, A9, 1749–1842.

Iman, R.L. and Hora, S.C. (1990). A robust measure of uncertainty importance for use in fault tree system analysis. *Risk Analysis*, **10**(3), 401–406.

Iman, R.L. (1992). Uncertainty and sensitivity analysis for computer modeling applications. *Reliability Technology*, **28**, 153–168.

Iman, R.L., Helton, J.C. and Campbell, J.E. (1981). An approach to sensitivity analysis of computer models, Part I. *Journal of Quality Technology*, **13**, 174–183.

Iman, R.L., Shortencarier, M.J. and Johnson, J.D. (1985). A FORTRAN 77 program and user'sguide for the calculation of partial correlation and standardized regression coefficients, Sandia National Laboratories, Albuquerque, NM.

Ishigami, T. and Homma, T. (1990). An importance quantification technique in uncertainty analysis for computer models, ISUMA '90. First International Symposium on Uncertainty Modelling and Analysis, University of Maryland, USA, p. 398.

Kalman, R. (1960). New approach to linear filtering and prediction problems. *ASME Trans., Journal Basic Eng.*, **82-D**, 35–45.

Kendall, M.G. and Stuart A., (1961). The Advanced Theory of Statistics, Vol. 2, Griffin, London.

Kennedy M.C. and O'Hagan A. (2001). Bayesian calibration of computer models, *J. Royal Statistical Society B.*, **63**, 425–464.

Kleijnen, J. (2007a). DASE: Design and analysis of simulation experiments. Springer Science + Business Media.

Kleijnen, J. (2007b). Kriging metamodeling in simulation: a review. CentER discussion paper.

Koda, M., Dogru, A.H. and Seinfeld, J.H. (1979). Sensitivity analysis of partial differential equations with application to reaction and diffusion processes. *Journal of Computational Physics*, **30**, 259–282.

Krykacz-Hausmann, B. (2001). Epistemic sensitivity analysis based on the concept of entropy. In: P. Prado and R. Bolado (Editors), Proceedings of SAMO2001, Madrid, pp. 31–35.

Krzykacz-Hausmann, B. (1990). Technical Report GRS-A-1700, Gesellschaft fuer Reaktor Sicherheit (GRS) MbH, Garching.

Kullback, S. and Leibler, R.A. (1951). One Information and Sufficiency. *The Annals of Mathematical Statistics*, **22**(1), 79–86.

Lence, B.J. and Takyi, A.K. (1992). Data requirements for seasonal discharge programs – an application of a regionalized sensitivity analysis. *Water Resources Research*, **28**(7), 1781–1789.

Li, G., Hu, J., Wang, S.-W., Georgopoulos, P.G., Schoendorf, J. and Rabitz, H. (2006). Random Sampling-High Dimensional Model Representation (RS-HDMR) and orthogonality of its different order component functions. *Journal of Physical Chemistry A*, **110**, 2474–2485.

Li, G., Wang, S.-W. and Rabitz, H. (2002). Practical approaches to construct RS-HDMR component functions. *Journal of Physical Chemistry*, **106**, 8721–8733.

Liu, H., Chen, W. and Sudjianto, A. (2004). Relative entropy based method for global and regional sensitivity analysis in probabilistic design, Proceedings of DETC'04 ASME

2004 International Design Engineering Technical Conferences and Computers and Information in Engineering Conference Salt Lake City, Utah.

Manache, G. and Melching, C.S. (2004). Sensitivity analysis of a water-quality model using Latin hypercube sampling, *Journal of Water Resources Planning and Management-Asce*, **130**, 232–242.

McCamley, F. and Rudel, R.K. (1995). Graphical sensitivity analysis for generalized stochastic dominance. *Journal of Agricultural and Resource Economics*, **20**(2), 403–403.

McKay, M.D., Beckman, R.J. and Conover, W.J. (1979). A comparison of three methods for selecting values of input variables in the analysis of output from a computer code., *Technometrics*, **21**, 239–245.

McKay, M.D. (1996). Variance-based methods for assessing uncertainty importance in NUREG-1150 analyses. LA-UR96-2695.

McRae, G.J., Tilden, J.W. and Seinfeld, J.H. (1982). Global sensitivity analysis – a computational implementation of the fourier amplitude sensitivity test (Fast). *Computers and Chemical Engineering*, **6**(1), 15–25.

Morris, M.D. (1991). Factorial sampling plans for preliminary computational experiments, *Technometrics*, **33**, 161–174.

Neter, J., Kutner, M.H., Nachtsheim, C.J. and Wasserman, W. (1996). Applied Linear Statsitical models. McGraw-Hill, New-York.

O'Hagan A. (2006). Bayesian analysis of computer code outputs: A tutorial. *Reliability Engineering and System Safety*, **91,** 1290–1300.

Oakley, J.E. and O'Hagan, A. (2004). Probabilistic sensitivity analysis of complex models: a Bayesian approach. *Journal Of The Royal Statistical Society Series B-Statistical Methodology*, **66**, 751–769.

Osidele, O.O. and Beck, M.B. (2004). Food web modelling for investigating ecosystem behaviour in large reservoirs of the south-eastern United States: lessons from Lake Lanier, Georgia. Ecological Modelling, **173**(2–3), 129–158.

Pappenberger, F. and Beven, K. (2004). Functional classification and evaluation of hydrographs based on multicomponent mapping. *International Journal of River Basin Management*, **2**(2), 89–100.

Pappenberger, F., Beven, K.J., Ratto, M. and Matgen, P. (2008). Multi-Method global sensitivity analysis of flood inundation models. *Advances in Water Resources*, **31**, 1–14.

Pappenberger, F., Beven, K., Frodsham, K., Romanowicz, R. and Matgen, P. (2007). Grasping the unavoidable subjectivity in calibration of flood inundation models: A vulnerability weighted approach, *Journal of Hydrology*, **333**, 275–287.

Pappenberger, F., Buizza, R., Bodis, K. and Ghelli, A. (2009). The skill of probabilistic forecasts under observational uncertainties within the Generalized Likelihood Uncertainty Estimation framework for hydrological applications, *Journal of Hydrometeorology*, **10**, 807–819.

Pappenberger, F., Iorgulescu, I. and Beven, K.J. (2006). Sensitivity Analysis based on Regional Splits (SARS – RT). *Environmental Modelling and Software*, **21**(7), 976–990.

Pappenberger, F., Matgen, P., Beven, K. J., Henry, J.-B., Pfister, L. and Fraipont de, P. (2006). Influence of uncertain boundary conditions and model structure on flood inundation predictions, *Advances in Water Resources*, **29**, 1430–1449.

Pappenberger, F., Ratto, M. and Beven, K.J. (submitted) Factor mapping to identify functional classed in flood inundation modelling.

Patil, S.R. and Frey, H.C. (2004). Comparison of sensitivity analysis methods based on applications to a food safety risk assessment model. *Risk Analysis*, **24**(3), 573–585.

Pearson, K. (1905). On th general theory of skew correlation and nonlinear regression. *Proceedings of the Royal Society London*, **71**, 303.

Rabitz, H. and Aliş, O.F. (2000). Managing the Tyranny of Parameters in Mathematical Modelling of Physical Systems [HDMR]. In: A. Saltelli, K. Chan and M. Scott (Editors), Sensitivity Analysis. John Wiley and Sons, New York.

Rabitz, H., Alis, O.F., Shorter, J. and Shim, K. (1999). Efficient input-output model representations. *Computer Physics Communications*, **117**(1–2), 11–20.

Ratto, M., Pagano, A., and Young, P.C. (2007). State Dependent Parameter meta-modelling and sensitivity analysis. *Computer Physics Communications*, **177**, 863–876.

Ratto, M., Tarantola, S., Saltelli, A. and Young, P.C. (2004). Accelerated estimation of sensitivity indices using State Dependent Parameter models. In: K. M. Hanson and F.M. Hemez (Editors), Sensitivity Analysis of Model Output, Proceedings of the 4th International Conference on Sensitivity Analysis of Model Output (SAMO 2004) Santa Fe, New Mexico, March 8–11, 2004., pages 61–70. http://library.lanl.gov/ccw/samo2004/.

Ratto, M., Tarantola, S., Saltelli, A. and Young, P.C. (2006). Improved and accelerated sensitivity analysis using State Dependent Parameter models. Technical Report EUR 22251 EN, ISBN 92-79-02036-6, Joint Research Centre, European Commission. http://webfarm.jrc.cec.eu.int/uasa/doc/mrat/eur/EUR22251EN.pdf.

Sacks J., Welch W.J., Mitchell T.J. and Wynn H.P. (1989). Design and analysis of computer experiments, *Statistical Science*, **4,** 409–435.

Saltelli, A. (2002). Making best use of model valuations to compute sensitivity indices. *Computer Physics Communications*, **145**, 280–297.

Saltelli A. Tarantola S., Campolongo, F. and Ratto, M. (2004). Sensitivity Analysis in Practice. A Guide to Assessing Scientific Models, John Wiley and Sons Publishers.

Saltelli, A. and Tarantola, S. (2002). On the relative importance of input factors in mathematical models: Safety assessment for nuclear waste disposal. *Journal of the American Statistical Association,* **97**(459), 702–709.

Saltelli, A., Andres, T.H. and Homma, T. (1993). Sensitivity Analysis of Model Output: An Investigation of New Techniques. *Computational Statistics and Data Analysis*, **15**, 211.

Saltelli, A., Chan, K. and Scott, M. (2000a). Sensitivity Analysis. John Wiley and Sons, New York.

Saltelli, A., Ratto, M., Tarantola, S. and Campolongo, F. (2005). Sensitivity analysis for chemical models. *Chemical Reviews*, **105**(7), 2811–2827.

Saltelli, A., Tarantola, S., Campolongo, F. and Ratto, M. (2004). Sensitivity Analysis in Practice – A Guide to Assessing Scientific Models. John Wiley and Sons, Chichester.

Saltelli, A., Ratto, M., Andres, T., Campolongo, F., Cariboni, J., Gatelli, D., Saisana, M., and Tarantola, S. (2007). Global sensitivity analysis. Gauging the worth of scientific models. Wiley.

Saltelli, A., Tarantola, S. and Campolongo, F. (2000b). Sensitivity analysis as an ingredient of modeling. *Statistical Science*, **15**(4), 377–395.

Saltelli, A., Tarantola, S. and Chan, K.P.S. (1999). A quantitative model-independent method for global sensitivity analysis of model output. *Technometrics*, **41**(1), 39–56.

Schaibly, J.H. and Shuler, K.E. (1973). Study of Sensitivity of Coupled Reaction Systems to Uncertainties in Rate Coefficients 2. Applications. *Journal of Chemical Physics*, **59**(8), 3879–3888.

Schweppe, F. (1965). Evaluation of likelihood functions for Gaussian signals. *IEEE Trans. on Information Theory*, **11**, 61–70.

Seibert, J. and McDonnell, J.J. (2002). Multi-criteria calibration of conceptual runoff models – the quest for improved dialogue between modeler and experimentalist. In: Q. Duan, S. Sorooshian, H. Gupta, A.N. Rousseau and D.S. Turcotte (Editors), Advances in Calibration of Watershed Models. American Geophysical Union, Washington.

Sobol, I.M. (1993). Sensitivity estimates for nonlinear mathematical models. *Mathematical Modelling and Computational Experiment*, **1**, 407–414.

Spear, R.C. and Hornberger, G.M. (1980). Eutrophication in Peel Inlet, II, Identification of critical uncertainties via Generalised Sensitivity Analysis. *Water Resource Research*, **14,** 43–49.

Spear, R.C., Grieb, T.M. and Shang, N. (1994). Parameter uncertainty and interaction in complex environmental-models. *Water Resources Research*, **30**(11), 3159–3169.

Storlie, C.B. and Helton, J.C. (2008a). Multiple predictor smoothing methods for sensitivity analysis: Description of techniques, *Reliability Engineering and System Safety*, **93**, 28–54.

Storlie, C.B. and Helton, J.C. (2008b). Multiple predictor smoothing methods for sensitivity analysis: Example results, *Reliability Engineering and System Safety*, **93**, 55–77.

Storlie, C. B., Bondell, H. D., Reich, B. J. and Zhang, H. H. (2009a). Surface estimation, variable selection, and the nonparametric oracle property. Statistica Sinica. In Press.

Storlie, C.B., Swiler, L.P., Helton, J.C. and Sallaberry, C.J. (2009b). Implementation and Evaluation of Nonparametric Regression Procedures for Sensitivity Analysis of Computationally Demanding Models, *Reliability Engineering and System Safety Reliability Engineering and System Safety*, **94**, 1735–1763

Tarantola S., Gatelli, D. and Mara, T. (2006). Random Balance Designs for the Estimation of First Order Global Sensitivity Indices, *Reliability Engineering and System Safety*, **91,** 717–727.

Valko, P. and Vajda, S. (1984). An extended ODE solver for sensitivity calculations. *Computers and Chemistry*, **8**, 255–271.

van Straten, G. and Keesman, J. (1991). Uncertainty propagation and speculation in projective forecasts of environmental change: a lake eutrophication example. *Journal of Forecasting*, 10, 163–190.

Wagener, T. (2003). Evaluation of catchment models, *Hydrological Processes*, **17**, 3375–3378.

Wagener, T. and Gupta, H. V. (2005). Model identification for hydrological forecasting under uncertainty, *Stochastic Environmental Research and Risk Assessment*, **19**, 378–387.

Wagener, T., Boyle, D. P., Lees, M. J., Wheater, H. S., Gupta, H. V. and Sorooshian, S. (2001). A framework for development and application of hydrological models, *Hydrology and Earth System Sciences*, **5**, 13–26.

Wagener, T., McIntyre, N., Lees, M.J., Wheater, H.S. and Gupta, H.V. (2003). Towards reduced uncertainty in conceptual rainfall-runoff modelling: dynamic identifiability analysis. *Hydrological Processes*, **17**(2), 455–476.

Wagener, T., Wheater, H.S. and Gupta, H.V. (2002). Advances in calibration of watershed models. In: Q., Duan, S., Sorooshian, H.V., Gupta, A. Rousseau, and R. Turcotte (Editors), AGU Monograph, Washington.

Wegman, E.J. (1990). Hyperdimensional data-analysis using parallel coordinates. *Journal of the American Statistical Association*, **85**(411), 664–675.

Weijers, S.R. and Vanrolleghem, P.A. (1997). A procedure for selecting best identifiable parameters in calibrating activated sludge model no.1 to full-scale plant data, *Water Science and Technology*, **36**, 69–79.

Young, P.C. and Ng, C.N. (1989). Variance intervention. *Journal of Forecasting*, **8**, 399–416.

Young, P.C. (1978). A general theory of modelling for badly defined dynamic systems. In *Modeling, Identification and Control in Environmental Systems*, G.C. Vansteenkiste (Editors), North Holland, Amsterdam, 103–135.

Young, P.C. (1999). Data-based mechanistic modelling, generalised sensitivity and dominant mode analysis. *Computer Physics Communications*, **117**(1–2), 113–129.

Young, P.C. (2000). Stochastic, dynamic modelling and signal processing: Time variable and state dependent parameter estimation. In W. J. Fitzgerald *et al.*, (Editor), *Nonlinear and Nonstationary Signal Processing*, pp. 74–114. Cambridge University Press, Cambridge.

Young, P.C. (2001) The identification and estimation of nonlinear stochastic systems. In A. I. Mees *et al.* (Editors), *Nonlinear Dynamics and Statistics*. Birkhauser, Boston.

Young, P.C., Parkinson, S. and Lees, M. (1996). Simplicity out of complexity in environmental modelling: Occam's razor revisited. *Journal of Applied Statistics*, **23**, 165–210.

Zeng, X.Q. and Yeung, D.S. (2003). A quantified sensitivity measure for multilayer perceptron to input perturbation. *Neural Computation*, **15**(1), 183–212.

Arnold, J.G., Williams, J.R., Srinivasan, R. and King, K.W. (1996). SWAT manual, USDA, Agricultural Research Service and Blackland Research Center, Texas, USA.

Reichert, P., Borchardt, D., Henze, M., Rauch, W., Shanahan, P., Somlyody, L. and Vanrolleghem, P.A. (2001b). River water quality model No. 1 (RWQM1). Scientific and Technical Report No. 12. IWA Publishing.

van Griensven, A. and Bauwens, W. (2001). Integral water quality modelling of catchments. *Water Science and Technology*, **43**(7), 321–328.

van Griensven, A. and Bauwens, W. (2003). Multi-objective auto-calibration for semi-distributed water quality models. *Water Resources Research*, **39**(10), 1348–1358.

8. APPENDIX

The appendix has to be used in conjunction with the text of each corresponding section above. It gives additional pointers which have been deemed too technical for the more general description.

Mathematical models are represented as a mapping $Y = f(X_1,\ldots,X_k)$, where X_i's are the uncertain input factors. The X_i's are treated as random variables characterised by specified distributions, implying that the output Y is also a random variable, with a probability distribution whose characterisation is the object of uncertainty analysis.

For illustration purposes we give a relatively large room to the description of the variance-based and Sobol' methods. This is due to their fundamental role in the sensitivity methods and because this allows discussing with simplicity and clarity most of the fundamental global sensitivity concepts. Moreover, variance-based sensitivity indices are considered as a term of reference to understand the meaning of other methodologies, like Morris (compared to total effect), Regionalised Sensitivity analysis and entropy-based measures (compared to main and total effects).

8.1. Screening methods

8.1.1. Morris

The elementary effect e_i of the x_i parameter is then defined as:

$$e_i = \frac{y^* - y}{\Delta} \tag{8.1}$$

This effect is quantitative, elementary and exclusive for the parameter i. The sensitivity of y to x_i depends on the values chosen for the remaining parameters, and this single result is only a sample of its sensitivity, so one can repeat this experiment for several sets of input parameters obtaining a sample from the distribution of elementary effects. The Morris sample is taken on a grid of p-levels within the k-dimensional hypercube of the input factors space. Sensitivity measures can be then calculated out of this sample. In the original Morris (1991) work, the average μ and the standard deviation σ of the sample of elementary effects is used. The mean μ assesses the overall influence of the factor on the output. The standard deviation σ estimates the ensemble of the factor's effects, whether nonlinear and/or due to interactions with other factors. Campolongo *et al.* (2007) introduced the average of the *absolute value* of elementary effects μ^* (see also Saltelli *et al.*, 2004, 2007), which turns out to be the only measure of interest for screening (and FF) purposes and to be an excellent proxy to the variance-based total effect.

8.2. Local methods

The general form of the systems that will be discussed is given by following equation:

$$\frac{dy}{dt} = f(y, x) \tag{8.2}$$

The sensitivity of a state variable y to a parameter x can be expressed as a sensitivity function (8.3). A state variable y is called sensitive to x if small changes in x produce significant changes in y. On the other hand, a variable y is called insensitive to x if changes in x produce insignificant changes in y.

$$S(t) = \frac{\partial y(t)}{\partial x} \tag{8.3}$$

This partial derivative can be analytically solved if the analytical solution of Equation (8.3) is known. Unfortunately, this is rarely the case and numerical methods have to be used in order to approximate the sensitivity function. Since sensitivity functions are used in several applications like optimal experimental design, model reduction, etc., it is of great importance that a correct approximation of the sensitivity function is calculated in order for the applications, in which the sensitivity functions are used, to produce accurate results. Local sensitivity analysis techniques evaluate this partial derivative at one specific set of parameter values, also called the nominal parameter set.

8.2.1. The finite difference method

The partial derivative defined in Equation (8.3) can be mathematically formulated by the equation given below (forward difference):

$$\frac{\partial y_i}{\partial x_j} = \lim_{\Delta x_j \to 0} \frac{y_i\left(t, x_j + \Delta x_j\right) - y_i\left(t, x_j\right)}{\Delta x_j} \tag{8.4}$$

This equation is only valid if an infinitesimal variation (perturbation) of the parameters, inputs or initial conditions x ($\Delta x_j \to 0$) is considered. Equation (8.4) shows that the application of the finite difference method requires the solution of the model (8.2) using the nominal value of the parameters y_i (t,x_j) and k

solutions of the equations using perturbed parameters $y_i(t, x_j + \Delta x_j)$. It should be noted that only one parameter is perturbed at a time while all others are kept at their nominal value. The sensitivities obtained actually belong to the $\left(x + \frac{\Delta x}{2}\right)$ parameter set because Equation (8.4) can also be seen as the average of the sensitivities of the model output y_i at x_j and Δx_j. Equation (8.4) can be modified into the central difference formula (8.5) which requires $2k$ model evaluations.

$$\frac{\partial y_i}{\partial \theta_j} \approx \frac{y_i\left(t, \theta_j + \Delta\theta_j\right) - y_i\left(t, \theta_j - \Delta\theta_j\right)}{2\Delta\theta_j} \tag{8.5}$$

The finite difference technique can be too computationally intensive, especially in cases where the number of parameters is large.

8.2.2. Direct method for sensitivity analysis

Atherton *et al.* (1975) developed the direct method for sensitivity analysis. Differentiation of Equation (8.2) with respect to x_j yields the following set of sensitivity differential equations. The solution of these equations results in the sensitivity functions:

$$\frac{d}{dt}\frac{\partial y}{\partial x_j} = \frac{\partial f}{\partial y}\frac{\partial y}{\partial x_j} + \frac{\partial f}{\partial x_j} \tag{8.6}$$

The term $\frac{\partial f}{\partial y}$ of equation (8.6) is recognized as the Jacobian J of the original system (8.2). Equations (8.2) and (8.6) are coupled through $\frac{\partial f}{\partial y}$ and $\frac{\partial f}{\partial x}$. This means that the solution of Equation (8.6) requires the knowledge of the solution of Equation (8.2) in all points where the ODE (ordinary differential equation) solver calculates the right-hand side of Equation (8.6). A link between these two sets of equations can be made in one of the following ways.

(i) Solve Equations (8.2) and (8.6) for every parameter x simultaneously, which requires the solution of $(k+1)n$ ODEs. The direct solution of this large system is inefficient. Although a solver has been built for these systems, based on the decomposed direct method (Valko and Vajda, 1984), it is not used very often.

(ii) Solve the couple of Equations (8.2) and (8.6), which requires the solution of $2n$ ODEs k times (for every parameter). This version is the simplest to code, but is the least economical and might cause numerical problems due to ill-conditioned Jacobian matrices.

(iii) The solution of Equations (8.2) and (8.6) can be decoupled. First, differential Equation (8.2) are solved and the results are stored in a table. Then Equation (8.6) are solved using the values stored in the table. If values are required at times without tabulated values, they are computed by interpolation.

8.2.3. Green function's method

The difficulties of solving the large sets of differential equations led to the development of the Green's function method, also called the variational method. This method makes use of the fact that the sensitivities can be expressed in integral, rather than differential form.

$$\frac{\partial y_t}{\partial x} = K(t,0)\frac{\partial y(t)}{\partial x}(0) + \int_0^t K(t,\tau)\frac{\partial f(\tau)}{\partial x}d\tau \tag{8.7}$$

In this equation, $K(t,\tau)$ is an $n \times n$ Green's function matrix or kernel, given by

$$\frac{d}{dt}K(t,\tau) - J(t)K(t,\tau) = 0 \; t > \tau \tag{8.8}$$

$$K(\tau,\tau) = I$$

There are a number of variations of the Green's function method and they differ among each other in the calculation of the matrix K. Probably the most used method is the GFM/AIM method (Analytically Integrated Magnus). This method approximates K by a matrix exponential. It is beyond the scope of this document to go into more detail on these solving techniques. In all Green's function methods, the numerical effort is proportional to the number of variables and not the number of parameters. So this method should be preferred when the number of parameters is large compared to the number of variables. When the number of variables is much larger than the number of parameters, direct differential methods should be used.

8.3. Global methods

8.3.1. Variance based methods

The two key measures are the main effect

$$V_i = V[E(Y|X_i)] \tag{8.9}$$

and the total effect

$$V_{Ti} = E[V(Y|\mathbf{X}_{-i})] \tag{8.10}$$

where X_{-i} indicates the array of all input factors except X_i, V and E denote variance and expectation operators.

As formalised by Sobol' (1993) the main effect indices are the first order terms in the decomposition of the unconditional variance, which for independent factors can be written as

$$V(Y) = \sum_i V_i + \sum_{\substack{i,j \\ i<j}} V_{ij} + \sum_{\substack{ijl \\ i,j \\ j<l}} V_{ijk} + \cdots + V_{123..k} \tag{8.11}$$

where

$$\begin{aligned} V_i &= V_{X_i}(E_{\mathbf{X}_{-i}}(Y|X_i)) \\ V_{ij} &= V_{X_iX_j}(E_{\mathbf{X}_{-ij}}(Y|X_i, X_j.)) - V_i - V_j \\ V_{ijl} &= V_{X_iX_iX_i}(E_{\mathbf{X}_{-ijl}}(Y|X_i, X_j, X_l.)) - V_{ij} - V_{jl} - V_{il} - V_i - V_j - V_l \\ &\ldots \end{aligned} \tag{8.12}$$

Note that in writing e.g. $V_{X_iX_j}(E_{\mathbf{X}_{-ij}}(Y|X_i, X_j.))$ we mean that the inner expectation is over all factors but X_i, X_j and the outer variance is over X_i, X_j.

All measures are usually normalised by the unconditional variance of Y, to obtain the sensitivity indices, scaled between 0 and 1:

$$\begin{aligned} S_i &= V_i/V(Y) \\ S_{Ti} &= V_{Ti}/V(Y) \end{aligned} \tag{8.13}$$

The S_i spectrum is sufficient to characterise the entire sensitivity pattern of Y only for additive models, for which

$$\sum_{i=1}^{k} S_i = 1 \tag{8.14}$$

Equation (8.14) tells us that for an additive model all the variance of the model Y can be explained in terms of first order effects. To make a trivial example, $Y = \sum_{i=1}^{k} \alpha_i X_i^2$ is non linear, and non-monotonic if the distribution function of the X_i's contains the zero. Yet this is an additive model, for which $V(Y) = \sum_i V_i$ and $\sum_i S_i = 1$. Models for which (8.14) does not hold are termed non-additive. For non-additive models $\sum_{j=1}^{r} (S_j) \leqslant 1$ and these models are characterised by the existence of so-called interaction effects, leading to the most general variance decomposition scheme (8.11) that becomes, after normalisation by V:

$$\sum_i S_i + \sum_i \sum_{j>i} S_{ij} + \sum_i \sum_{j>i} \sum_{l>j} S_{ijl} + \cdots S_{12\ldots k} = 1 \qquad (8.15)$$

The complete decomposition (8.15) comprises an array of $2^k - 1$ sensitivity terms, giving rise to the so-called 'curse of dimensionality', since the expression and its decomposition is neither cheap to compute, nor does it provide a succinct and easily readable portrait of the model characteristics. In this context, total indices provide the major part of the information needed to complement the main effects, with only k additional indices to be estimated. Total indices measure the overall effect of input factors, including both main effects and all possible interactions with any other input factor. Returning to the additivity of models, when a model is additive, we will have $S_i = S_{Ti}$ for all X_i's; while, in general, $S_i \leq S_{Ti}$ and the difference between main and total effects is due to all interaction terms involving X_i.

Let us consider Example 2. The total variance of the g-function is:

$$V = \prod_j \frac{1/3 + (1 + a_j)^2}{(1 + a_j)^2} - 1 \qquad (8.16)$$

and the main effects of the g-function are:

$$V_i = V[E(Y|X_i)] = \frac{(1/3 + (1 + a_j)^2)}{(1 + a_j)^2} - 1 \qquad (8.17)$$

So, we can compute the sensitivity indices $S_i = \frac{V_i}{V}$.

In order to compute total effects, one first computes the effect of all factors except X_i:

$$V_{-i} = V[E(Y|\mathbf{X}_{-i})] = \prod_{j \neq i} \frac{(1/3 + (1 + a_j)^2)}{(1 + a_j)^2} - 1 \qquad (8.18)$$

and then $S_{Ti} = 1 - \frac{V_{-i}}{V}$,

The analytical values of the main and total effects for Example 2 are hence easily computed:

Factor	S_i	S_{Ti}
X_1	0.7165	0.7871
X_2	0.1791	0.2422
X_3	0.0237	0.0343
X_4	0.0072	0.0105
$X_{5,6,7,8}$	0.0001	0.0001

One can also compute specific group effects $S_I^c = \frac{V_I^c}{V}$, where the members of the group are indexed by $I = i_1, \ldots, i_g$ I, using:

$$V_I^c = V[E(Y|\mathbf{X}_I)] = \prod_{j \in I} \frac{(1/3 + (1 + a_j)^2)}{(1 + a_j)^2} - 1$$

which equals the sum of all sensitivity indices containing *only* elements of I, and where c indicates that the effect is 'closed' over the elements of the group I.

The main effects and total effects are also strictly linked to two important sensitivity settings that are extremely relevant in the calibration context: *Factors prioritization* and *Factors fixing*.

FAST

When using FAST, S_i is computed by exploring the k-dimensional space of the input factors with a search curve defined by a set of parametric equations

$$x_i = G_i(\sin(\omega_i s)) \tag{8.19}$$

with $i = 1, 2, \ldots k$, where s is a scalar varying in $(-\infty, +\infty)$ and the ω_i are a set of different angular frequencies associated with each factor and the G_i are properly selected transformation function. Scanning Equation (8.19) for different values of s results in a curve in the k-dimensional hypercube whereby each dimension is explored with a different frequency ω_i. Fourier analysis allows then the computation of $V_{X_i}(E_{\mathbf{X}_{-i}}(Y|X_i.))$ based on the signal at ω_i and its harmonics. The implementation of the method requires care, mostly in avoiding interferences, based on accurate selection of the set of k frequencies ω_i (Koda *et al.*, 1979; McRae *et al.*, 1982). Extensions of the FAST method are described in Saltelli *et al.* (1999) and Tarantola *et al.* (2006).

In classic FAST only the main effect terms V_i are computed. Extended FAST (Saltelli *et al.*, 1999) allows the computation of higher order terms, in particular it allows to compute the entire set of main and total effects.

Monte Carlo based estimation and the work of Sobol'

Several Monte Carlo based estimates of the first order terms V_i have been proposed (Homma and Saltelli, 1996; Hora and Iman, 1986; Iman and Hora, 1990; Ishigami and Homma, 1990; Krzykacz-Hausmann, 1990; McKay, 1996; Saltelli *et al.*, 1993; Sobol, 1993). It will be sufficient here to consider the work of Sobol' (Sobol, 1993). Sobol' noted that the function itself can be decomposed into terms of increasing dimensionality, i.e.

$$Y = f(X_1, X_2, \dots, X_k) = f_0 + \sum_i f_i(X_i) + \sum_{\substack{ij \\ i<j}} f_{ij}\left(X_i, X_j\right) + \cdots + f_{12\dots k} \qquad (8.20)$$

There are infinite ways of decomposing f, but for independent factors there is a unique decomposition in orthogonal terms which ensures:

$$V_i = V(f_i) \qquad (8.21)$$

$$V_{ij} = V(f_{ij})$$

and, as described in Sobol' (1993)

$$\begin{aligned} f_0 &= E(Y) \\ f_i &= E(Y|X_i) - E(Y), \\ f_{ij} &= E\left(Y|X_i, X_j\right) - E(Y|X_i) - E\left(Y\middle|X_j\right) + E(Y) \end{aligned} \qquad (8.22)$$

and so on. The link between Equation (8.22) with Equations (8.11) and (8.15) is obvious.

The f_i's are called the main effects, f_{ij}'s are the second order interaction effects, and so on. This decomposition has a finite number of terms and is known as ANOVA-HDMR (HDMR, Sobol 1993; Rabitz *et al.,* 1999; Rabitz and Aliş, 2000), where HDMR stands for High Dimensional Model Representation and ANOVA refers to the analysis of variance from experimental design (Box *et al.*, 1978). A complete HDMR fully characterises the non linear relationship f.

Sobol' offered a Monte Carlo strategy to compute indices of any order, that is based on a Monte Carlo exploration of the input space. To make an example, to estimate V_i, the following algorithm is used

$$\tilde{V}_i = \frac{1}{N}\sum_{j=1}^{N} f\left(x_{j1}^a, x_{j2}^a, \Lambda\, x_{ji}^a, \Lambda\, x_{jk}^a\right) f\left(x_{j1}^b, x_{j2}^b, \Lambda\, x_{ji}^a, \Lambda\, x_{jk}^b\right) - \tilde{f}^2 \tag{8.23}$$

where N is the sample size of a MC simulation, k the number of independent factors, $\tilde{f} = \frac{1}{N}\sum_{j=1}^{N} f(x_{j1}^a, ..., x_{jk}^a)$ and the superscript a, b stand to indicate that different independent input matrices have been used:

$$\mathbf{A} = \begin{pmatrix} x_{11}^a & \Lambda & x_{1k}^a \\ \Lambda & \Lambda & \Lambda \\ x_{N1}^a & \Lambda & x_{Nk}^a \end{pmatrix}, \quad \mathbf{B} = \begin{pmatrix} x_{11}^b & \Lambda & x_{1k}^b \\ \Lambda & \Lambda & \Lambda \\ x_{N1}^b & \Lambda & x_{Nk}^b \end{pmatrix} \tag{8.24}$$

Sobol' noted that an important objective of the sensitivity analysis is to identify those factors that have no sensible effect on the output. To tackle the problem he rewrote the decomposition scheme for sets of factors as

$$V(Y) = V_{\mathbf{U}}(E_{\mathbf{Z}}(Y|\mathbf{U})) + V_{\mathbf{Z}}(E_{\mathbf{U}}(Y|\mathbf{Z})) + V_{\mathbf{UZ}} \tag{8.25}$$

where all factors in $\mathbf{X}$ have been partitioned in two sets, a trial set $\mathbf{U}$ of supposedly non influential factors and the remaining factors $\mathbf{Z}$. V_{UZ} is the pure interaction effect between the two sets and can be easily be computed by difference. If $V(Y) = V_{\mathbf{Z}}(E_{\mathbf{U}}(Y|\mathbf{Z}.))$, then one can conclude that the set $\mathbf{U}$ is truly non influential. Note that the condition of non influence implies $V_{\mathbf{U}}(E_{\mathbf{Z}}(Y|\mathbf{U}.)) + V_{\mathbf{UZ}} = 0$, and not simply$V_{\mathbf{U}}(E_{\mathbf{Z}}(Y|\mathbf{U}.)) = 0$. If the set $\mathbf{U}$ in (8.25) contains only one factor, X_i, then $\mathbf{Z} = \mathbf{X}_{-i}$. Hence (8.25) becomes

$$V(Y) = V_{X_i}(E_{\mathbf{X}_{-i}}(Y|X_i)) + V_{\mathbf{X}_{-i}}(E_{X_i}(Y|\mathbf{X}_{-i})) + V_{X_i\mathbf{X}_{-i}} \tag{8.26}$$

and the condition for X_i to be truly non influential is that $V_{X_i}(E_{\mathbf{X}_{-i}}(Y|X_i.)) + V_{X_i\mathbf{X}_{-i}} = 0$, which is the same as $V(Y) - V_{\mathbf{X}_{-i}}(E_{X_i}(Y|\mathbf{X}_{-i}.)) = 0$, or equivalently $E_{\mathbf{X}_{-i}}(V_{X_i}(Y|\mathbf{X}_{-i}.)) = 0$. In summary, if X_i is non influential, then $S_{Ti} = \frac{E_{\mathbf{X}_{-i}}(V_{X_i}(Y|\mathbf{X}_{-i}.))}{V(Y)} = 0$. S_{Ti} is the total sensitivity index of factor X_i. It is easy to prove (Saltelli *et al.*, 2004) that the condition S_{Ti} is necessary and sufficient for X_i to be non influential (factors fixing setting).

To estimate S_{Ti}, the following algorithm is used

$$\tilde{V}_{-i} = V_{\mathbf{X}_{-i}}(E_{X_i}(Y|\mathbf{X}_{-i})) \\ = \frac{1}{N}\sum_{j=1}^{N} f\left(x_{j1}^{a}, x_{j2}^{a}, \Lambda\, x_{ji}^{a}, \Lambda\, x_{jk}^{a}\right) f\left(x_{j1}^{a}, x_{j2}^{a}, \Lambda\, x_{ji}^{b}, \Lambda\, x_{jk}^{a}\right) - \tilde{f}^{2} \\ S_{Ti} = 1 - \frac{\tilde{V}_{-i}}{V(Y)} \tag{8.27}$$

Computing S_i and S_{Ti} for each factor X_i has a computational cost of $N \cdot (k+2)$ model realisations, with N of the order of 500–1,000 (see Saltelli *et al.*, 2005, 2007 for a discussion of computational cost of various types of sensitivity analysis).

In Figures 8.2 and 8.3 of Saltelli *et al.* (2000, pp. 179–180), the boxplots of the Sobol' estimates for Example 2 can be viewed at increasing sample size.

8.3.2. Regression analysis

In a regression analysis a relationship between the conditional expected variable y given some input variables x is established (see e.g. Iman *et al.*, 1985). For example a linear regression takes the form of:

$$Y = \alpha + \beta X + \gamma X^2 + \varepsilon \tag{8.28}$$

More general combinations are possible. The right-hand side in general compromises a linear combination of parameters (α, γ and β). X's are the model factors and Y is the model response. ε represents the error with the unpredicted or unexplained variation in the response variable and is assumed to be independent from X.

8.3.3. Regionalised sensitivity analysis

The distributions of the factors which are classified as NB and B can be compared by a Smirnov test (Saltelli *et al.*, 2004, p. 154) with the following hypothesis:

$$H_0 : f_m(X|B) = f_n(X|NB) \\ H_1 : f_m(X|B) \neq f_n(X|NB) \tag{8.29}$$

where $F_m(X|B)$ and $F_n(X|NB)$ are the sample marginal distribution functions for m behaviours and n nonbehaviours ($n+m = N$); $f_m(X|B)$ and $f_n(X|NB)$ are the respective marginal probability density functions. The Smirnov sensitivity test is a sensitivity measure that can be linked to variance-based main effects, but may also include some contribution of higher order, so we could say that it conveys information between a main and a total effect (see for example the exercises in Chapter 5 of Saltelli *et al.*, 2007).

8.3.4. Entropy

Krykacz-Hausmann (2001) argued that sensitivity should be measured as deviation of the factor distribution from being uniform (if the factor has been sampled from a uniform prior distribution). The author developed a sensitivity measure which is based on the Kullback-Leibler Entropy (Kullback and Leibler, 1951) and was extended by Liu *et al.* (2004).

Kullback-Leibler Entropy:

$$D_{KL}(pdf_1|pdf_0) = \int_{-\infty}^{\infty} pdf_1(y) \cdot \log\frac{pdf_1(y)}{pdf_0(y)}dy \tag{8.30}$$

D_{KL}: Kullback-Leibler Entropy
y: model response
pdf_0: pdf of the response $y = h(\mathbf{X})$
pdf_1: pdf of the response $y = h(\mathbf{X_1})$ due to changes in X_1

Liut *et al.* (2004) point out that this entropy shares many properties of a metric measure such as non-negativity, additive property and convexity despite not being a metric measure. The K-L entropy can be understood as a measure which expresses the lack of overlapping between probability density functions (within given limits). Liu *et al.* (2004) define a 'total' and 'main' effect indices in analogy to the Sobol method:

$$\begin{aligned} D_{KL_{x_i}}(pdf_1|pdf_0) &= \int_{-\infty}^{\infty} pdf_1(y(x_1,...,\overline{x_i},...,x_n))\left|\log\frac{pdf_1(y(x_1,...,\overline{x_i},...,x_n))}{pdf_0(y(x_1,...,x_i,...,x_n))}\right|dy \\ D_{KL_{\bar{x}_i}}(pdf_1|pdf_0) &= \int_{-\infty}^{\infty} pdf_1(y(x_1,...,\overline{x_i},...,x_n))\left|\log\frac{pdf_1(y(\overline{x_1},...,x_i,...,\overline{x_n}))}{pdf_0(y(x_1,...,x_i,...,x_n))}\right|dy \end{aligned} \tag{8.31}$$

where $\overline{x_i}$ means fixing this factor X_i. The larger $D_{KL_{x_i}}(pdf_1|pdf_0)$ is the more important is X_i. The smaller $D_{KL_{\bar{x}_i}}(pdf_1|pdf_0)$ the more important is the main effect of X_i.

The concept of entropy and sensitivity is also exploited (in a slightly different formulation) in methodologies such as the dynamic identifiability analysis (Wagener *et al.*, 2003) for rainfall-runoff models and its numerical expression by

Horritt (2006) for flood inundation models. Similarly to the Smirnov test, one can say that the entropy-based 'main' effect conveys information that is between variance-based main and total effect.

8.3.5. Sensitivity and model emulators

Regression analysis and the HDMR

There are strong links between regression analysis and the theory of variance based sensitivity analysis. Let $g(\cdot)$ be the generic function approximating the true model $Y = f(X_1, ..., X_k)$ and let us assume a quadratic loss function $E[(Y - g(\cdot))^2]$ as a measure of 'fit' for g. If we were to approximate f with a function of *one single* parameter X_i, which is the function $g_i^*(X_i)$ that provides minimum loss?

It is well known from standard school texts on statistics that the univariate function

$$g_i^* = E(Y|X_i) \tag{8.32}$$

i.e. the conditional expectation of Y given X_i, is the minimum loss approximation to f. The expression (8.32) tells us that at any p-location $X_i = x_{i,p}$, the value of $g*$ is obtained by integrating (averaging) Y over all the remaining $(X_1, ..., X_{i-1}, X_{i+1}, ..., X_k)$ input factors.

Equation (8.32) can be generalised to any subset of the input factors, by

$$g_i^* = E(Y|X_{i1}, ..., X_{il}). \tag{8.33}$$

Equations (8.32–8.33) are linked to the HDMR decomposition of f described in (8.20):

$$Y = f(X_1, X_2, ..., X_k) = f_0 + \sum_i f_i(X_i) + \sum_{\substack{ij \\ i<j}} f_{ij}(X_i, X_j) + \cdots + f_{12...k} \tag{8.34}$$

Recalling that the various terms of (8.34) are expressed as:

$$\begin{aligned} f_0 &= E(Y) \\ f_i(X_i) &= E(Y|X_i) - f_0 \\ f_{ij}(X_i, X_j) &= E(Y|X_i, X_j) - f_1 - f_2 - f_0 \end{aligned} \tag{8.35}$$

we explicitly highlight the link between the terms of the HDMR and the minimum loss approximating functions $g(\cdot)$. Each term of the decomposition tells the analyst how much Y moves around its mean level f_0 as a function of single input factors or groups of them.

Coming back to the emulation problem, we require that the smoothing function(s) of increasing order approximate and ultimately converge to $g_i^*(\cdot)$. This assures that, given a truncation of the full HDMR expansion, the emulator obtained is the optimal one in a least squares sense among all possible models of the same order of interaction.

cut-HDMR

The papers by H. Rabitz and co-workers on High Dimensional Model Representation (HDMR).

Rabitz *et al.* (1999) and Rabitz and Aliş (2000) can be seen as precursory applications of the emulation approach in SA. There, the emulator is given by the so-called cut-HDMR expansion, obtained by evaluating the function $Y = f(X_1, ..., X_k)$ on quadrature points falling on lines, planes and hyper-planes passing through a 'base' point $\mathbf{X}_0$ in the input factor space, and truncating the order of the hyper-planes at a value $l < k$. Then, all the required integrals are computed on the truncated cut-HDMR function. This helps to reduce the computational cost of the analysis with respect to variance based methods, albeit still not sufficiently to treat expensive models. Moreover, the cut-HDMR approach still depends significantly on the dimensionality of the problem k.

RS-HDMR

The same research group who worked on cut-HDMR presented the so-called Random Sampling-HDMR (RS-HDMR, Liu *et al.,* 2002; Liu *et al.,* 2006), where the HDMR terms are estimated by smoothing a Monte Carlo sample of the mapping $Y = f(X_1, ..., X_k)$ with orthogonal polynomials. Due to the curse of dimensionality problem, a truncated HDMR is estimated, usually up to order three, using polynomials of order up to 2–3:

$$\begin{aligned}
f_i(X_i) &\approx \sum_{r=1}^{k} \alpha_r^i \varphi_r^i(X_i) \\
f_{ij}(X_i, X_j) &\approx \sum_{p=1}^{l} \sum_{q=1}^{l'} \beta_{pq}^{ij} \varphi_p^i(X_i) \varphi_q^j(X_j) \\
f_{ijk}(X_i, X_j, X_k) &\approx \sum_{p=1}^{m} \sum_{q=1}^{m'} \sum_{r=1}^{m''} \gamma_{pqr}^{ijk} \varphi_p^i(X_i) \varphi_q^j(X_j) \varphi_r^k(X_k)
\end{aligned} \tag{8.36}$$

where k, l, l', m, m', m'' and integers, α_r^i, β_{pq}^{ij}, γ_{pqr}^{ijk} are constant coefficients to be determined from the MC sample and the polynomials $\{\varphi\}$ possess the weighted orthonormality properties (see Liu *et al.*, 2006 for more details).

RS-HDMR can be seen as the first implementation of the emulation concept within the sensitivity context. RS-HDMR proved to be good in all applications presented by the authors and allows to significantly reduce the computational cost of sensitivity analysis.

Some aspects are worth noting in the RS-HDMR framework: the problem of possible over-fitting and of the optimisation of the elements and the order of the polynomials to be used for the different terms in the RS-HDMR approximation. For example, some terms characterised by particularly nonlinear shapes, could require higher order polynomial, while other ones might simply be described by linear functions. In this sense, a combined use with other non-parametric methods would allow to get valuable information for a better accuracy of polynomial approximation. An important aspect of cut-HDMR and RS-HDMR is that they explicitly consider the curse of dimensionality problem and, instead of trying to smooth or fit the MC mapping with a generic k-dimensional function (e.g. using kernel regression), they fix the order of interaction terms that is possible to estimate in practice, in a limited sample MC exercise.

Bayesian emulator

In very short detail, a Gaussian Process (GP) emulator uses a linear model structure to define the prior mean for the emulator:

$$m(\cdot) = E\{f(\mathbf{X})|\beta\} = \mathbf{h}(\cdot)^T\beta \tag{8.37}$$

where $\mathbf{h}(\cdot) = (h_1(\cdot), h_2(\cdot), ..., h_p(\cdot))^T$ is a vector of p known functions over the domain of the input factors and β is a vector of unknown coefficients. Thus, the definition of $\mathbf{h}(\cdot)$ reflects our prior belief about the class of function that may approximate $f(\cdot)$. The prior uncertainty $e(\mathbf{X})$ is expressed with a multivariate normal distribution. Since $f(\cdot)$ is believed to be smooth, one can get an idea of the behaviour of $f(\cdot)$ at a location $\mathbf{X}'$, given the observation of $f(\cdot)$ at some other location $\mathbf{X}$. This is reflected by the prior covariance:

$$\text{cov}\{f(\mathbf{X}), f(\mathbf{X}')|\sigma^2\} = \sigma^2 c(\mathbf{X}, \mathbf{X}') \tag{8.38}$$

where $c(\mathbf{X}, \mathbf{X}')$ is a function that decreases with $|\mathbf{X} - \mathbf{X}'|$ and satisfies $c(\mathbf{X}, \mathbf{X}) = 1$ for all $\mathbf{X}$. In the end, the unknown mapping $f(\cdot)$ is expressed as:

$$f(\mathbf{X}) = \mathbf{h}(\mathbf{X})^T\beta + e(\mathbf{X}) \tag{8.39}$$

Given one set of Monte Carlo realisations (learning sample) of the model at hand, one can draw, from Bayesian updating, the posterior estimations of the unknown parameters $\hat{\beta}$ and $\hat{\sigma}^2$, together with the posterior estimate for the emulator $m^*(\cdot)$ and its covariance structure $c^*(\mathbf{X}, \mathbf{X}')$. From this, any desired posterior inference about single terms of the HDMR or about sensitivity indices can be drawn. More details on Gaussian emulator can be found in Kennedy and O'Hagan (2001) and Oakley and O'Hagan (2004).

SDP

The HDMR expansion truncated up to the third order can be viewed as a state dependent regression model:

$$Y_t - f_0 = \sum_i p_{i,t}(s_{i,t}) + \sum_{j>i} p_{ij,t}(s_{ij}) \sum_{l>j>i} p_{ijl,t}(s_{ijl}) + e_t \tag{8.40}$$

where each state dependent parameter $p_{I,t}(s_I)$, $I = i_1, \ldots, i_l$ depends on a state variable s_I that moves according to a generalised sorting strategy along the co-ordinates of the single factors of group of factors indexed by I. According to the generalised sorting strategy, the group of input factors of interest I is characterised by a low frequency spectrum (e.g. by some quasi-periodic pattern) while the remaining ones present a white spectrum. In this way, the estimation of the various HDMR terms reduces to the extraction of the low frequency component of the sorted output Y. To do so, the SDP's are modelled by one member of generalised random walk (GRW) class of non-stationary processes. For instance, the integrated random walk (IRW) process turns out to provide good results, since it ensures that the estimated SDP relationship has the smooth properties of a cubic spline. Given the IRW characterisation, the model (8.41) can be put into state space form as:

$$\begin{aligned} Y_t - f_0 &= \mathbf{1}_t^T \mathbf{p}_t + e_t \\ p_{i,t} &= p_{i,t-1} + d_{i,t-1} \\ d_{i,t} &= d_{i,t-1} + \eta_{i,t} \end{aligned} \tag{8.41}$$

where e_t (observation noise) and $\eta_{i,t}$ (system disturbances) are zero mean white noise inputs with variance σ^2 and $\sigma^2_{\eta_{i,t}}$ respectively. Given this formulation, SDP's are estimated using the recursive Kalman Filter (KF) and associated recursive Fixed Interval Smoothing (FIS) algorithm (Kalman, 1960; Bryson and Ho, 1969; Young, 1999). The recursive state estimation requires that each SDP is estimated in turn, each one with a different ordering of the data, within a backfitting procedure (Young, 2000; Young, 2001; see also Hastie and Tibshirani, 1990). At each backfitting iteration, the hyper-parameters associated with (8.41), namely the noise variances σ^2 and $\sigma^2_{\eta_{i,t}}$, are optimised by maximum likelihood, using prediction error decomposition (Schweppe, 1965).

A MATLAB Toolbox for the estimation of Smoothing Spline ANOVA models with Recursive algorithms can be found on http://eemc.jrc.ec.europa.eu/Software-SS_ANOVA_R.htm.

Splines

The spline approach is similar to the SDP method introduced above. A recent methodology estimates the relationship of the state dependent parameters over hermite cubic splines (Pappenberger and Stauch, 2006). For this, an interval of the random input variable X_i is divided into $m-1$ subsections resulting in m node locations, each with node values $Y_1 \ldots Y_m$. For each subinterval of the random variable X_i, the polynomials $pol(X_{i,m})$ are estimated as cubic *hermite* interpolating polynomials based on four interpolation conditions (see de Boor, 2001). The shape preserving algorithm by Fritsch and Carslon (1980) is applied and the values of the knots of the cubic spline optimised using a Levenberg-Marquart nonlinear least square minimisation (Stauch and Jarvis, in press). The optimisation procedure duplicates the backfitting algorithm for the SDP procedure.

The number of nodes and location has been fixed after several initial tests, but could be optimised (Bitterlich *et al.*, 2004). However, initial tests suggest that this is not always necessary.

Smoothing splines are also successfully applied for sensitivity analysis. In the framework of smoothing spline ANOVA models, Storlie et al (2009a) developed the so-called ACOSSO method, a model selection procedure for nonparametric models whose estimator is obtained via solving a regularization problem with a novel adaptive penalty on the sum of functional component norms. The method is very efficient in identifying ANOVA models up to third order and R-packages are available on the web page of the first author (see http://www.math.unm.edu/~storlie/).

Guidance Report I.4

Uncertainty analysis in model-based water management

Jens Christian Refsgaard, Jeroen P. van der Sluijs, Anker Lajer Højberg and Peter A. Vanrolleghem

Abstract

The main subject in this guidance report is methodologies and tools to characterize uncertainty and to assess the various sorts and sources of uncertainty, and the propagation of uncertainty through models to management information.

A terminology and typology of uncertainty is presented together with a framework for the modeling process, its interaction with the broader water management process and the role of uncertainty at different stages in the modeling processes. Brief reviews have been made of 14 different (partly complementary) methods commonly used in uncertainty assessment and characterization: data uncertainty engine (DUE), error propagation equations, expert elicitation, extended peer review, inverse modeling (parameter estimation), inverse modeling (predictive uncertainty), Monte Carlo analysis, multiple model simulation, NUSAP, quality assurance, scenario analysis, sensitivity analysis, stakeholder involvement and uncertainty matrix. The applicability of these methods has been mapped according to purpose of application, stage of the modeling process and source and type of uncertainty addressed. It is concluded that uncertainty assessment is not just something to be added after the completion of the modeling work. Instead uncertainty should be seen as a red thread throughout the modeling study starting from the very beginning, where the identification and characterization of all uncertainty sources should be performed jointly by the modeler, the water manager and the stakeholders.

This report focuses on uncertainty in the modeling process. As such it touches upon aspects of uncertainty related to the broader policy and public participation processes, but it does not intend to fully cover these broader aspects.

Keywords

Decision-making, mathematical modeling, uncertainty assessment, uncertainty propagation

1. WHY IS UNCERTAINTY ASSESSMENT IMPORTANT?

1.1. Uncertainty and risk in decision making

Integrated river basin management requires making a large number of decisions by operational agencies. A decision maker has to make decisions based on the available information. In most cases this information is deficient, incomplete

 Modelling Aspects of Water Framework Directive Implementation Volume 1. By Peter A. Vanrolleghem. ISBN: 9781843392231. Published by IWA Publishing, London, UK.

and characterised by uncertainties of different kinds. How should this affect the decision making process? With increased uncertainty the chance of taking wrong decisions increases. Can the decision maker accept this? What can he or she do to make decisions that anticipate that outcomes can differ from what was expected? How should knowledge about the various uncertainties that characterise the available information be used to make better decisions? Nearly all information is uncertain. The main subject in this guidance document is methodologies and tools to characterise uncertainty and to assess the various sorts and sources of uncertainty, and the propagation of uncertainty through models to management information.

As one option, the decision maker may decide to postpone making a decision. This will allow more effort (money) to be put into collecting additional data or increasing modelling efforts, so as to improve the quality of information and, thereby, reducing the probability of taking a wrong decision. Further work on data collection and modelling involves costs in itself, and delays in the decision may also involve additional costs as well. The typical question "Do we have enough understanding to responsibly make a decision responsibly?" is thus in a rational sense a question on whether the risks of wrong decisions caused by imperfectness of the information are acceptable, or whether it is advisable to improve the quality of the information by, for instance, further data collection/ modelling. One formal approach to address this question is through "value of information" analysis (VOI). Essentially, VOI encourages the prioritisation of research such that the expected value of additional information is compared against the opportunity costs of uncertainty, including the costs of making the wrong decision. This is sometimes difficult to do quantitatively in the environmental field because of problems associated with putting a value on natural resources. Nevertheless, even in the absence of explicit quantification, it is useful to remember that continuous attempts to reduce uncertainty by collecting additional information does not always make economic sense. Additional information—in the form of more data or more refined algorithms—almost always incurs costs. These costs must be balanced against the expected benefits of such information.

A note should be made here that higher quality of information does not automatically imply information with less uncertainty. It can also be information providing a richer insight in the sorts and magnitudes of uncertainty, so that the decision maker better understands what possible outcomes and risks to anticipate. This allows for contingent planning where options are kept open and flexible and where emergency risk management options can be prepared and kept on the shelf (compare to a fire extinguisher) so that they can be implemented immediately if that turns out to be necessary.

Another option open to the decision maker is to become risk-avers. This may result in incorporating a large safety margin or a 'tolerance', so that increased resources are spent on measures, such as clean-up of a site or water body, where there is a large probability that it may not be required to protect the water resources.

A decision maker can adopt one of the strategies developed in the field of decision theory if the ranges of outcomes for different options to choose from are known, but not the probabilities of each outcome. For instance, maximin is the strategy that chooses the option that has the best (that is: the least severe) worst case scenario. It makes sense if we have little to win and a great deal to loose, but it tends to prevent us from taking advantage of opportunities. Closely related to maximin is the difference principle: one society is better off than another if the worst-off members of the former do better than the worst-off in the latter. Maximin allows the most disadvantaged members of society to be harmed if the overall society benefits; the difference principle would forego an overall benefit to the society if it harmed the most disadvantaged members. The difference principle has been criticised for that it does not weigh limiting disadvantages to a subset of people against a possible increased average utility of society.

On the other extreme, maximax is the strategy that chooses the option that has the best best-case scenario. It is a risky strategy, often preferred by people with a risk seeking attitude that want to take advantage of opportunities and it makes sense if one has a great deal to gain and little to loose. Maximin can be seen as excessively pessimistic and maximax as excessively optimistic. An approach that attempts to balance between good and bad outcomes is the principle of insufficient reason: when we lack objective evidence to specify probabilities of outcomes, we should treat all outcomes as equally probable. (Resnik, 2003)

A widely advocated strategy is the Precautionary Principle if both the bounds on the outcomes and their probabilities are unknown. The Precautionary Principle grants greater benefit of doubt to the environment and to public health than to the activities that may be held to threaten these things (Stirling, 2003). Because the Precautionary Principle applies to those cases where serious adverse effects and surprises can occur with an unknown probability, it is rational to follow a better safe than sorry strategy. Failing to take precautionary measures in a timely manner could result in devastating and irreversible consequences (Harremoës *et al.*, 2001). Such consequences might have been avoided by proactive and anticipatory interventions whose costs are justifiable in comparison to the damages and losses that could occur.

The above illustrates that decision makers should act differently under different situations of uncertainty. However, they will only be able to do this on a rational basis when they know how uncertain the available information is and when they know how to incorporate this in their decision making. Uncertainty is a difficult concept, and there is a need to educate and assist the decision maker working in a situation where there is uncertainty. In this document we will describe methodologies and tools for uncertainty assessment that may be used in the decision making process, allowing decision makers to take rational decisions on how to act under a situation of uncertainty.

1.2. Water Framework Directive – requirements

The Water Framework Directive (WFD) provides a European environmental policy basis at the river basin scale. The river basin management and planning process prescribed in the WFD focuses on integrated management, involving all physical domains in water management, all sectors of water use, socio-economics and stakeholder participation. The planning process should move from a more rational-instrumental type of planning to an interactive planning with an open eye for the power of fundamental debate. The uncertainties present in such planning processes are judged of an increasing importance (EC, 2004a). An iterative planning process can deal with these uncertainties, for instance by revising the programmes of measures according to the circumstances (EC, 2004b). As such, the WFD poses new challenges to river basin managers. The traditional physical domain specific and sectoral approaches need to be combined and extended to fulfil the WFD requirements. In practise, the preparation of the river basin management plans prescribed in the WFD is in addition to these new challenges, influenced by uncertainties on the underlying data and modelling results.

A basic principle in EU environmental policy on which the WFD is based is "… to contribute to the pursuit of the objectives of preserving, protecting and improving the quality of the environment in prudent and rational use of natural resources, and to be based on the precautionary principle …[1]". Therefore, the holistic concept that is prescribed in the WFD with its integrated approach to natural resources and socio-economic issues requires that uncertainty be considered in the decision making process in order for it to become truly rational.

Uncertainty is addressed in several sections of the WFD document, (Blind and de Blois, 2003). For example, the WFD states with respect to monitoring that

[1]Directive 2000/60/EC, paragraph 11 in the introductory section.

"Estimates of the level of confidence and precision of the results provided by the monitoring programmes shall be given …". In addition, most of the WFD guidance documents, being more specific than the WFD document itself, explicitly emphasise that uncertainty analyses should be performed. For instance, the guidance document on the planning process (EC, 2004a) states: "Uncertainty can be defined as the occurrence of events that are beyond our control. Uncertainty is always an element in the planning process. It arises because the complexity of the many factors involved. In fact, meteorological, demographic, social, technical, and political conditions which will determine the planning process have behaviour patterns not always known with sufficient accuracy. Uncertainty arises mainly due to the stochastic nature of some key elements affecting these processes." Similarly, the WATECO document on economic analysis states that "Uncertainty on costs, effectiveness and time-lagged effects of measures needs to be dealt with throughout the economic analysis process, and more generally throughout the process of identifying measures and developing the river basin management plan". However, despite strong recommendations to consider uncertainty aspects, the guidance documents do not include recommendations on how to do so.

1.3. A motivating example

The problem is illustrated by an example from practise based on a study conducted by the County of Copenhagen (Copenhagen County, 2000; Refsgaard *et al.*, 2000). The County of Copenhagen is the authority responsible for water resources management in the county where the city of Copenhagen abstracts groundwater for most of its water supply. According to a new Water Supply Act the county had to prepare an action plan for protection of groundwater against pollution. As a first step, in 2000, the county asked five groups of Danish consulting firms to conduct studies of the aquifer's vulnerability to pollution in a 175 km^2 area west of Copenhagen, where the groundwater abstraction amounts to about 12 million m^3/year. The key question to be answered was: "which parts of this particular area are most vulnerable to pollution and need to be protected?" The five consultants were selected from among the most well reputed consulting firms in Denmark, and they were known to have different views and preferences on which methodologies are most suitable for assessing vulnerability. As the job was one of the first consultancy studies in a new major market for preparation of groundwater protection action plans it was considered a prestigious job to which the consultants generally allocated some of their most qualified professionals.

The five consultants used significantly different approaches. One consultant based his approach on annual fluctuations of piezometric heads assuming that the larger the fluctuations the more interaction between aquifer and surface water systems and, hence, the larger vulnerability. Several consultants used the DRASTIC multi-criteria method (Aller *et al.*, 1987), but modified it in different ways by changing weights and adding new, mainly geochemically oriented, criteria. One consultant based his approach on advanced hydrological modelling of both groundwater and surface water systems using the MIKE SHE code. Two other consultants used simpler groundwater modelling approaches. The three consultants applying modelling used simulated recharge as inputs to their respective DRASTIC approach. Thus, the five consultants used five different conceptual models to describe the possibility of groundwater pollution in the area. In addition, their different interpretations and interpolations made from a common data base resulted in significantly different figures; for example, for areal means of precipitation and evapotranspiration and the thickness of various geological layers (Refsgaard *et al.*, 2000). Due to lack of concentration data in the aquifer system the methods could, mostly, not be tested against field data and their use could therefore be characterised as non-documented extrapolation. Such lack of rigorous validation tests is common in studies dealing with aquifer vulnerability towards pollution.

The conclusions of the five consultants regarding vulnerability to nitrate pollution are shown in Figure 1.1. It is seen that the five estimates differ substantially from each other. In the present case, no data exist to validate the model predictions, because the five models have been used to make extrapolations towards unobservable futures. Thus, it is not possible, from existing field data, to tell which of the five model estimates are more reliable. The differences in prediction originate from two main sources: (i) data and parameter uncertainty and (ii) conceptual uncertainty. However, despite the significant data and parameter uncertainty, the main cause of the differences lies in the different conceptual models that were used by the five consultants.

Usually a water manager commissions only one study and bases his decisions on the conclusions from that study. The uniqueness of the present study was that five consultants were asked to answer the same question on the basis of the same data. In this respect the differences between the five best estimates are striking and clearly do not provide a sound basis for deciding anything about which areas should be protected. A worrying question, which is left unanswered, is whether the basis for decisions is just as poor in the many other cases where only a single conceptual model has been used and where, subsequently, a lot of money has been used to prepare and implement action plans.

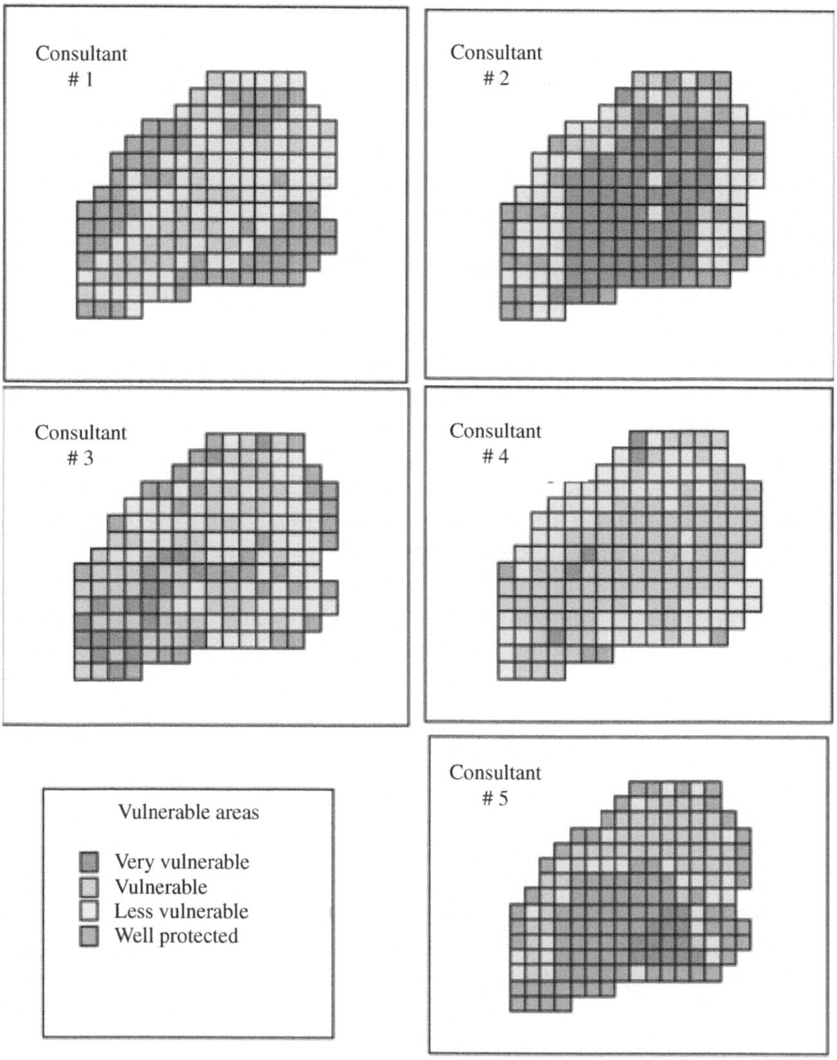

Figure 1.1. Model predictions on aquifer vulnerability towards nitrate pollution for a 175 km^2 area west of Copenhagen (Copenhagen County, 2000)

In this case the uncertainty was so large that a basis for making a rationale decision did not exist. The conceptual uncertainty was discovered by chance, and the level of the uncertainty was a large, and uncomfortable, surprise to the water manager.

1.4. Context and objective of this document

Models describing water flows, water quality, ecology and economy are being developed and used in increasing number and variety. With the requirements imposed by the WFD the trend in recent years to base water management decisions to a larger extent on model studies and to use more sophisticated models is likely to be reinforced. It is important to note that the modelling studies typically do not address the entire water resources management decision process, such as the WFD process, but rather support certain elements of the process.

The role of modelling as part of the WFD process may be illustrated schematically as in Figure 1.2. The inner circle in Figure 1.2 depicts a simplified version of the WFD planning process (EC, 2004a) with the main elements:

- *Identification* including assessment of present status, analysis of impacts and pressures and establishment of environmental objectives. Here modelling may be useful for example for supporting assessments of what are the reference conditions, for assessment of the pressures and what are the impacts of the various pressures in combination to monitoring data and expert judgement (EC, 2004b).
- *Designing* including the set up and analysis of programme of measures designed to be able in a cost effective way to reach the environmental objectives. Here modelling will typically be used for supporting assessments of the effects and costs of various measures under consideration.
- *Implementing* the measures. Here on-line modelling in some cases may support the operational decisions to be made.
- *Evaluation* of the effects of the measures on the environment. Here modelling may support the monitoring in order to extract maximum information from the monitoring data, e.g. by indicating errors and inadequacies in the data and by filtering out the effects of climate variability.

This main WFD process is a participatory process with important elements of public participation.

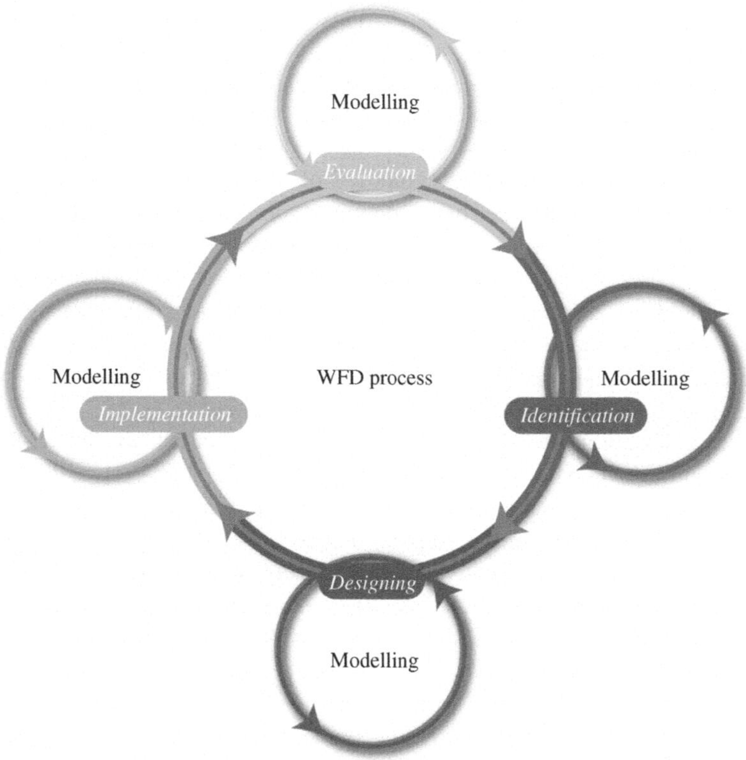

Figure 1.2. The role of modelling in the water resources management process within the context of the Water Framework Directive (WFD)

Modelling can be used as a tool at various stages of the WFD process, as illustrated by the four smaller circles in Figure 1.2. The typical steps and elements of a modelling process are illustrated in Figure 1.3 and briefly described in Chapter 2. The most important interactions between the modelling process and the main planning process are:

- The modelling process starts with a thorough framing of the problem to be addressed and definition of modelling objectives and requirements for the modelling study (Step 1 in Figure 1.3). Water managers and stakeholders dominate this step, which basically is identical to part of the broader planning process. A participatory based assessment of the most important sources of uncertainty for the decision (WFD) process should be used as a basis for prioritising the elements of the modelling study. The uncertainty assessments made at this stage will typically be qualitative.

- The main modelling itself is composed of Steps 2, 3 and 4 of Figure 1.3. Here the link with the main planning process consists of dialogue, reviews and discussions of preliminary results involving water managers and, according to decisions made at the start of the modelling, some stakeholders. As the modelling process proceeds uncertainty assessments are typically made more quantitative (rather than qualitative) and the uncertainties assessed by the modeller are confronted with the water manager and stakeholder expectations to accuracy.
- The finalisation of the modelling study (equivalent to the last step in Figure 1.3), typically including scenario simulations. Here the water managers and the stakeholders again have a dominant role. The decisions made at the outcome of this step on the basis of modelling results are made in the context of the main planning process. Uncertainty assessment of model predictions is a crucial aspect of the modelling results and should be communicated in a way that is accessible for the stakeholders in the further WFD process.

The objective of this document is to provide guidance on uncertainty related to modelling in water resources management. Thus, the document does not focus on uncertainty related to the broader policy and public participation processes. The target audience for the document is professionals involved in modelling. This includes modellers themselves as well as the persons in the water manager's and stakeholders' organisations designated to interact with the modeller in the modelling process.

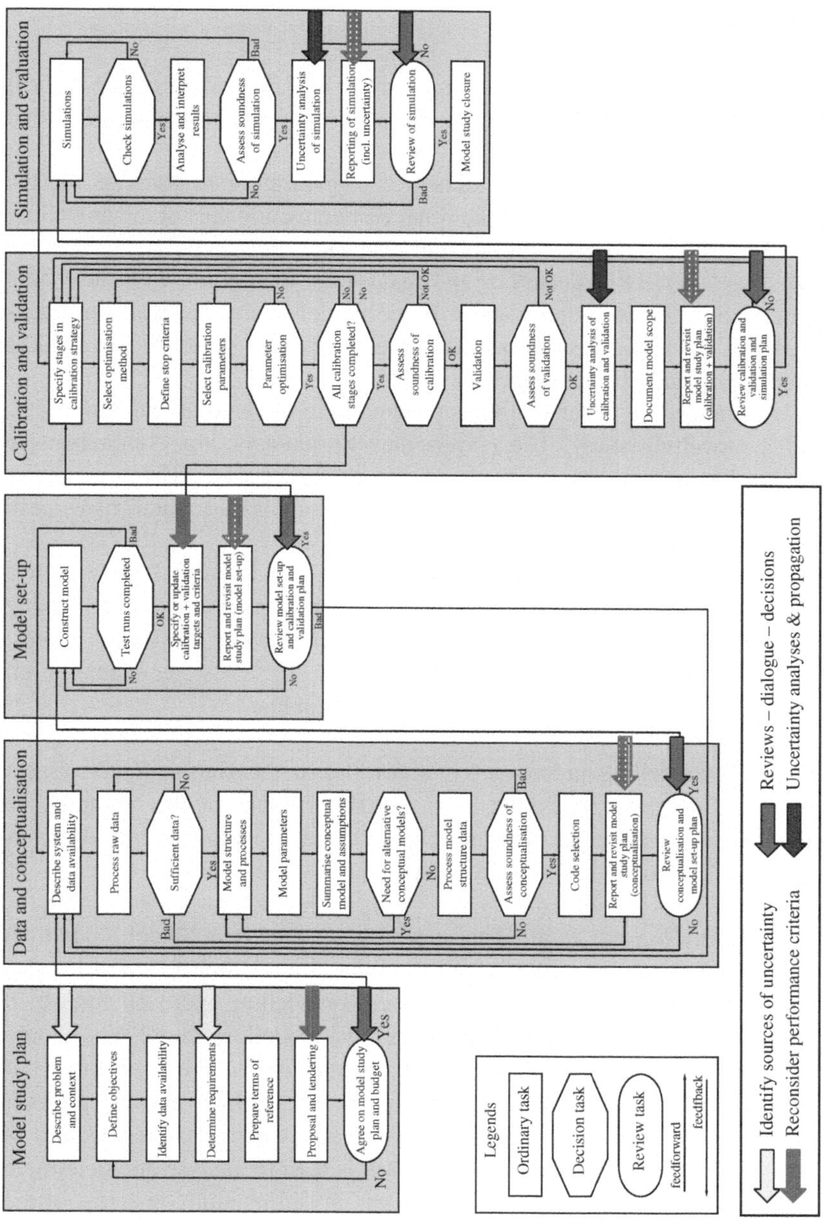

Figure 1.3. The five modelling steps and 45 tasks in the HarmoniQuA modelling protocol (Refsgaard *et al.*, 2004)

2. WHEN IS UNCERTAINTY ASSESSMENT REQUIRED?

2.1. The modelling process

A modelling study will involve several phases and several actors. A typical WFD modelling study will involve the following four different types of actors:

- *The water manager*, i.e. the person or organisation responsible for the management or protection of the water resources, and thus of the modelling study and the outcome (the problem owner).
- *The modeller*, i.e. a person or an organisation that works with the model conducting the modelling study. If the modeller and the water manager belongs to different organisations, their roles will typically be denoted consultant and client, respectively.
- *The reviewer*, i.e. a person that is conducting some kind of external review of a modelling study. The review may be more or less comprehensive depending on the requirements of the particular case. The reviewer is typically appointed by the water manager to support the water manager to match the modelling capability of the modeller.
- *The stakeholders/public*, i.e. a stakeholder is an interested party with a stake in the water management issue, either in exploiting or protecting the resource. Stakeholders include the following different groups: (i) competent water resource authority (typically the water manager, cf. above); (ii) interest groups; and (iii) general public.

The WFD modelling process may, according to the HarmoniQuA project (Refsgaard *et al.*, 2004; Scholten *et al.*, 2004, first part of this volume), be decomposed into five major steps which again are decomposed into 45 tasks (Figure 1.3). The contents of the five steps are:

- *Step 1* (*Model Study Plan*). This step aims to agree on a Model Study Plan comprising answers to the questions: Why is modelling required for this particular model study? What is the overall modelling approach and which work should be carried out? Who will do the modelling work? Who should do the technical reviews? Which stakeholders/public should be involved and to what degree? What are the resources available for the project? The water manager needs to describe the problem and its context as well as the available data. A very important task is then to analyse and determine what are the various requirements of the modelling study in terms of the expected accuracy of modelling results. The acceptable level of accuracy will vary from case to case and must be seen in a socio-economic context. It should, therefore, be defined through a dialogue between the modeller, water manager and stakeholders/public. In this respect an analysis of the key sources of uncertainty is crucial in order to focus the study on the elements that produce most information of relevance to the problem at hand.

- *Step 2* (*Data and Conceptualisation*). In this step the modeller should gather all the relevant knowledge about the study basin and develop an overview of the processes and their interactions in order to conceptualise how the system should be modelled in sufficient detail to meet the requirements specified in the Model Study Plan. Consideration must be given to the spatial and temporal detail required of a model, to the system dynamics, to the boundary conditions and to how the model parameters can be determined from the available data. The need to model certain processes in alternative ways or to differing levels of detail in order to enable assessments of model structure uncertainty should be evaluated. The availability of existing computer codes that can address the model requirements should also be addressed.
- *Step 3* (*Model Set-up*). Model Set-up implies transforming the conceptual model into a site-specific model that can be run in the selected model code. A major task in Model Set-up is the processing of data in order to prepare the input files necessary for executing the model. Usually, the model is run within a Graphical User Interface (GUI) where many tasks have been automated. The GUI speeds up the generation of input files, but it does not guarantee that the input files are error free. The modeller performs this work.
- *Step 4* (*Calibration and Validation*). This step is concerned with the process of analysing the model that was constructed during the previous step, first by calibrating the model, and then by validating its performance against independent field data. Finally, the reliability of model simulations for the intended domain of applicability is assessed through uncertainty analyses. The results are described so that the scope of model use and its associated limitations are documented and made explicit. The modeller performs this work.
- *Step 5* (*Simulation and Evaluation*). In this step the modeller uses the calibrated and validated model to make simulations to meet the objectives and requirements of the model study. Depending on the objectives of the study, these simulations may result in specific results that can be used in subsequent decision making (e.g. for planning or design purposes) or to improve understanding (e.g. of the hydrological/ecological regime of the

study area). It is important to carry out suitable uncertainty assessments of the model predictions in order to arrive at a robust decision. As with the other steps, the quality of the results needs to be assessed through internal and external reviews.

Each of the last four steps is concluded with a reporting task followed by a review task. The review tasks include dialogues between water manager, modeller, reviewer and, often, stakeholders/public. The protocol includes many feedback possibilities (Figure 1.3).

2.2. Uncertainty aspects

Uncertainty aspects are important throughout the modelling process. Thus, uncertainty is considered explicitly in 14 out of the 45 tasks in Figure 1.3. But uncertainty is treated in different ways at different stages of the process. The four main actions of dealing with uncertainty may be characterised as:

A. Identify and characterise sources of uncertainty (Figure 1.3 )

The various sources of uncertainty need to be identified and characterised in connection with the tasks *Describe Problem* and Context and *Determine Requirements* in Step1 *Model Study Plan* (yellow arrows in Figure 1.3). This should be done by the water manager but typically after a dialogue with relevant stakeholders. Depending on the framing of the model study some of these uncertainties may be located as external non-controllable sources. It is crucial that uncertainty is considered explicitly so early in the definition phase of the model study. Here uncertainties are seldom quantified. It is also at this early stage that the first analyses are made on the acceptable level of uncertainty and the expected model performance.

B. Modeller reconsiders uncertainty and performance criteria (Figure 1.3 )

The modeller has to reconsider several times during the modelling process if the accuracy performance criteria formulated by the water manager are realistic for the particular model study. Most importantly (full green arrows in Figure 1.3), the modeller has to do this in connection with preparation of a proposal, and when, after having made the first model test run, s/he re-evaluates the performance criteria before the model validation tests in the next step. Furthermore (dotted green arrows in Figure 1.3) in connection with reporting at the end of each step, the modeller has the opportunity to reconsider whether the originally promised performance criteria are still realistic, given the new information produced during the previous tasks in the respective step.

*C. Reviews – dialogue – decisions (*Figure 1.3 *)*

The last task in each step is a dialogue or decision task where a dialogue between water manager and modeller takes place. Often independent reviews are conducted as a basis for the decision and stakeholders and/or the general public are involved in the dialogue. As part of this dialogue, uncertainty aspects become important, e.g. when discussing whether there are sufficient data to proceed with the modelling, or whether the uncertainty of the model simulations are at a level where the results can be expected to be useful. The reviews and the stakeholder dialogues are also important platforms for a reflection on whether the assumptions made in the model are realistic and on how the study outcome may be influenced by the implicit and explicit assumptions made in the model. In many cases, more than one assumption is scientifically tenable. If such assumptions influence the model outcome, then the ignorance regarding which assumption is the best assumption can be an important source of uncertainty.

*D. Uncertainty assessment and propagation (*Figure 1.3 *)*

Towards the end of the step Calibration and Validation and the step Simulation and Evaluation there are two tasks dealing exclusively with uncertainty assessment. In the first (Uncertainty Analysis of Calibration and Validation) an assessment is made of the model uncertainty related to simulations in the validation test cases. This is used for evaluating possible biases in model simulations and assessing if the model performance is good enough compared to the agreed accuracy requirements. In the second task (Uncertainty Analysis of Simulation) the uncertainties in the problem framing (the context) and the management scenarios are also taken into account.

The uncertainty assessment and propagation tasks (item D above) are the traditional uncertainty tasks often conducted in connection with model studies. These tasks are often comprehensive and may involve a lot of model calculations and are often limited to quantitative uncertainty. These tasks, which are both located towards the end of the modelling process, are very important. However, it is equally important to introduce uncertainty in the introductory phase of a model study. Therefore, the identification and characterisation of all uncertainty sources recommended in the task Determine Requirements under Step 1 (item A above) is crucial. The uncertainty aspects mentioned under items B and C are "less heavy" and may be seen as a follow up to item A.

3. WHAT IS UNCERTAINTY?

3.1. Definitions

Uncertainty and associated terms such as error, risk and ignorance are defined and interpreted differently by different authors, see Walker *et al.* (2003) for a review. The different definitions reflect the underlying scientific philosophical way of thinking and therefore typically vary among different scientific disciplines. In addition they vary depending on their purpose. Some are rather generic, such as Funtowicz and Ravetz (1990), while others apply more specifically to model based water management, such as Beck (1987).

In this document we will use the terminology of Klauer and Brown (2003) that has emerged after discussions between social scientists and natural scientists specifically aiming at applications in model based water management. More details and discussion on the definitions given below can be found in Klauer and Brown (2003). By doing so we adopt a subjective interpretation of uncertainty in which the *degree of confidence* that a decision maker has about possible outcomes and/or probabilities of these outcomes is the central focus. For reasons of completeness and comparison we will also briefly sketch the objective interpretation of each form of uncertainty we discuss.

Uncertainty

The notion of uncertainty includes both subjective and objective aspects. Becoming confident or establishing lack of confidence is an act of subjective judgement about the validity of some information. However, the judgement might be supported and informed by the evaluation of 'objective' facts and other forms of evidence.

Definition (Uncertainty): A person is uncertain if s/he lacks confidence about the specific outcomes of an event or action. Reasons for this lack of confidence might include a judgement of the information as incomplete, blurred, inaccurate or potentially false or might reflect intrinsic limits to the deterministic predictability of complex systems or of stochastic processes.

Similarly, a person is certain if s/he is confident about the outcome of an event. It is possible that a person feels certain but has misjudged the situation (i.e. s/he is wrong).

Example: A person may be uncertain about the exact value of a river discharge value due to uncertainty related to the instruments used for measurements, representativeness of measurements and the method of transforming measurements (of often secondary variables) to discharge. Two different people may have different perceptions of the magnitude of this uncertainty.

Note that other authors define the term uncertainty not as a property (state of confidence) of the decision maker but as a property (state of perfection) of the total body of knowledge or information that is available at the moment of judgement. Uncertainty is then seen as an expression of the various forms of imperfection of the available information and depends on the state-of-the-art of scientific knowledge on the problem at the moment that the decision needs to be made (assuming that the decision maker has access to the state-of-the-art knowledge).

Ignorance

Awareness of the information on the potential outcomes of an event is a precondition for any grade of certainty or uncertainty. There are different, gradual stages of awareness starting from awareness of the existence of information to a deep understanding of the information.

Definition (ignorance): A person is ignorant with respect to an event if s/he is unaware of the (potential) outcomes of that event or of the event itself.

Examples: Oestrogens were not suspected of being harmful to fishes until a few decades ago. Clayey till was supposed to be virtually impermeable for pesticides and other contaminants and, therefore, provide good protection of groundwater until it was discovered 10–20 years ago that some clayey layers contain fractures through which pollutants can be transported very rapidly.

Note that ignorance refers to unawareness of the entire scientific community about potential outcomes of an event or side effects of an activity in the interpretation that sees ignorance as a property of the state of scientific knowledge rather than as the state of the individual decision maker.

Risk situations

If uncertainty is recognised as being an important issue, then the most common strategy to cope with this is to use probabilities. However, the use of probabilities presupposes a number of things about the available representation. First, it assumes that all potential outcomes of the event are known. In other words, that the event is properly characterised by the set of potential outcomes. Secondly, it assumes that the probabilities of each outcome are also known. We will call such a situation a risk situation.

Definition (risk situation): A risk situation is a person's representation of an event, where s/he assumes to know all potential outcomes as well as the probabilities of each outcome.

In some disciplines risk is defined as being equivalent to probability, while in others it is defined as damage multiplied by probability. Our definition is compatible with the latter (but not the first) of these.

Example: A water manager has to decide whether to implement a certain measure; for example, cleaning a polluted aquifer or to make additional field measurements to improve the data basis and, thereby, reduce the uncertainty involved (i.e. two possible outcomes). The water manager believes s/he can calculate the uncertainty (in terms of a probability distribution function) for how polluted the aquifer is and how much this uncertainty will change in case of new data. At the same time s/he knows the costs of making a wrong decision and the costs of the additional field program.

Precaution

The Precautionary Principle (PP) has become an underlying rationale over the past decades for the satisfactory and ethically justified management of uncertain risks to public health, society or environment. The PP aims to protect humans and the environment against uncertain risks of human action by means of pre-damage control (anticipatory measures). The PP is to supplement, but not necessarily replace, other management strategies that fall short of being able to handle large scale scientific uncertainty and ignorance. It is incorporated in a large and increasing number of international treaties and declarations in the fields of, inter alia, sustainable development, environmental protection, health, trade, and food safety. The PP is on its way to becoming a widely accepted part of international law. In its basic form, the Precautionary Principle states that actions to protect human health and the environment against possible danger of severe and irreversible damage, need not wait for rigorous scientific proof (Weiss, 2003).

Definition (precautionary principle): In order to protect the environment, the precautionary approach shall be widely applied by States according to their capabilities. The lack of full scientific certainty shall not be used as a reason for postponing cost-effective measures to prevent environmental degradation where there are threats of serious or irreversible damage.

The triple negative notion in this definition stemming from the Rio Declaration (1992), that the absence of rigorous proof of harm does not justify inaction, is perceived as being weak. It forces the consideration of proactive intervention but does not require such intervention. Other definitions are stronger and put the burden of proof on the proponent of an action to show that it does not pose a danger of environmental harm. For examples, the Wingspread Statement on the Precautionary Principle defines it as follows: *"When an activity raises threats of harm to human health or the environment, precautionary measures should be taken even if cause and effect relationships are not fully established scientifically...[The] proponent of the activity, rather than the public, should bear the burden of proof."*

3.2. Taxonomy of imperfect knowledge

There are many different decision situations, with different possibilities for characterising our confidence in the available information or in other words the uncertainty. A first distinction is between ignorance as a lack of awareness about imperfect knowledge and uncertainty as a state of confidence about knowledge. Our state of confidence may range from being certain to admitting that we know nothing (of use), and uncertainty may be expressed at a number of levels in between. Regardless of our confidence in what we know, ignorance implies that we can still be wrong ('in error'). In this respect Brown (2004) has defined a taxonomy of imperfect knowledge as illustrated in Figure 3.1.

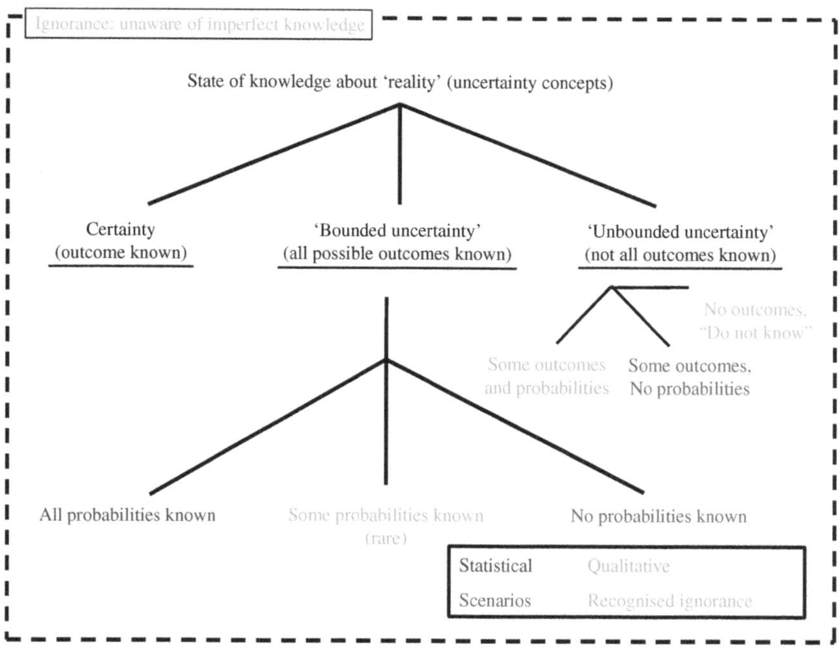

Figure 3.1. Taxonomy of imperfect knowledge resulting in different uncertainty situations (Brown, 2004)

It is useful in evaluating scientific uncertainty to distinguish between uncertainty about the 'outcomes' or scenarios, as possible states of 'reality' (mechanisms, events, observations), and uncertainty in terms of 'probability' (chance, likelihood, plausibility) for each outcome to occur. If one throws a perfect dice, the outcome is uncertain, but the 'draw' of a perfect dice is certain: we know precisely the probability for each of the 6 outcomes, each being 1/6. This is what we mean by 'uncertainty in terms of probability'. However, the estimates for the probability of each outcome can also be uncertain. If a model study says: "there is a 30% probability that this area will flood two times in the next year", there is not only 'uncertainty in terms of probability' but also uncertainty regarding whether the estimate of 30% is a reliable estimate.

Secondly, it is useful to distinguish between bounded uncertainty, where all possible outcomes are deemed 'known' (they can be distinct or indistinct) and unbounded uncertainty, where some or all possible outcomes are deemed unknown. Since quantitative probabilities require 'all possible outcomes' of an uncertain event and each of their individual probabilities to be known, they can only be defined for 'bounded uncertainties'. If probabilities cannot be quantified in any undisputed way, we often can still qualify the available body of evidence for the possibility of various outcomes. Inspired by legal practices, Weiss (2003a, 2003b, + personal communication 2004) developed the following 12 point subjective scale for qualifying evidence that can be used for this purpose:

- Impossible
- Hunch
- Reasonable suspicion
- Reasonable belief
- Reasonable indication
- Preponderance of the evidence
- Substantial and credible
- Clear indication
- Clear showing
- Clear and convincing
- Beyond a reasonable doubt
- Beyond any doubt

If outcomes but no probabilities are known we have to rely on 'scenario analysis'.

The bounded uncertainty where all probabilities are assumed known (the blue case in Figure 3.1) is often denoted 'statistical uncertainty' (e.g. Walker *et al.*, 2003). This is the case that is traditionally addressed in model based uncertainty assessments. It is important to note that this case only constitutes one of many of the decision situations outlined in Figure 3.1, and, in many situations, the main uncertainty in a decision situation cannot be characterised quantitatively.

3.3. Sources of uncertainty

Walker *et al.* (2003) describes the uncertainty as manifesting itself at different locations in the model based water management process. These locations, or sources, may be characterised as follows:

- *Context*, i.e. at the boundaries of the system to be modelled. The model context is typically determined at the initial stage of the study where the problem is identified and the focus of the model study selected as a confined part of the overall problem. This includes, for example, the external economic, environmental, political, social and technological circumstances that form the context of problem.
- *Input uncertainty* in terms of external driving forces (within or outside the control of the water manager) and system data that drive the model such as land use maps, pollution sources and climate data.
- *Model structure uncertainty* is the conceptual uncertainty due to incomplete understanding and simplified descriptions of processes as compared to nature.
- *Parameter uncertainty*, i.e. the uncertainties related to parameter values.
- *Model technical uncertainty* is the uncertainty arising from computer implementation of the model, e.g. due to numerical approximations and bugs in the software.
- *Model output uncertainty*, i.e. the total uncertainty on the model simulations taken all the above sources into account, e.g. by uncertainty propagation.

3.4. Nature of uncertainty

Walker *et al.* (2003) explain that the *Nature* of uncertainty can be categorised into:

- *Epistemic uncertainty*, i.e. the uncertainty due to imperfect knowledge.
- *Stochastic uncertainty*, i.e. uncertainty due to inherent variability, e.g. climate variability.

Epistemic uncertainty is reducible by more studies: e.g. comprising research and data collection. Stochastic uncertainty is non-reducible.

Often the uncertainty on a certain event includes both epistemic and stochastic uncertainty. An example is the uncertainty of the 100 year flood at a given site. This flood event can be estimated: e.g. by use of standard flood frequency analysis on the basis of existing flow data. The (*epistemic*) uncertainty may be reduced by improving the data analysis, by making additional monitoring (longer time series) or by a deepening our understanding of how the modelled system works. Note that this does not imply that data collection and further research automatically leads to a reduction of epistemic uncertainty. More research sometimes increases epistemic uncertainty in the short term. For instance, the recognised epistemic ignorance has temporary increased if new data show that our mechanistic understanding of the system (as represented in a model structure) cannot be right and science has not yet discovered what mechanisms or internal system feedbacks, or other factors were overlooked. The epistemic uncertainty is reduced only after we have discovered what was wrong about our earlier mechanistic understanding of the system. However, no matter how perfect both the data collection and the mechanistic understanding of the system are, and, no matter for how long historical data time series exist, there will always be some (*stochastic*) uncertainty inherent to the natural system, related to the stochastic and chaotic nature of several natural phenomena, such as weather. Perfect knowledge on these phenomena cannot give us a deterministic prediction, but would have the form of a perfect characterisation of the natural variability; for example, a probability density function for rainfall in a month of the year.

3.5. The uncertainty matrix

The uncertainty matrix in Figure 3.2 can be used as a tool to get an overview of the various sources of uncertainty in a modelling study. The matrix is modified after Walker *et al.* (2003) in such a way that it matches Figure 3.1 and so that the taxonomy now gives ‘uncertainty type’ in descriptions that indicate in what terms uncertainty can best be described. The vertical axis identifies the location or source of uncertainty while the horizontal axis covers the level and nature of uncertainty.

It is noticed that the matrix is in reality three-dimensional (source, type, nature). Thus, the categories Type and Nature are not mutually exclusive, and it may be argued that the matrix should be modified in such a way that the two uncertainties within Nature (epistemic and variability) should become subcells within the Type categories. This is not done for graphical reasons.

An inventory can be made of where the uncertainties are located and how they can be characterised by filling out all the cells in the matrix.

Source of uncertainty		Taxonomy (types of uncertainty)				Nature	
		Statistical uncertainty	Scenario uncertainty	Qualitative uncertainty	Recognised ignorance	Epistemic uncertainty	Stochastic uncertainty
Context	Natural, technological, economic, social, political						
Inputs	System data						
	Driving forces						
Model	Model structure						
	Technical						
	Parameters						
Model outputs							

Figure 3.2. The uncertainty matrix (modified after Walker *et al.*, 2003)

4. METHODOLOGIES FOR UNCERTAINTY ASSESSMENT

In general, transparency and reporting are essential for a good uncertainty assessment. Many methodologies and tools suitable for supporting uncertainty assessment have been developed and reported in the scientific literature. No methodology is applicable to address all the different aspects of uncertainty assessment. In this chapter some important methods are briefly described in one-page summaries:

- Data Uncertainty
- Error Propagation Equations
- Expert Elicitation
- Extended Peer Review (review by stakeholders)
- Inverse modelling (parameter estimation)
- Inverse modelling (predictive uncertainty)
- Monte Carlo Analysis
- Multiple Model Simulation
- NUSAP
- Quality Assurance
- Scenario Analysis
- Sensitivity Analysis
- Stakeholder Involvement
- Uncertainty Matrix

References to more detailed descriptions and to supporting software tools are provided in the summary descriptions. For several of the methodologies more extensive descriptions are available in the RIVM/MNP Tool Catalogue, that served as a starting point for the overview presented here (van der Sluijs *et al.*, 2004).

The methods can roughly be divided in three groups that differ slightly in purpose:

(1) *Methods to characterise and prioritise uncertainty:*
This includes methods for handling data uncertainty, methods of expert elicitation, parameter estimation through inverse modelling, sensitivity analysis, the NUSAP method and the uncertainty matrix.
(2) *Methods aiming to increase the quality of information:*
This includes procedures for quality assurance, extended peer review and stakeholder involvement.
(3) *Methods to quantify and propagate uncertainty in model calculations to produce uncertainty in model outcome:*

This includes the error propagation equations, Monte Carlo analysis, inverse modelling (parameter estimation and predictive uncertainty), multiple model simulation, various forms of sensitivity analysis and scenario analysis.

The methods are not necessarily valid for only one of the listed groups. Sensitivity analysis may, for instance, be used both to identify the importance of a given uncertainty source at an early stage in the modelling process, and again at a later stage to quantify the uncertainty with respect to the model results. Each of the methods will be reviewed in the following sections.

4.1. Data uncertainty

Description

Uncertainty in data may be described in 13 uncertainty categories depending on how data varies in time and space (Brown, 2004; Brown *et al.*, 2005).

Space-time variability	Measurement scale			
	Continuous numerical	Discrete numerical	Categorical	Narrative
Constant in space and time	**A1**	**A2**	**A3**	**4**
Varies in time, not in space	**B1**	**B2**	**B3**	
Varies in space, not in time	**C1**	**C2**	**C3**	
Varies in time and space	**D1**	**D2**	**D3**	

Each data category is associated with a range of uncertainty models, for which more specific probability density functions (pdfs) may be developed with different simplifying assumptions (e.g. Gaussian; second-order stationarity; degree of temporal and spatial autocorrelation). Furthermore, correlation in time and space is characterised by correlogram/variogram functions. Categorical data (3) differ from numerical data (1, 2), because the categories are not measured on a numerical scale.

A software tool for supporting the assessment of data uncertainty within the above framework is being developed within the EU FP5 project HarmoniRiB. This tool will become publicly available by 2006.

Resources required

Assessment of uncertainty requires a basic understanding of uncertainty concepts such as probability distribution and correlogram functions and their relation to the scale of measurement.

Uncertainty assessment deduced from specific information on the individual sources of uncertainty or from data analysis is a laborious and difficult task requiring substantial resources. This approach requires a specific knowledge on the various sources of uncertainty such as instrument accuracy (http://www.harmonirib.com) transformation functions from variable actually measured (e.g. water table) to variable of interest (e.g. discharge), aggregation in time and/or space, representativeness of sampling, etc.

Assessments based on expert judgements and literature values from similar settings are often the only feasible way in practice. Such assessments may be supported by guidelines for assessing uncertainty in various types of data originating from meteorology, soil physics and geochemistry, hydrogeology, land cover, topography, discharge, surface water quality, ecology and socio-economics (Van Loon and Refsgaard, 2005). This report has been prepared on the basis of literature reviews.

Strengths and limitations

+ Important input when assessing uncertainty of model output
+ Useful feedback information to design of monitoring programmes
− May require a lot of work
− Complex issue with many possibilities to make theoretically inconsistent assessments. Especially the correlation structure and its link with the scale of support may be difficult to understand

References

http://www.harmonirib.com
http://www.swift-wfd.com

Brown, J.D. (2004). Knowledge, uncertainty and physical geography: towards the development of methodologies for questioning belief. *Transactions of the Institute of British Geographers*, **29**(3), 367–381.

Brown, J.D., Heuvelink, G.B.M. and Refsgaard, J.C. (2005). An integrated methodology for recording uncertainties about environmental data. *Water Science and Technology*, **52**(6), 153–160.

Refsgaard, J.C., Nilsson, B., Brown, J., Klauer, B., Moore, R., Bech, T., Vurro, M., Blind, M., Castilla, G., Tsanis, I. and Biza, P. (2005). Harmonised techniques and representative river basin data for assessment and use of uncertainty information in integrated water management (HarmoniRiB). *Environmental Science and Policy*, **8**(3), 267–277.

Van Loon, E. and Refsgaard, J.C. (eds.) (2005). Guidelines for assessing data uncertainty in hydrological studies. First draft version prepared September 2004. Final version to be published beginning of 2005 on http://www.harmonirib.com.

4.2. Error propagation equations

Description

The error propagation equations (e.g. Mandel, 1984, Bevington and Robinson, 1992) are widely used in the experimental and measurement sciences to estimate error propagation in calculations. The error propagation equations are valid only if the following conditions are met:

- The uncertainties have Gaussian (normal) distributions;
- The uncertainties are relatively small: the standard deviation divided by the mean value being less than 0.3;
- The uncertainties have no significant covariance.

The error propagation equations for the most common operators are (σ is the standard deviation):

Addition and Subtraction: $z = x + y + \ldots$ or $z = x - y - \ldots$

$$\sigma_z = \sqrt{(\sigma_x^2) + (\sigma_y^2) + \ldots}$$

Multiplication by an exact number: $z = c\,x$

$$\sigma_z = c\sigma_x$$

Multiplication and Division: $z = x\,y$ or $z = x/y$

$$\frac{\sigma_z}{z} = \sqrt{\left(\frac{\sigma_x}{x}\right)^2 + \left(\frac{\sigma_y}{y}\right)^2 + \ldots}$$

The method can be extended to allow non-Gaussian distributions and to allow for covariances (see e.g.: http://www.itl.nist.gov/div898/handbook/mpc/section5/mpc55.htm

Resources required

The error propagation equations require no specific hardware or software and can typically be applied on the back of the envelope or on an ordinary scientific calculator, or using a spreadsheet.

Most of the time will be consumed by quantifying the uncertainties in the parameters and inputs, which can be derived from statistics if available or otherwise can for instance be obtained by means of expert elicitation.

Strengths and limitations

+ Requires very little resources and skills (but the choice of the aggregation level for the analysis is an important issue that does require skills)
+ Quick (but can be dirty)
– Has a limited domain of applicability (e.g. near-linearity assumption)
– The basic error propagation equations cannot cope well with distributions with shapes other than normal (but the method can be extended to account for other distributions).
– Leads to a tendency to assume that all distributions are normal, even in cases where knowledge of the shape is absent and, hence, a uniform distribution would be a better reflection of the state of knowledge.
– Cannot easily be applied in complex calculations.

References

Bevington, P.R. and Robinson, D.K. (1992). Data Reduction and Error Analysis for the Physical Sciences, WCB/McGraw-Hill Boston USA, p. 328.

IPCC (2000). Good Practice Guidance and Uncertainty Management in National Greenhouse Gas Inventories, IPCC.

Ku, H. (1966). Notes on the Use of Propagation of Error Formulas, J Research of National Bureau of Standards-C. Engineering and Instrumentation, **70C**(4), 263–273.

Mandel, J. (1984). The Statistical Analysis of Experimental Data, Dover Publications New York, USA, p. 410.

4.3. Expert elicitation

Description

Expert elicitation is a structured process to elicit subjective judgements from experts. It is widely used in quantitative risk analysis to quantify uncertainties in cases where there are no or too few direct empirical data available to infer on uncertainty. Usually the subjective judgement is represented as a 'subjective' probability density function (PDF) reflecting the expert's degree of belief.

Expert elicitation in the context of uncertainty quantification aims at a credible and traceable account of specifying probabilistic information regarding uncertainty, in a structured and documented way. Typically it is applied in situations where there is scarce or insufficient empirical material for a direct quantification of uncertainty, and where it is relevant to obtain inscrutable and defensible results (Hora, 1992).

Several elicitation protocols have been developed amongst which the much-used Stanford/SRI Protocol is the first (Spetzler and von Holstein, 1975; see also Morgan and Henrion, 1990; Chapters 6 and 7).

Expert elicitation typically involves the following steps: (1) Identify and select experts; (2) Explain to the expert the nature of the problem and the elicitation procedure. Create awareness of biases in subjective judgements and explore these. (3) Clearly define the quantity to be assessed and chose a scale and unit familiar to the expert. (4) Discuss the state of knowledge on the quantity at hand (strengths and weaknesses in available data, knowledge gaps, qualitative uncertainties). (5) Elicit extremes of the distribution. (6) Assess these extremes: could the range be broader than stated? (7) Further elicit and specify the distribution (shape and percentiles or characterising parameters). (8) Verify with the expert that the distribution that you constructed from the expert's responses correctly represents the expert's beliefs. (9) Decide whether or not to aggregate the distributions elicited from different experts (this only makes sense if the experts had the same mental models of the quantity for which a distribution was elicited).

Resources required

Typically performing a formal expert elicitation is a time and resource intensive activity. The whole process of setting up a study, selecting experts, preparing elicitation questions, performing expert training, expert meetings, interviews, analyses, writing rationales, documentation etc. can easily stretch over months or years. The choice of whether to perform a formal or a more informal elicitation (NCRP, 1996) depends on the price one is willing to pay for more inscrutable and defensible results, and will be influenced by the relevance and controversies regarding the problem area.

One needs to have good interviewing skills and a reasonable understanding of the field under consideration. A good understanding of biases in subjective judgements by experts is required to avoid these biases to the maximum extent possible. Skills are needed to draft a good questionnaire or template for the elicitation. Training in elicitation techniques may be needed.

Strengths and limitations

+ It has the potential to make use of all available knowledge including knowledge that cannot be easily formalised otherwise.
+ It can easily include views of sceptics and reveals the level of expert disagreement on certain estimates.
– The fraction of experts holding a given view is not proportional to the probability of that view being correct.
– One may safely average estimates of model parameters, but if the expert's models were incommensurate, one may not average models (Keith, 1996).
– If differences in expert opinion are irresolvable, weighing and combining the individual estimates of distributions is only valid if weighted with competence of the experts regarding making the estimate. There is no good way to measure competence.
– The results are sensitive to the selection of the experts whose estimates are gathered.

References

Ayyub, B.M. (2001). *Elicitation of Expert Opinions for Uncertainty and Risks*, CRC Press, Florida.

Hora, S.C. (1992). Acquisition of expert judgement: examples from risk assessment. *Journal of Energy Engineering*, 118, 136–148.

Meyer, M.A. and Booker, J.M. (1991). Eliciting and Analyzing Expert Judgement: A practical Guide, Academic Press, London.

Morgan, M.G. and Henrion, M. (1990). Uncertainty, A Guide to Dealing with Uncertainty in Quantitative Risk and Policy Analysis, Cambridge University Press.

Spetzler, C.S. and von Holstein, S. (1975). Probability encoding in decision analysis. Management Science, 22(3). http://legacy.ncsu.edu/classes/ce456001/www/Background1.html

4.4. Extended peer review (review by stakeholders)

Description

Extended peer review is the involvement of non-scientific actors in the quality assurance processes of knowledge production and assessment for policy making and risk management. Extended peer review can include all stakeholders engaged in the management of the problem at hand. Stakeholders are those actors who are directly or indirectly affected by an issue and who could affect the outcome of a decision making process regarding that issue or are affected by it.

Stakeholders' reasoning, observation and imagination are not bounded by scientific rationality. This can be beneficial when tackling ill-structured, complex problems. Consequently, the knowledge and perspectives of the stakeholders can bring in valuable new views on the problem and relevant information on that problem. The latter is known as "extended facts". Stakeholders can contribute to the quality of knowledge in a number of ways. These include improvement of the quality of the problem formulation and the questions addressed by the scientists; the contribution of knowledge on local conditions which may help determine which data are strong and relevant or which response options are feasible; providing personal observations which may lead to new foci for empirical research addressing dimensions of the problem that were previously overlooked; criticism of assumptions made by the scientist, which may lead to changes towards assumptions that better match real life conditions; and, creative thinking of mechanisms and scenarios through which projected environmental and hydrologic changes may affect different sectors of society.

Resources required

Extended peer review requires well-developed communication and group moderation skills, along with a good understanding of public perceptions of risks and of science in general. Didactic skill is also required to help stakeholders to understand the sometimes complex and abstract concepts used in scientific assessments. Involving social scientists in the design and implementation of extended peer review processes is recommended.

Strengths and limitations

+ Allows for the use of extra knowledge from non scientific sources
+ Increases the level of public accountability in knowledge production
+ Promotes a development from knowledge consumption towards knowledge co-production

- Scientific and non-scientific participants are often not reciprocally accountable
- Public tends to get co-opted according to dominant view
- May reproduce power asymmetries.

References

Funtowicz, S.O. and Ravetz, J.R. (1996). Risk Management, post-normal science, and extended peer communities. In: Hood C and Jones DKC, Accident and Design, Contemporary debates in Risk Management. UCL Press, pp. 172–182.

Healy, S. (1999). Extended peer communities and the ascendance of post-normal politics. *Futures*, **31**, 655–669.

Fuller, S. (2003). On the need to extend peer review: A reply to Kihara. *Social Epistemology*, **17**(1), 74–78

Kihara, H. (2003). The extension of peer review, How should it or should not be done? *Social Epistemology*, **17**(1), 65–73.

Pellizzoni, L. (1999). Reflexive modernization and beyond: knowledge and value in the politics of environment and technology. *Theory, Culture & Society* **16**(4), 99–125.

4.5. Inverse modelling (parameter estimation)

Description

Parameter values are often optimised through inverse modelling. This is also denoted as automatic calibration (Duan *et al.*, 1994; Hill, 1998; Doherty, 2003). An optimal parameter set is sought "automatically" by minimising an objective function, often defined as the summed squared deviation between the calibration targets (field data) and their simulated counterparts.

Most inversion techniques have the benefit that they in addition to optimal parameter values also produce calibration statistics in terms of parameter- and observation sensitivities, parameter correlation and parameter uncertainties. As the model calibration is based on a single model (with one possible model structure), errors in the model structure will be propagated to the model parameter uncertainties. The estimated parameter uncertainties are thus uncertainties for the effective model parameter given both the model structure and available observations (number, location and types of observations). This also means that estimated parameter uncertainties will not compensate adequately for the model structure uncertainty, when the model is used for prediction of conditions beyond the calibration base (e.g. when calibrating on groundwater flow and subsequently using the model to simulate groundwater transport and concentrations). Unless the model structure uncertainty is somehow considered the resulting simulation uncertainty will be significantly underestimated.

A different approach to autocalibration has been developed by Beven (Beven and Binley, 1992; Beven, 2002) in the Generalised Uncertainty Likelihood Estimation (GLUE) method. Beven argues that one optimal set of parameter values does not exist (equifinality concepts), and that many likely sets therefore should be included in model simulations. Thus modelling is carried out in a Monte Carlo framework, where a large number of realisations of parameter sets are sampled from a broad a priori space of possible parameter values. By adopting threshold values for acceptable simulation accuracy, and rejecting parameter combinations for non-behavioural models, it is then possible to estimate the space of plausible parameter values and hence make inferences about the parameter uncertainty.

Resources required

Many software tools support inverse modelling, which often requires only moderate resources and skills to operate. Good modelling skills and comprehensive experience are, however, required to understand and analyse the results properly.

Universal optimisation routines can be downloaded as freeware, e.g. PEST (http://www.sspa.com/pest) and UCODE (http://water.usgs.gov/software/ucode.html).

Strengths and limitations

+ Gives an objective estimate of parameter uncertainty based on information in available field data
+ Most methods gives information on correlation between parameters
– Most methods rely on assumptions, e.g. linearity as well as random and normally distributes residuals
– Parameter uncertainties are computed for the specific model structure and available observations. The uncertainties may thus not be valid for conditions beyond the calibration base.

References

Beven, K. and Binley, A.M. (1992). The future role of distributed models: model calibration and predictive uncertainty. *Hydrological Processes*, **6**, 279–298. (See also http://www.es.lancs.ac.uk/hfdg/glue.html for further information on GLUE)

Beven, K. (2002). Towards a coherent philosophy for modelling the environment. *Proceedings of the Royal Society of London A*, **458**(2026), 2465–2484.

Doherty, J. (2003). Ground water model calibration using pilot points and regularization. *Ground Water*, **41**(2), 170–177. (See also http://www.sspa.com/pest for further information/download of PEST).

Duan, Q., Sorooshian, S. and Gupta, V.K. (1994). Optimal use of the SCE-UA global optimization method for calibrating watershed models. *Journal of Hydrology*, **158**, 265–284.

Hill, M.C. (1998). Methods and guidelines for effective model calibration. U.S. Geological Survey, Water-Resources Investigations Report 98-4005. Denver CO (document available at http://pubs.water.usgs.gov/wri984005/).

4.6. Inverse modelling (predictive uncertainty)

Description

In addition to parameter estimation some of the inverse optimisation routines include the ability to estimate predictive uncertainties. The method by which the predictive uncertainty is derived varies among the inversion routines. But common to many of the local optimisation routines based on non-linear regression, is that the prediction of interest is treated as an observation, and the regression algorithm is then used to quantify the effect of the parameter uncertainty on this "observation". Some methods rely on a semi-analytical solution in which the regression algorithm is used to compute either a predictive uncertainty interval for the output variable or uncertainty in the difference between a reference case and a scenario simulation. Other methods use the regression to seek the maximum or minimum value of the prediction under the constraint that the model must be calibrated at an acceptable level, which is defined by some predefined acceptance level of the objective function.

Resources required

If inverse optimisation is used for calibration, the extra resources required to run the predictive mode is very limited. Some methods may need a considerable number of model evaluation and thereby long execution times.

Strenghts and limitations

+ Gives an objective estimate of the predictive uncertainty given the applied model structure
− Most methods rely on assumptions, e.g. linearity as well as random and normally distributes residuals
− The predictive uncertainty are often limited to data types that can be defined as observations, this means that uncertainties on variables that are interpolated or extrapolated compared to the available field data can not be quantified by these methods.

References

Doherty, J. (2003). Ground water model calibration using pilot points and regularization. *Ground Water*, **41**(2), 170–177.

Hill, M.C. (1998). Methods and guidelines for effective model calibration. U.S. Geological Survey, Water-Resources Investigations Report 98-4005. Denver CO (document available at http://pubs.water.usgs.gov/wri984005/).

Tonkin, M.J., Hill, M.C. and Doherty, J.E. (2004). MODFLOW-2000, the U.S. geological survey modular ground-water model – Documentation of mod-predict for predictions, prediction sensitivity analysis, and enhanced analysis of model fit. Open-File Report 03-385 (document available at http://water.usgs.gov/nrp/gwsoftware/modpredict/modpredict.html).

4.7. Monte Carlo Analysis

Description

Monte Carlo Simulation is a statistical technique for stochastic model-calculations and analysis of error propagation in calculations. Its purpose is to trace out the structure of the distributions of model output. In it's simplest form this distribution is mapped by calculating the deterministic results (realisations) for a large number of random draws from the individual distribution functions of input data and parameters of the model. Advanced sampling methods have been designed such as Latin Hyper Cube sampling to reduce the required number of model runs needed to get sufficient information about the distribution in the outcome (mainly to save computation time). Latin Hyper Cube sampling makes use of stratification in the sampling of individual parameters. As in random Monte Carlo sampling, pre-existing information about correlations between input variables can be incorporated. Monte Carlo analysis requires the analyst to specify probability distributions of all inputs and parameters, and the correlations between them. Both probability distributions and correlations are usually poorly known. Ignoring correlations and co-variance in input distributions may lead to substantial under- or over-estimation of uncertainty in model outcome.

Monte Carlo analysis is a skill, and when it is used without following some basic principles of good practice it may lead to meaningless results. A good primer for responsible use of Monte Carlo Analysis is, for instance, the EPA "Guiding Principles for Monte Carlo Analysis" (EPA, 1997).

Most Monte Carlo analysis software offers the possibility to determine the relative contribution of uncertainty in each parameter to the uncertainty in a model output, e.g. by sensitivity charts, and can be used for a sophisticated analysis of trends in the presence of uncertainty.

Resources required

A number of commercial and free software packages (@Risk, Crystal Bal, SimLab) are available to do Monte Carlo analysis. Another commercial package is Analytica (http://www.lumina.com), which is a quantitative modelling environment with built-in Monte Carlo algorithms. Packages such as Crystal Ball are very easy to learn. If you are familiar with Excel it takes less than one hour to get proficient with Crystal Ball. It takes more time to get proficient with SimLab and requires more skills because one has to interface SimLab with one's own model.

Performing Monte Carlo Analysis on complex models with long computation times and many variables requires major computer resources and computation time. Software solutions to combine PCs in a network to create a super computer capacity can be an outcome (e.g. the Super Muse software from the US-EPA: http://www.epa.gov/athens/research/modeling/supermuse/supermuse.html)

Strengths and limitations

+ Provides comprehensive insight in to how specified uncertainty in inputs propagates through a model.
+ Forces analysts to consider uncertainty and interdependencies among different inputs explicitly.
+ Is capable of coping with any conceivable shape of PDF and can account for correlations.
+ Can be used in 2-dimensional models to assess variability and epistemological uncertainty separately.
– Monte Carlo assessment is limited to quantifiable uncertainties.
– One may not have any reasonable basis on which to ascribe a probability distribution to parameters.
– May take large run-time for computational intensive models. This can partly be remedied by using more efficient sampling techniques (e.g. Latin Hypercube Sampling).
– The interpretation of a probability distribution of the model output by decision makers is not always straightforward. There is no single rule arising out of such a distribution that can guide decision-makers.

References

Crystal Ball (2000). User Manual Decision Engineering Inc., Denver.

EPA (1997). Risk Assessment Forum, Guiding Principles for Monte Carlo Analysis, EPA/630/R-97/001. (http://www.epa.gov/NCEA/pdfs/montcarl.pdf)

Palisade Corporation (2000). Guide to Using @RISK – Risk Analysis and Simulation Add-in for Microsoft Excel, Version 4, March 2000.
Saltelli, A., Chan, K. and Scott, M. (2000). Sensitivity Analysis, John Wiley & Sons publishers, Probability and Statistics Series.
Saltelli, A., Tarantola, S., Campolongo, F. and Ratto, M. (2004). Sensitivity Analysis in Practice: A Guide to Assessing Scientific Models, John Wiley & Sons publishers.
Vose, D. (2000). Risk Analysis – A quantitative guide, 2nd edition. John Wiley & Sons, Ltd. Chichester.

4.8. Multiple model simulation

Description

Multiple Model Simulation is a strategy to address uncertainty about model structure. Instead of doing an assessment using a single model, the assessment is carried out using different models. For instance, this can be realised by having alternative model codes with different process descriptions (Linkov and Burmistrov, 2003; Butts *et al.*, 2004) or, in the groundwater case, by having different conceptual models based on different geological interpretations (Troldborg, 2000, Selroos *et al.*, 2001).

The strategy of applying several alternative models based on codes with different model structures is also common in climate change modelling. Thus in its description of uncertainty related to model predictions of both present and future climates the IPCC Special Report on Emission Scenarios (IPCC, 200x) bases its evaluation on scenarios of many (up to 35) different models.

The same strategy is followed in the Dialogue Model (Visser *et al.,* 2000). Dialogue simulates the cause effect chain of climate change, using mono-disciplinary sub-models for each step in the chain. The chain starts with scenarios for economic growth, energy demand, fuel mix etc., leading to emissions of greenhouse gasses, leading to changes in atmospheric composition, leading to radiative forcing of the climate, leading to climate change, leading to impacts of climate change on societies and ecosystems. Rather than picking one main-stream mono-disciplinary sub-model for each step, Dialogue uses multiple models for each step (for instance, three different carbon cycle models, five different GCM model-outcomes, etc.), representing the major part of the spectrum of expert opinion in each discipline. This multiple model approach facilitates the inclusion of new alternative models in each step to accommodate new scientific ideas on the structure of a given sub-model.

Refsgaard *et al.* (2006) present a new framework for dealing with uncertainty due to model structure error, based on alternative conceptual models and assessment of their pedigree and adequacy.

Resources required

Multiple model simulation requires modelling skills to design and implement the various model structures. Varying model structure is a much more complex task than varying parameters and there are no ready to use software solutions available. The required resources will further depend on the complexity of the model and the required time for each model run. If models developed and run by different modelling groups are used, a substantial part of the resources will be required for co-ordination and project management.

Strengths and limitations

+ The effects of alternative model structures can be analysed explicitly.
+ It makes it possible to include expert knowledge on plausible model structures.
+ It substantially reduces the chances that the assessment overlooks important aspects of the problem compared to the use of single models only. In that sense it may reduce surprises.
− We cannot be sure whether we have adequately sampled the relevant space of plausible models. Important plausible model structures could be overlooked.
− Often we do not know the plausibility and even less the probability of each conceivable model structure.
− The expert knowledge on which the formulations of the alternative conceptual models have to be based has an unavoidable subjective element.

References

Butts, M.B., Payne, J.T., Kristensen, M. and Madsen, M. (2004). An evaluation of the impact of model structure on hydrological modelling uncertainty for streamflow simulation. *Journal of Hydrology*, **298**, 242–266.

Linkov, I. and Burmistrov, D. (2003). Model uncertainty and choices made by the modelers: lessons learned from the international atomic energy agency model intercomparisons. *Risk Analysis*, **23**(6), 1297–1308.

Nakicenovic, N., Alcamo, J., Davis, G., de Vries, B., Fenhann, J., Gaffin, S., Gregory, K., Grübler, A., Jung, T.Y., Kram, T., La Rovere, E.L., Michaelis, L., Mori, S., Morita, T., Pepper, W., Pitcher, H., Price, L., Riahi, K., Roehrl, A., Rogner, H.-H., Sankovski, A., Schlesinger, M., Shukla, P., Smith, S., Swart, R., van Rooijen, S., Victor, N. and Dadi, Z. (2000). IPCC Special Report on Emissions Scenarios. Cambridge University Press, Cambridge, United Kingdom and New York, NY, USA, 599 pp.

Refsgaard, J.C., van der Sluijs, J.P., Brown, J. and van der Keur, P. (2006). A framework for dealing with uncertainty due to model structure error. *Adv. Water Res.*, **29**(11), 1586–1597.

Selroos, J.O., Walker, D.D., Strom, A., Gylling, B. and Follin, S. (2001). Comparison of alternative modelling approaches for groundwater flow in fractured rock. *Journal of Hydrology*, **257**, 174–188.

Troldborg, L. (2000). Effects of geological complexity on groundwater age prediction. Poster session 62C, AGU. *EOS Transactions*, **81**(48), F435.

Visser, H., Folkert, R.J.M., Hoekstra, J. and De Wolff, J.J. (2000). Identifying key sources of uncertainty in climate change projections. *Climatic Change*, **45**, 421–457.

4.9. NUSAP

Description

The NUSAP system for multidimensional uncertainty assessment (Funtowicz and Ravetz, 1990) aims to provide an analysis and diagnosis of uncertainty in science for policy. The basic idea is to qualify quantities using the five qualifiers of the NUSAP acronym: Numeral, Unit, Spread, Assessment, and Pedigree. NUSAP complements quantitative analysis with expert judgement of reliability (Assessment) and systematic multi-criteria evaluation of the different phases of production of a given knowledge base (Pedigree). Pedigree criteria can be: proxy representation, empirical basis, methodological rigor, theoretical understanding, and degree of validation. Pedigree assessment can be further extended to also address societal dimensions of uncertainty, using criteria addressing different types of value weighting, quality of problem frames etc. NUSAP provides insight on two independent properties related to uncertainty in numbers, namely spread and strength. Spread expresses inexactness whereas strength expresses the methodological and epistemological limitations of the underlying knowledge base. The two metrics can be combined in a Diagnostic Diagram mapping strength of for instance model parameters and sensitivity of model outcome to spread in these model parameters. Neither spread alone nor strength alone is a sufficient measure for quality. Robustness of model output to parameter strength could be good even if parameter strength is low, if the spread in that parameter has a negligible effect on model outputs. In this situation our ignorance of the true value of the parameter has no immediate consequences. Alternatively, model outputs can be robust against parameter spread even if its relative contribution to the total spread in the model is high provided that parameter strength is also high. In the latter case, the uncertainty in the model outcome adequately reflects the inherent irreducible uncertainty in the system represented by the model. Uncertainty then is a property of the modelled system and does not stem from imperfect knowledge on that system. Mapping components of the knowledge base in a diagnostic diagram thus reveals the weakest spots and helps in the setting of priorities for improvement.

Resources required

Resources required for assessing the Spread qualifier depend on the method chosen (some form of Sensitivity Analysis or Monte Carlo analysis usually in combination with expert elicitation will be needed).

For the assessment of Pedigree, many resources (pedigree matrices, pedigree calculator, kite diagram maker, elicitation protocol and questionnaires) are freely available from http://www.nusap.net. Basic skills of Expert Elicitation are required.

Basic skills for facilitating structured group discussions are needed if one uses an expert workshop. In addition, skills are needed to arrive at a balanced composition of the workshop audience to minimise biases.

Time required per expert elicitation in a one to one interview depends on the number of parameters and the complexity of the case. Typically, it may vary between 1 and 5 hours. A substantial amount of time may be needed for a good preparation of the elicitation interviews.

Recommended length for a NUSAP expert elicitation workshop is between one and one and a half days.

Strengths and limitations

+ Identifies both quantitative and qualitative uncertainty in quantitative information and enables them to be displayed in a standardised and self-explanatory way.
+ Promotes criticism by clients and users of all sorts, both expert and lay, and will thereby support extended peer review processes.
+ It is flexible in its use and can be used on different levels of comprehensiveness: from a 'back of the envelope' sketch based on self elicitation to a comprehensive and sophisticated procedure involving structured, informed, in-depth group discussions on a parameter by parameter format.
– The scoring of pedigree criteria is to a large extent based on subjective judgements. Therefore, outcomes may be sensitive to the selection of experts involved in the scoring.
– It is hard to apply the method to complex models with large numbers of parameters.

References

Www.nusap.net

Funtowicz, S.O. and Ravetz, J.R. (1990). Uncertainty and Quality in Science for Policy. Dordrecht: Kluwer.

Craye, M., van der Sluijs, J.P. and Funtowicz, S.O. (2004). A reflexive approach to dealing with uncertainties in environmental health risk science and policy. *International Journal for Risk Assessment and Management*, **5**(2), 216–236.

Van der Sluijs, J.P., Craye, M., Funtowicz, S.O., Kloprogge, P., Ravetz, J.R. and Risbey, J. (2005). Combining quantitative and qualitative measures of uncertainty in model based environmental assessment: the NUSAP system. *Risk Analysis*, **25**(2), 481–492.

Van der Sluijs, J.P., Risbey, J. and Ravetz, J.R. (2005). Uncertainty assessment of VOC emissions from paint in The Netherlands, environmental monitoring and assessment, **105**(1–3), 229–259.

4.10. Quality assurance

Description

Quality assurance (QA) may be defined as protocols and guidelines to support the proper application of models. Important aims of QA are to ensure the use of best practise and to ensure that the expected accuracy and model performance are in accordance with the project objectives. Uncertainty and QA are intimately linked as uncertainty plays a very important role throughout the modelling process.

Many QA guidelines exist such as Middlemis (2000), Van Waveren *et al.* (1999) and Scholten *et al.* (2007). Most guidelines have been prepared for a single modelling domain, such as groundwater and for use in a particular, confined, market such as a single country (Refsgaard, 2002; Refsgaard *et al.*, 2005).

Key elements of QA procedures include:

- Definition of the purpose of the modelling study, including translation of the water manager and stakeholder needs to preliminary performance criteria.
- Establishment of performance criteria. The accuracy of the model predictions has to be established via a trade off between the benefits of improving the accuracy in terms of less uncertainty on the management decisions and the costs of improving the accuracy through additional model studies and/or collection of additional field data.
- Reviews carried out by independent auditors with subsequent consultation between the modeller, the water managers and possibly the stakeholders at different phases of the modelling project. Important aspects of the reviews are to ensure that good practise has been followed and to evaluate if the uncertainty assessments made by the modeller are credible and adequate.
- Performance of model validation tests, i.e. testing of model performance against independent data that have not been used for calibration in order to assess the accuracy and credibility of the model simulations for situations comparable to those where it is intended to be used.

The HarmoniQuA project has developed a comprehensive set of QA guidelines for multiple modelling domains combined with a supporting software tool, MoST (downloadable via www.harmoniqua.org). In addition to the software tool, the HarmoniQuA guidelines are unique in their dedication aspects, namely that different tasks and responsibilities are described for different users, different modelling domains and different levels of modelling job complexity.

Resources required

For QA to become successful it is required that both the modeller and the water manager are motivated and active in supporting its use. The water manager has a particular responsibility, because they have the power to request and pay for adequate QA in modelling studies. Therefore, QA guidelines can only be expected to be used in practice, if the water manager prescribes their use. In this respect, it is essential that the water manager has the technical capacity to organise the QA process.

Strengths and limitations

+ Improves the chances that best practise is used.
+ Improves the quality of the work and the credibility of the results.
+ Makes it possible to include stakeholders into the modelling process in a formalised framework, either as information, consultation or involvement.
+ Improves the transparency and reproducibility of the modelling work.
– To be done properly it requires resources and adds to the cost of a modelling study.
– May become a 'rubber stamp' and generate a false credibility if not designed and performed thoroughly.

References

Middlemis, H. (2000). Murray-Darling Basin Commission. Groundwater flow modelling guideline. Project no. 125, Aquaterra Consulting Pty Ltd., South Perth. Western Australia.

Refsgaard, J.C., Henriksen, H.J., Harrar, W.G., Scholten, H. and Kassahun, A. (2005). Quality assurance in model based water management – review of existing practice and outline of new approaches. *Environmental Modelling and Software*, **20**(10), 1201–1215.

Refsgaard, J.C. (Ed) (2002). State-of-the-Art Report on Quality Assurance in modelling related to river basin management. Report from the EU research project HarmoniQuA. www.harmoniqua.org. 18 chapters, 182 pp. Geological Survey of Denmark and Greenland, Copenhagen.

Scholten, H.S., Kassahun, A., Refsgaard, J.C., Kargas, T., Gavadinas, C. and Beulens, A.J.M. (2007). The HarmoniQuA methodology to support multidisciplinary model-based water management. *Environmental Modelling and Software*, **22**(5), 743–759.

Van Waveren, R.H., Groot, S., Scholten, H., Van Geer, F., Wösten, H., Koeze, R. and Noort, J. (1999). Good Modelling Practice Handbook, STOWA, Utrecht, The Netherlands (in Dutch). English version from http://www.landandwater.tudelft.nl/Downloads/GMP-uk.pdf.www.harmoniqua.org from where the supporting tool MoST can be downloaded.

4.11. Scenario analysis

Description

Scenario Analysis aims to describe logical and internally consistent sequences of events to explore how the future may, could or should evolve from the past and present. The future is inherently uncertain. Different alternative futures can be explored through scenario analysis. As such, scenario analysis is also a tool to deal explicitly with different assumptions about the future. A definition is "Scenarios are descriptions of journeys to possible futures. They reflect different assumptions about how current trends will unfold, how critical uncertainties will play out and what new factors will come into play" (UNEP, 2002). Another definition is: "A scenario is a description of the present state of a social and or natural system (or a part of it), of possible and desirable future states of that system along with sequences of events that could lead from the present state to these future states." Other definitions also include the purposes of the use of scenarios.

Different types of scenarios can be distinguished. For instance, Alcamo (2001) discerns baseline vs. policy scenarios, exploratory vs. anticipatory scenarios and qualitative vs. quantitative scenarios.

Baseline scenarios present the future state of society and environment in which no (additional) environmental policies do exist or have a discernible influence on society or the environment. Policy scenarios depict the future effects of environmental protection policies. Exploratory scenarios start in the present and explore possible trends into the future. Anticipatory scenarios start with a prescribed vision of the future and then work backwards in time to visualise how this future could emerge. Qualitative scenarios describe possible futures in the form of narrative texts or so-called "story-lines". Quantitative scenarios provide tables and figures incorporating numerical data often generated by sophisticated models.

Finally, scenarios can be surprise-free or trend scenarios, that extend foreseen developments, on the one hand or include surprises and exploring the extremes (e.g. best case/worst case) on the other hand.

Resources required

Scenario analysis requires creativity and ability to think outside the scope of the familiar and the present. Furthermore, it requires insight in dynamics, relationships, in synergies of systems and their environment and thus it requires a broad knowledge of the field involved. Therefore, scenarios analysis should take place in an interdisciplinary team.

In the case of a quantitative approach, computer models or spreadsheets or other software are needed to run/visualise scenarios. Access to relevant data is important in order to be able to construct the scenarios. In the case of a qualitative approach, input has to be collected from experts, stakeholders or users in workshops with stakeholders to be able to develop storylines. Basic skills for facilitating groups are required. Both approaches are time and resource consuming.

Strengths and limitations

+ Scenarios are often the only way to deal with the unknown future;
+ Assumptions about future developments are made transparent and documented;
+ Gives insight in key factors that determine future developments;
+ Creates awareness on alternative development paths, risks, and opportunities and possibilities for policies or decision-making;
– The analysis is limited to those aspects of reality that can be quantified (quantitative scenarios);
– Difficult to test underlying assumptions (qualitative scenarios);
– Frequently scenarios do not go beyond trend extrapolation (quantitative scenarios) and are surprise-free;
– Frequently models used contain only one view, which will make the outcomes narrow in scope, thus not doing justice to the wish to explore fundamentally different futures.

References

Alcamo, J. (2001). Scenarios as tools for international environmental assessments. Environmental issues report. Experts corner report. Prospects and Scenarios No. 5, European Environment Agency, Copenhagen.

Van Der Heijden, K. (1996). Scenarios: The Art of Strategic Conversation, John Wiley & Sons; ISBN: 0471966398.

UNEP (2002). Global Environmental Outlook 3: Past, present and future perspectives, Earthscan.

4.12. Sensitivity analysis

Description

Sensitivity analysis (SA) is the study of how the variation in the output of a model (numerical or otherwise) can be qualitatively or quantitatively apportioned to different sources of variation, and of how the outputs of a given model depend upon the information fed into it (Saltelli *et al.*, 2000, 2004).

Depending on the complexity of a model's output space SA methods may range from the simple to the relatively complex. If a model's output space is linear or approximates a hyperplane, SA may be conducted through a straightforward application of differential analysis. This is typically done by taking partial derivatives of the output with respect to one input, holding all other inputs constant.

If a model's output space is nonlinear (or does not approximate a hyperlplane) then the assumptions for differential analysis do not hold. Differential analysis may be conducted, but the analyst should be aware that the results may apply only to a narrow range of the output space. For this reason, differential analysis in this situation is referred to as Local SA.

If the analyst wishes to conduct Global SA (i.e. SA across the model's entire output space) for nonlinear (non-hyperplaner) models, then other analytical methods should be used. These include such methods as Monte Carlo Analysis, Morris' One-at-a-time method and various variance based methods such as Fourier Amplitude Sensivity Test (FAST).

Resources required

Sensitivity Analysis (SA) requires basic computer skills and basic knowledge of statistical concepts. Software for SA is available both as freeware (such as SIMLAB) and commercial. SIMLAB: http://www.jrc.cec.eu.int/uasa/prj-sa-soft.asp and http://sensitivity-analysis.jrc.cec.eu.int/default2.asp?page=SIMLAB

Commercial packages for Monte Carlo analysis such as @Risk (http://www.palisade.com) and Crystal Ball (http://www.decisioneering.com/crystal_ball) include options for sensitivity analysis. @Risk and Crystal Ball are designed as fully integrated MS-Excel add-in programs with their own toolbar and menus.

The precise requirements depend on the complexity of the model to which one applies the Sensitivity Analysis and the number of factors one wants to include in the analysis. It may be necessary to program an interface between the model and the Sensitivity Analysis software.

Strengths and limitations

+ Gives insight in the potential influences of all sorts of changes in inputs.
+ Helps discrimination across parameters according to importance for the accuracy of the outcome.
+ Software for Sensitivity Analysis is freely available.
+ SA can help in evaluating the fit-for-purpose of a model.
− Has a tendency to yield an overload of information.
− SA does not require one to assess how likely it is that specific values of the parameters will actually occur.
− Commercial software packages for SA can be used with minimal knowledge on the sampling and calculations techniques itself, which allows incompetent use.
− In practice SA often takes the model structure and system boundaries for granted.

References

Saltelli, A., Chan, K. and Scott, M. (2000). Sensitivity Analysis, John Wiley & Sons publishers, Probability and Statistics Series.

Saltelli, A., Tarantola, S., Campolongo, F. and Ratto, M. (2004). Sensitivity Analysis in Practice: A Guide to Assessing Scientific Models, John Wiley & Sons Publishers.

Campolongo, F., Tarantola, S. and Saltelli, A. (1999). Tackling quantitatively large dimensionality problems. *Computer Physics Communication*, **117**, 75–85.

Morris, M.D. (1991). Factorial sampling plans for preliminary computational experiments. *Technometrics*, **33**(2). http://sensitivity-analysis.jrc.cec.eu.int/default.htm

4.13. Stakeholder involvement

Description

Stakeholder involvement in not only the decision making process but also in knowledge production and knowledge use, can help to assess and manage complex (environmental) problems in a better way. This potential can be tapped in three ways:

(1) By enabling them to articulate issues of concern and to improve the problem framing for research and policy;
(2) By utilising their own (non scientific) knowledge and observations and their capacity to invent new options; and
(3) By involving them actively in the quality control of the operational knowledge that is co-produced (extended peer review).

By engaging stakeholders into the risk management process it is likely that they will not only contribute to future risk management decisions but also support their implementation.

We have another concern as well when addressing complex environmental policy issues in a participatory way. The scientific soundness of the presumed causal mechanisms underlying alternative problem definitions needs to be warranted if we want to base our policies on scientifically sound underpinnings. Although there is more than one legitimate interpretation of the science, this plurality of perspectives does not deny the special competence of scientists. It does mean that there is a mixing and blending of skills, partly technical and partly personal, of all those engaged that can enrich the comprehension of the whole.

The RIVM/MNP Guidance for Uncertainty Assessment and Communication (Van der Sluijs *et al.*, 2004) has a useful section on stakeholder involvement, including an instrument for discourse analysis.

Stakeholders can also be used in environmental monitoring, as is done for instance in the European Phenological Data Platform for Climatological Applications and other initiatives where citizens provide observations on effects of climate changes on nature (e.g. first day in the year that a bird species is observed).

The HarmoniCOP project has developed a typology to characterise tools to support the public participation process in relation to implementation of the Water Framework Directive (Maurel, 2003).

Resources required

Stakeholder involvement in knowledge production requires very good communication and deliberation skills and the involvement of social scientists is recommended. It also puts high requirements on the infrastructure through which stakeholders can contribute their knowledge and on the quality assurance of the knowledge provided, especially if this infrastructure has to function during a long period of time (as with environmental monitoring systems).

Strengths and limitations

+ Increases the level of public accountability in knowledge production.
+ May increase the public support for the knowledge base that is co-produced.
– There are many unresolved challenges regarding the quality control of knowledge produced by non scientific actors.

References

Kloprogge, P. and van der Sluijs, J.P. (2006). The inclusion of stakeholder knowledge and perspectives in integrated assessment of climate change. *Climatic Change*, **75**(3), 359–389.

Maurel, P. (Ed.) (2003). Public participation and the European Water Framework Directive. Role of Information and Communication Tools. Workpackage 3 Report of the HarmoniCOP Project. Cemagref, Montpellier, www.harmonicop.info.

Osidele, O.O. and Beck, M.B. (2002). Integrating Stakeholder Imagination with Scientific Theory: A Case Study of Lake Lanier, USA, *Proceedings of the IEMSS*, 31–36

Stern, P.C. and Fineberg, H.V. (Eds.) (1996). Understanding risk, informing decisions in a democratic society, Washington DC: National Research Council, National Academy Press.

van der Sluijs, J.P., Risbey, J.S., Kloprogge, P., Ravetz, J.R., Funtowicz, S.O., Quintana, S.C., Guimarães Pereira, Â., De Marchi, B., Petersen, A.C., Janssen, P.H.M., Hoppe, R. and Huijs, S.W.F. (2003). RIVM/MNP Guidance for Uncertainty Assessment and Communication: Detailed Guidance Utrecht University & RIVM (available from www.nusap.net).

Yosie, F.T. and Herbst, T.D. (1998). Using stakeholder processes in environmental decision-making. An evaluation of lessons learned, key issues and future challenges, Ruder Finn Washington, ICF Inc., http://www.gdrc.org/decision/nr98ab01.pdf

4.14. Uncertainty matrix

Description

The uncertainty matrix (Walker *et al.*, 2003; Janssen *et al.*, 2003) can be used to identify and prioritise the most important uncertainties in a given model study. The matrix shown below is an example of a project specific adaptation of the more general uncertainty matrix shown in Figure 3.2, above.

Source of uncertainty	Type of uncertainty				Importance	
	Statistical uncertainty	Scenario uncertainty	Qualitative uncertainty	Recognised ignorance	Weighting	(uncertainty x weight)
Problem context						
– future agritultural practise		Medium	Medium	Medium	Large	Medium
– future climate		Medium	Medium	Large	Medium	Medium
Input data						
– catchment data	Medium			Small	Large	Medium
– nitrate load from agriculture	Small			Small	Large	Small
Parameter uncertainty						
– water quantity	Small			Small	Medium	Small
– water quality	Medium			Medium	Medium	Small
Model structure (conceptual)						
– geology		Large	Large	Medium	Large	Large
– nitrate reduction in underground		Medium	Medium	Large	Large	Large
Model technical uncertainty						
– numerical approximation	Small			Small	Medium	Small
– bugs in software				Medium	Medium	Small
					SUM:	

For a specific application the different sources of uncertainty are listed in the rows and the type of uncertainty associated to each source is noted and characterised. This may be done either quantitatively or, as shown in the above figure, qualitatively. The importance of each source may then be characterised by a weighting depending on its impact on the modelling study in question. The Sum of uncertainty may then be assessed, e.g. by use of the error propagation equation (Section 4.2).

It may not be possible to identify all sources of uncertainty and/or assigning correct weightings from the project start. The matrix may thus be used interactively by adding or reassigning weights during the modelling process as more insight into the system is gained.

Resources required

The uncertainty matrix can be used as a heuristic instrument and screening tool. One can draft a gross list of uncertainties to be addressed and give a priority (e.g. high, medium, low) to each cell by going over all the cells of the matrix systematically and reflecting on the question whether that cell may be relevant to include in an uncertainty assessment of the case at hand. This can be done by an individual researcher; in a dialogue in a research team; or in a dialogue between a research team and the client; or, in a wider dialogue between the research team, the client and a group of stakeholder representatives.

The resources required will depend on the way the matrix is used in the process. The broader the dialogue is in which the tool is used, the more resources will be required.

Strengths and limitations

+ Forces the analyst to systematically consider a broad range of possible sources and types of uncertainty. It thereby reduces the chance that important uncertainties are overlooked.
+ Can be used quick and dirty/back of the envelope or more sophisticated and comprehensive.
+ Facilitates and structures a dialogue on uncertainty, both within a research team and in communication with clients, peers and stakeholders.
– Strongly relies on expert judgement, the result is likely to be sensitive to the composition of the group of analysts involved in applying the matrix.
– Yields mainly qualitative insights, has limited value if it is not complemented by some form of quantitative analysis.

References

Walker, W.E., Harremoës, P., Rotmans, J., Van der Sluijs, J.P., Van Asselt, M.B.A., Janssen, P. and Krayer von Krauss, M.P. (2003). Defining uncertainty a conceptual basis for uncertainty management in model-based decision support, *Integrated Assessment*, **4**(1), 5–17.

Janssen, P.H.M., Petersen, A.C., van der Sluijs, J.P., Risbey, J.S. and Ravetz, J.R. (2003). RIVM/MNP Guidance for Uncertainty Assessment and Communication: Quickscan Hints & Actions List. RIVM/MNP, ISBN 90-6960-105-2, 2003 (available from www.nusap.net).

5. HOW TO SELECT THE APPROPRIATE METHODOLOGY FOR UNCERTAINTY ASSESSMENT

5.1. Introduction

Some of the more important types of methodologies and associated tools that may be applied for assessing uncertainties are listed and briefly described in Chapter 4. The next question is which methodology should be selected for a particular purpose and task of the modelling process.

The selection of an adequate methodology depends on:

- Where in the modelling process the analysis should be carried out, cf. Chapter 2.
- The type, nature and source of uncertainty, cf. Chapter 3.
- The priority that addressing each of the identified sources of uncertainty has, according to their importance for the decision-making process ("policy relevance").
- The available resources and level of ambition with respect to completeness of the analysis.

Guidance regarding the selection of appropriate methodologies/tools for different purposes is provided in the following two sections.

5.2 Methodologies according to modelling process and level of ambition

Table 5.1 provides a list of applicable methodologies that are considered to be adequate for the different tasks/steps in the modelling process described in Chapter 2 and Figure 1.3. Furthermore, it includes hints for which methodologies are more suitable for comprehensive analysis with relatively large economic resources for the study and which methodologies correspond to a lower level of ambition (denoted as "basic" in the table).

Table 5.1. Suitable methodologies to deal with uncertainty at various stages of a modelling process

Type of uncertainty aspect	Task in the modelling process (cf. Fig. 1.2)	Level of ambition/available resources	
		Basic	Comprehensive
Identify and characterise sources of uncertainty	Describe Problem and Context (Step 1)	UM	EPE, SI, UM
	Determine Requirements (Step 1)		
Modeller reconsiders uncertainty and performance criteria	Prepare and Evaluate Tender (Step 1)	(Update of) UM + Common sense	
	Reporting (Step 2)		
	Reporting (Step 3)		
	Reporting (Step 4)		
	Reporting (Step 5)		
	Specify or Update Calibration and Validation Targets and Criteria (Step 3)		
Reviews-dialogue-decisions	Agree on Model Study Plan and Budget		
	Review of Step 2	QA	EPR, QA
	Review of Step 3	QA	EPR, QA
	Review of Step 4	QA	EPR, QA
	Review of Step 5	QA	EPR, QA
Uncertainty assessment and propagation	Uncertainty Analysis of Calibration and Validation (Step 4)	DA, EPE, SA	DA, EPE, EE, IN-PA, IN-UN, MCA, MMS, NUSAP, SA
	Uncertainty Analysis of Simulation (Step 5)	DA, EPE, SA	DA, EPE, EE, IN-UN, MCA, MMS, NUSAP, SC,SA, SI

Abbreviations of methodologies:

DA Data Uncertainty
EPE Error Propagation Equations
EE Expert Elicitation
EPR Extended Peer Review (review by stakeholders)
IN-PA Inverse modelling (parameter estimation)
IN-UN Inverse modelling (predictive uncertainty)
MCA Monte Carlo Analysis
MMS Multiple Model Simulation
NUSAP NUSAP

QA	Quality Assurance
SC	Scenario Analysis
SA	Sensitivity Analysis
SI	Stakeholder Involvement
UM	Uncertainty Matrix

5.3. Methodologies according to source and type of uncertainty

Table 5.2 provides a list of applicable methodologies for addressing uncertainty of different types and originating from different sources. Note that the nature of uncertainty (epistemic or stochastic) has been omitted as compared to the uncertainty matrix in Figure 3.1. The reason for this is that this is a third dimension and that each of the cells below may be divided into reducible (epistemic) and irreducible (stochastic) uncertainty.

6. ILLUSTRATIVE CASES

The following cases aim to illustrate which uncertainty methods typically might be selected in different types of modelling studies. With reference to Figure 1.2 above the first two cases address model application to design programme of measures, while the last two cases deals with implementation and monitoring, respectively. The main difference between the two first cases lies in the level of ambition/available resources, cf. Table 5.1. The cases are hypothetical.

6.1. Case 1: Designing measures – nutrient load/comprehensive modelling

This case resembles some of the problems typically encountered in some central and northern European river basins with eutrophication of surface water bodies caused by nutrient loads primarily from agriculture but also to some extent from industry and urban areas. The conditions can furthermore be characterised by:

- There is a good amount of high quality data.
- Modelling studies have regularly been carried out during the past 10 years for various water management purposes.
- The river basin authority (water manager) has a large and highly professional staff with comprehensive experience in handling modelling projects.
- Some of the key stakeholders are very well organised and have professional support (e.g. agricultural advisory service and green NGOs) that are able to evaluate and challenge any possible scientific weaknesses of modelling studies.

Table 5.2. Correspondence of the methodologies with the source and types of uncertainty distinguished in the uncertainty taxonomy (inspired by van der Sluijs *et al.*, 2004)

Source of uncertainty		Taxonomy (types of uncertainty)			
		Statistical uncertainty	Scenario uncertainty	Qualitative uncertainty	Recognised ignorance
Context	Natural, technological, economic, social, political	EE	EE, SC, SI	EE, EPR, NUSAP, SI, UM	EE, EPR, NUSAP, SI, UM
Inputs	System data	DA, EPE, EE, MCA, SA	DA, EE, SC	DA, EE	DA, EE
	Driving forces	DA, EPE, EE, MCA, SA	DA, EE, SC	DA, EE, EPR	DA, EE, EPR
Model	Model structure	EE, MMS, QA	EE, MMS, SC, QA	EE, NUSAP, QA	EA, NUSAP, QA
	Technical	QA	QA	QA	QA
	Parameters	IN-PA, SA, QA	IN-PA, SA, QA	QA	QA
Model outputs		EPE, EE, IN-UN, MCA, MMS, SA	EE, IN-UN, MMS, SA	EE, NUSAP	EE, NUSAP

Abbreviations of methodologies:

DA Data Uncertainty
EPE Error Propagation Equations
EE Expert Elicitation
EPR Extended Peer Review (review by stakeholders)
IN-PA Inverse modelling (parameter estimation)
IN-UN Inverse modelling (predictive uncertainty)
MCA Monte Carlo Analysis
MMS Multiple Model Simulation
NUSAP NUSAP
QA Quality Assurance
SC Scenario Analysis
SA Sensitivity Analysis
SI Stakeholder Involvement
UM Uncertainty Matrix

In this case the river basin authority has to select the most appropriate measure or combination of several measures required in order to achieving the objectives of good ecological status. The possible measures may include some with various kind of restrictions on agricultural management, some focussing on removal of nutrients from industrial and urban point sources and some resulting in restoration of nature areas, e.g. with more frequent inundation and associated restrictions on land use.

A typical selection of uncertainty methods is illustrated in Table 6.1.

6.2. Case 2: Designing measures – water scarcity/basic modelling

This case resembles some of the problems typically encountered in some Mediterranean river basins with water scarcity and competing demand from different stakeholders for urban water supply, irrigation and nature. The conditions can furthermore be characterised by:

- There are relatively few data.
- Previously modelling studies have only been carried out for research purposes.
- The river basin authority (water manager) has a professional staff with little experience in handling modelling projects.
- The key stakeholders are very well organised but they do not have access to professional support that can assist them in evaluating and challenging the possible scientific weaknesses of modelling studies.

In this case the river basin authority has to select the most appropriate measure or combination of several measures required to solve the water scarcity problem and at the same time achieving the objectives of good ecological status. The possible measures may include some on water savings and demand management and some involving additional supply (e.g. from desalination, from a new reservoir or from new groundwater abstraction).

A typical selection of uncertainty methods is illustrated in Table 6.2.

6.3. Case 3: Implementation – Real-time forecasting (of Case 2)

This case can be seen as a continuation of Case 2 above. Assume that the measures designed in Case 2 has included a reservoir with a condition on a certain downstream flow regime to be maintained, e.g. in terms of required minimum flows and/or water quality criteria. In order to ensure the downstream flow regime and at the same time minimise the release from the reservoir real-time forecasts on the future flows (up to a week) is performed.

Table 6.1. Example of methods selected in Case 1: Design measures – nutrient load/comprehensive modelling

Modelling process			Link to decision (WFD) process
Modelling Step (Fig. 1.3)	**Uncertainty method**	**Content**	
Step 1 Framing of modelling study	UM	Identification and characterisation of sources of uncertainty used as a basis for prioritising what should be done in the modelling study	Stakeholders and water manager are key players in the WFD process and are continuously involved in this modelling step translating the needs in the decision process to modelling objectives and requirements
	SI	Stakeholders express their interests and wishes with respect to objectives of modelling and requirements to accuracy	
Steps 2, 3, 4 Conceptualise, set up and document model	UM	Update of uncertainty matrix	These steps in the modelling process are driven by the modeller. Stakeholders and water managers follow the modelling process through reports and meetings. The link to the WFD process is ensured by a formalised dialogue in connection with reviews at the end of each step, where decisions on approval of past work and possible adjustments of future project works are made
	DA	Assessment of data uncertainty done by modeller	
	SA	Sensitivity analysis to identify most important parameters for calibration	
	IN-PA	Assessment of uncertainty of model parameter values	
	MCA	Monte Carlo analysis to investigate whether the model results obtained during calibration and validation possibly can be explained by the assessed uncertainty on input data and parameter values	
	QA	Peer review of modelling work done by an external modelling expert	
	EPR	Stakeholder reviews of technical elements with special emphasis on the relevance and quality of the modelling study evaluated on the basis of the knowledge and interests possessed by the stakeholders	

(continued)

Table 6.1. *Continued*

Modelling process			**Link to decision (WFD) process**
Modelling Step (Fig. 1.3)	**Uncertainty method**	**Content**	
Step 5 Model application	EE	Expert elicitation on likely elements in scenarios for future developments	The main part of the work in this final step is done by the modeller, but in a close dialogue with stakeholders and the water manager, who carry the results of the modelling study back into the WFD decision process
	SC	Scenario analysis	
	MCA	Comprehensive Monte Carlo analysis with propagation of uncertainties on data and parameter values to assess uncertainty of model outcome	
	MMS	Formulation of multiple model structures to encapsulate the uncertainty on process equations, hydrogeological interpretations, etc.	
	QA	Peer review of modelling work done by an external modelling expert	
	EPR	Stakeholder reviews of technical elements with special emphasis on the relevance and quality of the scenario analysis and uncertainty assessments	

Table 6.2. Example of methods selected in Case 2: Design measures – water scarcity/basic modelling

Modelling process			Link to decision (WFD) process
Modelling Step (Fig. 1.3)	**Uncertainty method**	**Content**	
Step 1 Framing of modelling study	UM	Identification and characterisation of sources of uncertainty used as a basis for prioritising what should be done in the modelling study	Stakeholders and water manager are key players in the WFD process and are continuously involved in this modelling step translating the needs in the decision process to modelling objectives and requirements
Steps 2, 3, 4 Conceptualise, set up and document model	UM	Update of the uncertainty matrix	These steps in the modelling process are driven by the modeller. Stakeholders and water managers follow the modelling process through reports and meetings. The link to the WFD process is ensured by a formalised dialogue in connection with reviews at the end of each step, where decisions on approval of past work and possible adjustments of future project works are made
	DA	Assessment of data uncertainty done by modeller	
	SA	Sensitivity analysis to investigate whether the model results obtained during calibration and validation possibly can be explained by reasonable uncertainty assumptions on input data and parameter values	
	QA	Peer review of modelling work done by an external modelling expert	
Step 5 Model application	SC	Scenario analysis	The main part of the work in this final step is done by the modeller, but in a close dialogue with stakeholders and the water manager, who carry the results of the modelling study back into the WFD decision process
	SA	Sensitivity analysis to investigate the robustness of the model outputs to uncertainty in input data and model parameters	
	QA	Peer review of modelling work done by an external modelling expert	

Table 6.3. Example of methods selected in Case 3: Implementation – real-time forecasting (of Case 2)

Modelling process			Link to decision (WFD) process
Modelling Step (Fig. 1.3)	**Uncertainty method**	**Content**	
Step 1 Framing of modelling study	UM	Identification and characterisation of sources of uncertainty used as a basis for prioritising what should be done in the modelling study	Stakeholders and water manager are key players in the WFD process and are continuously involved in this modelling step translating the needs in the decision process to modelling objectives and requirements
Steps 2, 3, 4 Conceptualise, set up and document model	UM	Update of the uncertainty matrix	These steps in the modelling process are driven by the modeller. Stakeholders and water managers follow the modelling process through reports and meetings. The link to the WFD process is ensured by a formalised dialogue in connection with reviews at the end of each step, where decisions on approval of past work and possible adjustments of future project works are made
	DA	Assessment of data uncertainty done by modeller	
	SA	Sensitivity analysis to investigate whether the model results obtained during calibration and validation of the new model elements (new on-line stations, data assimilation algorithm) possibly can be explained by reasonable uncertainty assumptions on input data and parameter values	
	QA	Peer review of modelling work done by an external modelling expert	
Step 5 Model application	DA	Assessment of data uncertainty done by modeller	The main part of the work in this final step is done by the modeller, but in a close dialogue with stakeholders and the water manager, who carry the results of the modelling study back into the WFD decision process
	MCA	Assessment of uncertainty of forecasted flows/water quality. This may for instance be done by Ensemble Kalman Filtering, which includes Monte Carlo techniques	
	QA	Peer review of modelling work done by an external modelling expert	

Table 6.4. Example of methods selected in Case 4: Evaluation – post project appraisal (of Case 1)

Modelling process			Link to decision (WFD) process
Modelling Step (Fig. 1.3)	**Uncertainty method**	**Content**	
Step 1 Framing of modelling study	UM	Identification and characterisation of sources of uncertainty used as a basis for prioritising what should be done in the modelling study	Stakeholders and water manager are key players in the WFD process and are continuously involved in this modelling step translating the needs in the decision process to modelling objectives and requirements.
Steps 2, 3, 4 Conceptualise, set up and document model			These steps are not performed in this case, because the model from Case 1 is re-used.
Step 5 Model application	DA	Assessment of data uncertainty done by modeller	The main part of the work in this final step is done by the modeller, but in a close dialogue with stakeholders and the water manager, who carry the results of the modelling study back into the WFD decision process.
	SA	Sensitivity analysis to investigate the robustness of the model outputs to uncertainty in input data and model parameters	
	QA	Peer review of modelling work done by an external modelling expert	
	EPR	Stakeholder reviews of technical elements with special emphasis on the relevance and quality of the scenario analysis and uncertainty assessments	

In this case the hydrological model setup and applied during Case 2 will typically be extended and used in real-time operation. Most of the model can be used directly, but it needs to be tailored to run operationally in a real-time environment, implying a modification to use input data from only on-line stations and development of an updating routine to enable feed-back from real-time flow data.

A typical selection of uncertainty methods is illustrated in Table 6.3.

6.4. Case 4: Evaluation – Post project appraisal (of Case 1)

This case can be seen as a continuation of Case 1 above. Assume that the measures designed in Case 1 have been implemented and that a monitoring programme has been established to monitor to which extent the measures have had the desired effect. In case of nutrient loads from agriculture, this is known to show large temporal variations generated by climate variability. Thus winters with more rainfall and higher temperatures will give larger nutrient load than winters with less rainfall and/or lower temperatures. This temporal variability is usually much larger than the change in average load that is the objective of the designed programme of measures. Such climatically generated variance in a monitoring time series appears as noise if a statistical test is performed to identify a possible trend or a step change in the data.

In this case a dynamic model driven by climate input may be used to simulate the variation generated by the climate alone. By filtering out part of the climatically generated noise it may then be possible to identify changes in nutrient loads caused by the WFD measures (Refsgaard *et al.*, 1989; Lørup *et al.*, 1998). As the basic model has been established in Case 1 this model application can be seen as a post project appraisal. It will therefore typically not be required to start from scratch, but rather to re-use the already calibrated/validated model – just with new input data.

A typical selection of uncertainty methods is illustrated in Table 6.4. It is noted that, due to the re-use of an existing model, the modelling Steps 2, 3 and 4 are not performed here.

7. REFERENCES

The references below do not include the references in 14 tables in Chapter 4.

Aller, L.T., Bennet, T., Lehr, J.H. and Petty, R.J. (1987). DRASTIC: A Standardized System for Evaluating Ground Water Pollution Potential Using Hydrogeologic Setting, U.S. EPA Robert S. Kerr Environmental Research Laboratory, EPA/600/287/035, Ada, OK; 1987.

Beck, M.B. (1987). Water quality modelling: a review of the analysis of uncertainty. *Water Resources Research*, **23**(8), 1393–1442.

Blind, M. and de Blois, C. (2003). The Water Framework Directive and its Guidance Documents – Review of data aspects. Chapter 5 in Refsgaard, J.C. and Nilsson, B. (Eds.) Requirements Report. Geological Survey of Denmark and Greenland, Copenhagen. Available on http://www.harmonirib.com.

Brown, J.D. (2004). Knowledge, uncertainty and physical geography: towards the development of methodologies for questioning belief. *Transactions of the Institute of British Geographers*, **29**, 367–381.

Copenhagen County (2000). Pilot project on establishment of methodology for zonation of groundwater vulnerability. Proceedings from seminar on groundwater zonation, November 7, 2000. County of Copenhagen. (In Danish).

EC (2004a). Common Implementation Strategy for the Water Framework Directive (2000/60/EC) Guidance Document No 11, Planning Processes. European Communities. Available on http://forum.europa.eu.int/Public/irc/env/wfd/library?l=/framework_directive/guidance_documents/0publishedsguidancesdocu&vm=detailed&sb=Title

EC (2004b). Common Implementation Strategy for the Water Framework Directive (2000/60/EC) Guidance Document No 3, pressures and impactes, IMPRESS. Available on http://forum.europa.eu.int/Public/irc/env/wfd/library?l=/framework_directive/guidance_documents/0publishedsguidancesdocu&vm=detailed&sb=Title

Harremoës, P., Gee, D., MacGarvin, M., Stirling, A., Keys, J., Wynne, B. and Guedes Vaz, S. (Eds.) (2001). Late lessons from early warnings: The precautionary principle1896–2000, Environmental issue report no. 22. Copenhagen. European Environment Agency.

Funtowicz, S.O. and Ravetz, J. (1990). Uncertainty and Quality in Science for Policy. Kluwer Academic Publishers, Dordrecht.

Healy, S. (1999). Extended peer communities and the ascendance of post-normal politics. *Futures*, **31**, 655–669.

Klauer, B. and Brown, J.D. (2003). Conceptualising imperfect knowledge in public decision making: ignorance, uncertainty, error and 'risk situations'. *Environmental Research, Engineering and Management*, **27**, 124–128.

Lørup, J.K., Refsgaard, J.C. and Mazvimavi, D. (1998). Assessing the effect of land use change on catchment runoff by combined use of statistical tests and hydrological modelling: case studies from Zimbabwe. *Journal of Hydrology*, **205**, 147–163.

Refsgaard, J.C., Alley, W.M. and Vuglinsky, V.S. (1989). Methodology for Distinguishing Between Man's Influence and Climatic Effects on the Hydrological Cycle. IHP-III Project 6.3. Technical Documents in Hydrology, UNESCO.

Refsgaard, J.C., Hansen, L.K. and Vahman, M. (2000). Groundwater zonation in Copenhagen County – Intercomparision of thematic results from different consultants. Seminar on groundwater zonation, November 7, 2000. County of Copenhagen. (In Danish).

Refsgaard, J.C., Henriksen, H.J., Harrar, W.G., Scholten, H. and Kassahun, A. (2004). Quality Assurance in Model Based Water Management – Review of Existing Practise and Outline of new Approaches. *Environmental Modelling & Software*, **20**, 1201–1215.

Resnik, D.B. (2003). Is the precautionary principle unscientific? *Studies in History and Philosophy of Biological and Biomedical Sciences*, **34**, 329–344.

Rio Declaration on Environment and Development (1992). In Brown Weiss, Edith *et al.* (1999), International Environmental Law: Basic Instruments and References, 1992–1999. Ardsley NY: Transnational Press.

Scholten, H., Refsgaard, J.C. and Kassahun, A. (2004). Structuring Multidisciplinary Knowledge for Model Based Water Management; The HarmoniQuA Approach. In: Pahl, C., Schmidt, S., Jakeman,s T. (Eds.), iEMSs 2004 International Congress: "Complexity and Integrated Resources Management". International Environmental Modelling and Software Society, Osnabrück, Germany, June 2004.

Stirling, A. (2003). The Precautionary Approach To Risk Appraisal, NWMO Background Paper, Sussex, 29 pp.

Van der Sluijs, J.P., Janssen, P.H.M., Petersen, A.C., Kloprogge, P., Risbey, J.S., Tuinstra, W. and Ravetz, J.R. (2004). RIVM/MNP Guidance for Uncertainty Assessment and Communication Tool Catalogue for Uncertainty Assessment. Utrecht University. (Downloadable from http://www.nusap.net/sections.php?op=viewarticle&artid=17)

Walker, W.E., Harremoës, P., Rotmans, J., Van der Sluijs, J.P., Van Asselt, M.B.A., Janssen, P. and Krayer von Krauss, M.P. (2003). Defining uncertainty a conceptual basis for uncertainty management in model-based decision support. *Integrated Assessment*, **4**(1), 5–17.

Weiss, C. (2003). Expressing scientific uncertainty. *Law, Probability and Risk*, **2**, 25–46.

Weiss, C. (2003). Scientific Uncertainty and Science Based Precaution. *International Environmental Agreements: Politics, Law and Economics*, **3**, 137–166.